Developmental and Cellular Skeletal Biology

Developmental and Cellular Skeletal Biology

BRIAN K. HALL

Department of Biology
Dalhousie University
Halifax, Nova Scotia
Canada

ACADEMIC PRESS

NEW YORK SAN FRANCISCO LONDON 1978

A Subsidiary of Harcourt Brace Jovanovich, Publishers

ACADEMIC PRESS, INC.
111 Fifth Avenue, New York, New York 10003

United Kingdom Edition published by
ACADEMIC PRESS, INC. (LONDON) LTD.
24/28 Oval Road, London NW1 7DX

Library of Congress Cataloging in Publication Data

Hall, Brian K., Date
Developmental and cellular skeletal biology.

Bibliography: p.
Includes index.
1. Bone. 2. Bone--Growth. I. Title.
[DNLM: 1. Skeleton. WE101 H174d]
QM569.H28 596'.01'852 78-208
ISBN 0-12-318950-0

PRINTED IN THE UNITED STATES OF AMERICA

Contents

3 THE ORIGIN OF SKELETAL CELL TYPES

4 LOCATION OF THE SKELETON WITHIN THE EMBRYO

5 INITIATION OF CENTERS OF SKELETOGENESIS

9 INITIATION OF SKELETAL GROWTH

Preface

The skeleton has fascinated man ever since it was realized that, aside from one or several sets of genes, bare bones are his only bequest to posterity. But the skeleton is more than an articulated set of bare bones: Its three-dimensional conformation establishes the basis of our physical appearance; its formation and rate of differentiation determine our shape and size at birth; its postnatal growth orders us among our contemporaries and sets our final stature, while its decline in later life is among the primary causes for loss of the swiftness and agility of youth. Not surprisingly, the skeleton is a central focus of many biomedical disciplines and investigations.

For the developmental or cell biologist, the skeleton provides an excellent model for studies in cell differentiation, morphogenesis, polarized growth, epithelial–mesenchymal interactions, programmed cell death, and the role of the extracellular matrix. The skeleton supplies the geneticist with a permanent record of the vicissitudes of its growth, whereby the phenotypic expression of genetic abnormalities can be studied. The orthopedic surgeon earns a livelihood from the correction of these abnormalities, while the orthodontist corrects the position of teeth displaced consequent to alveolar bone dysfunction. Physiologists, biochemists, and nutritionists all are concerned with the skeleton's store of calcium and phosphorus and its response to vitamins and hormones; the hematologist, on the other hand, finds that the skeleton houses the progenitors of the blood cells. Pathologists endeavor to understand the disease states that result from abnormalities in skeletal cellular differentiation or function; surgeons want to prevent formation of skeletal tissues in the wounds that bear witness to their work. Veterinarians, physical anthropologists, radiographers, forensic scientists—indeed, the list could go on.

All of these individuals work in specialized fields, each with its own literature, jargon, and mode of operation. And while they share a common interest in the skeleton as an *organ,* practitioners in a given area are rarely exposed to the viewpoints and advances in the other fields. Working from the premise that knowledge of the cell biology, development, and growth of the skeleton is basic to *all* these specialties, this book is intended as a review of what is known of how the skeleton arises, differentiates, and grows. It is my hope that this overview of the status of these aspects of skeletal biology will appeal to all those specialist groups whose members claim the skeleton as "their organ," to research workers entering the fascinating and diverse field of skeletal biology, and to graduate and senior undergraduate students interested in cell biology, development, and/or growth. To this end, the present work utilizes examples and literature from several of these fields. Of course, venturing outside one's proscribed specialty in this way is hazardous, and errors of omission and of commission are almost inevitable. Although I have profited greatly from discussions and correspondence with colleagues, especially Drs. W. A. Beresford, A. I. Caplan, W. A. Elmer, A. W. Ham, M. A. Hardy Fallding, A. Y. Friedenstein, R. J. Goss, E. J. Kollar, R. A. Kosher, C. S. LeLièvre, P. F. A. Maderson, A. H. Melcher, W. J. Moore, M. L. Moss, P. Person, R. L. Searls, G. Strudel, P. V. Thorogood and M. S. Tyler, I take full responsibility for any errors that remain.

My own interest in the skeleton was kindled by the late P. D. F. Murray, whose 1936 monograph *Bones, A Study of the Development and Structure of the Vertebrate Skeleton* still remains one of the most lucid, and paradoxically, one of the most modern treatments of the developing skeleton. My own research and that of my honors and graduate students have been supported by the National Research Council of Canada (grant no. A5056) and by the Research Development Fund of Dalhousie University. I am indebted to Sharon Brunt, who has provided expert technical assistance for this research, created an unfailingly cheerful atmosphere in the laboratory, and spent many hours proofreading and correcting the present manuscript. Finally, I owe an immeasurable debt to my wife June, who spent countless hours with Derek and Imogen so that I could lose myself in the intricacies of the developing skeleton. If this monograph were to have a dedication it would be "In spite of Derek and Imogen."

Brian K. Hall

1

Types of Skeletal Tissues

I. INTRODUCTION

Of the four classes of mineralized tissues, namely, bone, cartilage, enamel, and dentine, the first and second are also skeletal tissues. For completeness, a fifth "catchall" category could also be added. This category would include the tissues which, on the basis of one criterion or another, are intermediate between two of the above tissues. Examples of these would be tissues intermediate between bone and cartilage or between enamel and dentine. The usefulness of these tissues in understanding the development of skeletal tissues will become apparent as our discussion progresses.

These four broad classes have been further subdivided. The various subdivisions proposed reflect the interests of the skeletal biologist and the scope of his or her field. For example, the embryologist wants to subdivide on the basis of developmental process; the anatomist on the basis of structure; the pathologist on the basis of deviation from the norm, and so on. A brief review of the four classes of mineralized tissues follows, with special attention paid to the two skeletal tissues.

II. BONE

Bone is a vascularized, supporting skeletal tissue (although it may arise ectopically outside the skeleton), which is deposited by osteoblasts and by osteocytes, and removed, and hence remodeled, by osteoclasts and by osteocytes. Glycosaminoglycans and collagen of type $[\alpha 1(\mathrm{I})]_2\alpha 2$ (type I collagen) comprise its extracellular matrix, which is permeated by canals and impregnated with hydroxyapatite. Bone functions to support the body; it acts as a site for attachment of ligaments and muscles, as a storehouse for calcium and phosphorus as well as for the hemopoietic tissues of the adult, and as a major site for the metabolic regulation of mineral homeostasis. Bone is found only in vertebrates.

A. Cellular Bone

Bone is classified on the basis of developmental processes as either endochondral (developing by the replacement of a cartilaginous model) or intramembranous (developing by the replacement of a fibrous or fibrocellular model). Although these terms primarily apply to the process of ossification, often they are used to specify the bones that result from these processes. Cellular vs. acellular bone; cancellous (woven) vs. lamellar bone; coarse-fiber vs. fine-fiber bone—these are the classifications of the histologist. [The reader is referred to H. M. Smith (1947), Ørvig (1967), Gardner (1971), Hancox (1972a), Pritchard (1972a), Ham (1974), and Patterson (1977) for in-depth discussions of bone type and bone development.] To illustrate the locations within the body where particular types of bone might be expected, the following list is provided. (a) Bone with coarse bundles of parallel fibers: at the sites of attachment of tendons and ligaments in both birds and mammals and in the ossified tendons of birds; (b) bone with coarse bundles of woven fibers: in the fetal mammal and in the early stages of fracture repair; (c) bone with fine bundles of parallel fibers: in the long bones of birds and young mammals, and around blood vessels in ossified tendons; (d) bone with fine, lamellated fibers: in the adult mammal; and (e) bone with both coarse and fine fibrous bundles: near attachment sites of tendons and ligaments and where coarse bundles are removed and replaced.

The processes that produce these various types of bone differ. Coarse, woven, bone is deposited rapidly, and as a result, the osteocytes and fibers are haphazardly arranged. Replacement by lamellar bone is considerably more orderly and predictably progressive, with the formation of primary and then secondary osteons. The primary osteon has a central canal of diameter less than 100 μ; it lacks a cement line, while having two or more central blood vessels. Interstitial lamellae are absent. The secondary osteon has a larger central canal, is limited externally by a cement line, has one central blood vessel, and is wedged between interstitial lamellae. The lamellae may be concentrically arranged, as in a haversian system; they may be circumferential, as in near-surface bone, or interstitial, as in remnants of old osteons. The life span of osteons, and the time required to produce them, vary from species to species. In a 2-year-old cat, it takes 50 days to make an osteon, while in a 45-year-old man, the process takes 100 days. And, since the average life span of that osteon will be 15 years, only 0.05% of the skeleton is turned over per day. Nor is the rate at which an osteon mineralizes uniform. While 70% of the mineralization occurs within one to two days of deposition of the uncalcified osteoid, the remainder can take many months. These rate differences become important when assessing pathological states, particularly in metabolic bone diseases, as has been emphasized by Bordier *et al.* (1969), Baylink *et al.* (1972), and by Fornasier (1977).

The standard histology textbook gives the impression that all mammalian bone is fine lamellar bone containing numerous secondary osteons. In fact, this picture is true only for human bone. Enlow and Brown (1958), Enlow (1966a), and Singh *et al.* (1974) have provided surveys of mammalian bone, with the latter authors attempting a quantification based on the number and size of primary longitudinal canals, and on the number of lacunae and empty lacunae. Necrotic, acellular areas, avascular areas, and areas lacking primary or secondary osteons, are all common and normal; these, moreover, can vary from bone to bone within an individual, with age for a given bone, and between individuals. One side of a bone may be highly vascularized and the other side may be avascular. A highly sophisticated knowledge of the microenvironment in which bone and bones develop is necessary before we can interpret this diversity of expression of the differentiated state of the osteocyte. One of the aims of this book is to marshal and analyze some of that evidence.

In the past, it has been argued that the specialized histology of bone represents either an adaptation to the mechanical stresses placed upon it, or an adaptation to the metabolic requirements for calcium and/or phosphorus. Recently, de Ricqlès (1973, 1974a,b) has related bone histology to the pattern of growth and general metabolism exhibited by the particular species or bone. These notions have achieved some prominence in the discussions on the possible warm-bloodedness of the dinosaurs (Bennett and Dalzell, 1973; Desmond, 1976). De Ricqlès contrasts the bones of the slow-growing salamander, which might achieve a body weight of 20 g after four years, with those of many mammals that achieve weights of several hundred kg after two years. He has correlated types of periosteal bone (which comprises most of the bulk of the long bones) with vascularity, species growth rate, rate of mineralization, fiber organization (a function of their rate of deposition), and periodicity of bone deposition. His studies merit the close attention of all those interested in the functional significance of bone histology.

The most recently deposited, unmineralized, metabolically active bone (*osteoid,* a term and concept developed by Virchow, 1853) is found adjacent to either periosteal or endosteal bone surfaces. These surfaces are lined by formative cells (osteoblasts), by resorptive cells (osteoclasts), and by precursor cells (osteoprogenitor cells). These bone surfaces are of prime importance in metabolic function, in reaction to vitamins and hormones, and in initiation of pathological change (Neuman, 1969; Ramp, 1975). Owen (1970, 1971) has studied the dynamics of the cells on these surfaces. Recent ultrastructural evidence indicates that junctions exist between adjacent osteocytes, between osteocytes and osteoblasts, and between osteoblasts (Holtrop and Weinger, 1971; Furseth, 1973; Weinger and Holtrop, 1974; Stanka, 1975). As a result, rapid transport throughout the bone can be achieved—one need only witness the appearance of radioisotope within cortical bone just minutes after intraperitoneal injection. Endosteal surfaces are especially important in bones that exhibit secondary remodeling. In 40 cm^3 of the human pelvis there is 80 cm^2 of periosteal surface, but 1600 cm^2 of endosteal surface. How these active cell layers are generated and maintained will be discussed in Chapters 6 and 7.

B. Acellular Bone

When bone is found to be acellular, it is usually regarded as degenerated, pathological, or the artifactual result of poor histological processing. [In fact, Stinson (1975a,b) has gone so far as to resurrect the medieval notion that *all* bone is dead and that osteocytes are mythical artifacts!] In compression-induced arthritis in rabbits, acellular areas occur in both the articular cartilage and the underlying subchondral bone (Gritzka *et al.*, 1973). Caisson disease of bone represents the development of avascular necrosis as a result of diving or tunneling under increased pressure, the degree of necrosis being closely correlated with diving time, depth of dive, etc. (Jaffe, 1972; Ohta and Matsunaga, 1974; Walder, 1976). These examples of acellularity obviously result from abnormal circumstances. However, the comparative histological studies mentioned in Section A indicate that large areas of acellular or avascular bone can be part of the normal histology of mammalian bone; hence, the sampling problem and the difficulty in assigning causation encountered by the paleontologist, forensic scientist, and anthropologist [see Enlow (1966a,b) and Wells (1973) for discussions].

There are, however, two major groups of vertebrates in which bone acellularity is the rule and not the exception. These are the modern teleost fishes and the Agnatha, the jawless vertebrates of the Ordovician period. The existence of this acellular bone is not well known. Because of this, and because it provides a link between evolutionary and developmental studies on the one hand, and contrasts normality with pathology on the other, acellular bone will be discussed at some length.

It was Kölliker (1859) who first described acellular bone in the teleosts. He called it *osteoid* (a term now reserved for recently deposited, unmineralized, cellular bone) and postulated that the ancestors of the teleosts possessed cellular bone, i.e., that acellularity was a secondary condition. I shall return to this evolutionary question in Chapter 2.

Moss (1961a) and Enlow and Brown (1956, 1957, 1958) provide the bulk of the histological studies available on acellular bone in the teleosts. Because acellular bone is found in various orders of higher teleosts that inhabit both fresh and salt water, it has been postulated that the development of acellular bone is not correlated with retention of calcium or phosphorus, since neither of these elements is in limited supply in aquatic environments (Moss, 1961a, 1963, 1965). However, the availability of the calcium and phosphorus stored in acellular bone varies from species to species (Simmons, 1971). When Simmons and Marshall (1970), examining ^{45}Ca uptake in the acellular bone of the toadfish (*Opsanus tau*), found little, they therefore concluded that little osteogenic activity was taking place. However, when administered intraperitoneally to the pike (*Esox lucius*), tetracycline is rapidly incorporated (within 3 hr) into the acellular bone, a finding that indicates both osteogenesis and metabolic activity (Meunier and Boivin, 1974). In the killifish (*Fundulus kansae*), 80% of the calcium is stored in the acellular bone in diffusible form and can be mobilized in response to seasonal needs or under conditions of induced stress, such as hypophysectomy (Brehe and Fleming, 1976). Calcium from skeletons of species having cellular bone, e.g., the American eel (*Anguilla rostrata*), is similarly mobile (Fenwick, 1974). The cells respond to exogenous calcitonin or pituitary extract by diminished bone resorption, as do the osteoblasts and osteocytes of mammalian bone (Lopez, 1970a,b; Lopez and Martelly-Bagot, 1971; Lopez and Deville, 1973). The availability of the calcium stored within the acellular bone might be explained by the presence of vascular canals, whose perivascular connective tissue cells might mobilize the calcium (Moss, 1963). However, the basis of calcium utilization remains to be elucidated, and might well provide useful information for the treatment or prevention of pathological necrosis of cellular bone.

Although it is especially interesting, the development of acellular bone during ontogeny has received little study. It was noted that the teleosts, which now have acellular bone, arose from ancestors that possessed cellular bone. Similarly, during teleost ontogeny, acellular bone arises secondarily from cellular tissues. Again, Moss (1964a,b) provides the little information that is available. He has shown that acellular bone may arise by osteogenesis within the osteoprogenitor cells of the periosteum, by osteogenesis within cells of tendons (the so-called tendinous osteogenesis), or by metaplastic transformation of cartilage into bone. In each case—and these various types of osteogenesis may be found within one individual at different skeletal sites—the extracellular matrix calcifies, either trapping the osteocytes that become pyknotic, or leaving the osteocytes on the surface. The entrapped pyknotic osteocytes calcify [much as in mineralization of invertebrate cartilages or in lignification of plants (Person and Philpott, 1963)] and an acellular tissue results (Moss, 1961b, 1963). A far more detailed analysis of the development of such bones still is necessary.

Another approach to the study of the development and physiology of acellular bone has been the exploration of its ability to repair fractures. A comparison of the responses to a fracture stimulus by cellular and acellular bone was investigated by Moss (1962a). The fractured acellular opercular bones and lower jaws of the cichlid (*Tilapia macrocephala*) produced a callus of calcified cartilage and bone. Fibroblasts from periskeletal connective tissue modulated to form osteo- and chondroprogenitor cells, with these cells forming the cartilage and bone of the callus. Under acalcemic conditions, the ability to form this callus was diminished, whereas under similar conditions species with cellular bone, such as the goldfish (*Carassius auratus*), were capable of initiating fracture repair. The loss of bone cells during ontogeny is not, then, an impediment to the repair of fractures of those same bones, for an ability to modulate adjacent cells into the skeletogenic series has been developed. (The concept of races of cells with skeletogenic potential will be explored in Chapter 3.) Once again, the study of acellular bone might provide valuable insights into this basic developmental question.

Moss (1962b) also utilized implantation of acellular and cellular bones to ectopic sites as a means of assessing the varying potentials of the two bone types. Bones were implanted either subcutaneously or into defects in the femora or crania of adult rats. In the subcutaneous sites, both cellular and acellular bone produced an immune response, and were resorbed and removed by host cells. In the intraskeletal sites, both bone types were initially incorporated into the host bones, and were then slowly removed by resorption. The extracellular matrix of the acellular bone possesses a species specificity, which can elicit a rejection response.

One finds acellular bone in the oddest places. In sharks, the basal tissue supporting the teeth is acellular bone (Moss, 1970; Kemp, 1977), even though elasmobranchii are traditionally classified as having a purely cartilaginous skeleton. The fact that they are capable of producing bone is a good example of the ability of the progenitor cells of the skeleton to retain, over very long periods of evolutionary time, the ability to modulate to either cartilage or bone. Moss (1977) has discussed possible inhibition of osteoblast activity by calcified cartilage in other parts of the shark skeleton.

Acellular bone [so-called aspidin(e)] is one of the skeletal tissues found in the earliest fossil vertebrates (see Chapter 2). The oft-debated views on the relationship between cellular and acellular bone during evolution take on an element of rationality when viewed in light of the knowledge of the ontogenetic development of acellular bone from cellular tissues. It is difficult to imagine how the acellular bone of the Agnatha could have arisen other than by development from a cellular tissue. The important question linking the development and evolution of this tissue is why the osteoblasts and osteocytes do not persist in the tissue as they develop. Moss (1968a–c), Hall (1975a), Maderson (1975), Schaeffer (1977), and Patterson (1977) have addressed this interplay between the knowledge of developmental processes and evolutionary mechanisms.

III. CARTILAGE

Cartilage is an avascular, supporting, and articular skeletal tissue (although like bone, it may arise ectopically outside the skeleton), deposited by both chondroblasts and by chondrocytes, and removed by chondroclasts. Its extracellular matrix, primarily composed of glycosaminoglycans, contains a smaller collagen component of type $[\alpha 1(\text{II})]_3$ (type II collagen). Cartilage may or may not exist as a mineralized tissue. Cartilage functions as the primary embryonic skeletal tissue in many parts of the embryo and as the articular tissue at joints on both endochondral and membrane bones (in the latter case, the cartilage is known as secondary cartilage). Cartilage is found in both vertebrates and invertebrates.

A. Vertebrate Cartilage

Although cartilage is subdivided into types primarily according to histological criteria, the cartilage that provides the model for endochondral bones can be classed as primary cartilage, and that which arises on membrane bones as secondary cartilage. Most cartilage develops from mesoderm, but some types, notably Meckel's cartilage, parts of the chondrocranium, and the visceral cartilages, are of neural crest (ectomesenchymal) origin.

On histological grounds, cartilage is considered to be *hyaline* if the extracellular matrix is composed of predominantly glycosaminoglycan; it is termed *elastic* if elastic fibers occur in the extracellular matrix, or *fibrous* (fibrocartilage) if there is an increased collagenous fiber content in the matrix. Hyaline cartilage is found in the embryonic models of endochondral bones and in the larynx; elastic cartilage in the pinna, larynx, and epiglottis; and fibrous cartilage where ligaments and tendons attach to bone, in the intraarticular discs of the joints, and as articular cartilage at joint surfaces. While cartilage is normally avascular, nonchondrified channels, which may carry blood vessels some distance into the body of the cartilage, may be present within the cartilaginous matrix (Novak, 1964; Moss-Salentijn, 1975). And, in contrast to the solely appositional growth of bone, the growth of cartilage is both interstitial and appositional. It is thought that this feature explains its success as an embryonic skeletal tissue. The resistance of cartilage to compression enhances its usefulness in growing organisms and makes it an ideal tissue for joint surfaces. For basic references on cartilage, the reader is referred to Fell (1925) and Gardner (1971) for the cytology of chondrogenesis in avian and human embryos, respectively; to Ham (1974) for the histology of cartilage; to Godman and Porter (1960) for the classic ultrastructural analysis of cartilage, and to Serafini-Fracassini and Smith (1974) for the biochemistry and physical structure of cartilage.

Historically, the distinction between cartilage and bone has been known at least since the time of Aristotle (384–322 B.C.), who recognized and separated the Chondrichthyes from the Osteichthyes on the basis of the presence of a cartilaginous or an osseous skeleton. In fact, up until the 18th century, it was thought that cartilage transformed into bone. In 1736, Robert Nesbitt set himself to "shew the ancient and common notion of all bones being originally cartilaginous to be a vulgar error," and that the "bony particles in foetuses begin to be deposited or to shoot either between membranes or within cartilages." In 1848, W. S. Sharpey refined this, proposing his Theory of Substitution, viz., that bone replaces cartilage in endochondral bones.

Today, the difficulty of distinguishing cartilage from bone arises in at least three contexts. The task of the paleohistologist is the identification of these tissues, and to determine their relationships during the evolution of the vertebrates. The pathologist, when examining the skeleton, often finds skeletal tissues that are not classifiable into any one category, and there are classes of tissues in the nonpathological skeleton that are apparently intermediate between cartilage and bone.

When discussing bone, I used acellular bone to illustrate the usefulness of studies of those classes of skeletal tissues that are not often discussed in textbooks, but that yield valuable insights into developmental processes. I shall do the same with cartilage, using some of the invertebrate cartilages as examples.

B. Invertebrate Cartilages

Interest in invertebrate cartilage has been rekindled over the last eighteen years, primarily through the work of Person. Historically, the existence of invertebrate cartilage has been known since the early 1800's. The recognition that many invertebrates possessed a cartilagelike (chondroid, chordoid, mucoid) supporting tissue structurally similar to the supporting and parenchymal tissues of plants was one of the generalities that led Schwann and Schleiden to the formulation of the Cell Theory in 1838–1839. The studies of the 19th and early 20th centuries have been exhaustively summarized and reviewed by Schaffer (1930), who, however, did not regard these tissues as true cartilages. Their "immature" histology, combined with a scant extracellular matrix, the inability to detect glycosaminoglycans and collagen in their matrices—both of which are diagnostic of vertebrate cartilage—and their failure to mineralize or to calcify, led Schaffer to regard them as chondroid tissues. By "chondroid" he meant cartilagelike, and he classified such tissues into grades of greater or lesser resemblance to cartilage (without implying any phylogenetic trends or affiliations).

Schaffer's view prevailed, and skeletal biologists subsequently turned to other problems. Following his 1930 review, no research publications on invertebrate cartilages appeared until 1959. However, since 1959, information on the light and electron microscopy, the biochemical characterization of the extracellular matrical products, the histochemistry, the mineralization, and some results on the regenerative ability of invertebrate cartilage have been reported. Little, if any, information on the development of the cartilages is available. Person and Philpott (1969a) and Person, in the symposium proceedings edited by Slavkin (1972), have reviewed the first decade of the modern work and have presented the arguments in favor of regarding these tissues as true cartilages. These arguments include the possession of cells in lacunae in an extracellular matrix containing chondroitin sulfate A or C [chondroitin 4- or 6-sulfate; see Jeanloz (1960) for a discussion of the nomenclature of the glycosaminoglycans (acid mucopolysaccharides)] and collagen, producing a tissue providing mechanical support and a site for the attachment of muscles and ligaments. Other pertinent reviews are those of Mathews (1968, 1975), who has reviewed the composition of invertebrate collagen, and Kobayashi (1971) and Anderson (1976b), who have surveyed the distribution and composition of glycosaminoglycans and glycoproteins. Like that of vertebrate cartilage, the collagen of invertebrate cartilage is composed of three identical chains—perhaps this provides an argument for the primitiveness of this collagen type during evolution?

The nature of invertebrate cartilages probably can be reviewed best by summarizing the available information on the several well-studied examples. These are the odontophoral cartilages of the whelk, *Busycon canaliculatum* (Mollusca), the branchial (gill book) cartilage of the horseshoe crab, *Limulus polyphemus* (Arthropoda), the cranial cartilage of the squid, *Loligo pealii*, and that of the octopus, *Octopus vulgaris* (Mollusca), as well as the tentacular cartilages of the polychaetes, *Eudistylia polymorpha* and *Sabella melanostigma* (Annelida).

1. *Busycon canaliculatum* (Whelk)

This marine snail has a large odontophoral cartilage consisting of hypertrophic cells in capsules, which are separated by scant amounts of extracellular matrix (Person, 1960; Person and Philpott, 1963). While it is true that invertebrate cartilage of this histological type presented one of the stumbling blocks to the acceptance of these tissues as true cartilages, there are in fact a number of vertebrate cartilages characterized by very scant amounts of extracellular matrix. Examples would include the xiphisternal cartilages, secondary cartilages, and the callus cartilage found in the repair of fractured bone.

The odontophoral cartilage lacks chondroitin sulfate, having instead a polyglucose sulfate akin to chitin (Lash and Whitehouse, 1960). (The glycosaminoglycans of invertebrate cartilages tend to be oversulfated in comparison to those from vertebrate cartilages.) The connective tissues of the echinoderms have a similar polyfucose sulfate, but also possess chondroitin sulfate (Katzman and Jeanloz, 1969).

Collagen fibers with a 640 Å periodicity and a typical X-ray diffraction pattern are found in the extracellular matrix (Person and Philpott, 1963). *Busycon* cartilage is unusual in containing myoglobin within its cells (Lash, 1959; Person *et al.*, 1959; Person and Philpott, 1963). Recall that the blood pigment in molluscs is hemocyanin. *Busycon* cartilage was also the first vertebrate or invertebrate cartilage in which cytochrome oxidase was found (Person *et al.*, 1959). (The low utilization of oxygen by vertebrate cartilage, a fact paradoxically coupled with both high levels of dehydrogenases requiring cytochrome oxidase, and the considerable synthesis of sugars requiring oxygen, prompted Person to examine invertebrate cartilages in the first place.)

Members of a number of genera of boring prosobranch Mollusca (*Erosalpinx*, *Eupleura*) will, after amputation of the proboscis, regenerate a new proboscis complete with new odontophoral cartilage. In so doing, a cap (blastema?) forms, which, like that in vertebrate limb development or in the regeneration of an amphibian limb, gives rise to both the cartilage and the muscle of the new appendage (Carriker *et al.*, 1972).

2. *Limulus polyphemus* (Horseshoe Crab)

The gill books of *Limulus* contain a cartilaginous endoskeleton whose histological organization is similar to that just described for the odontophore cartilage of *Busycon* (Person, 1960; Person and Philpott, 1963; Cowden, 1967).

The scant extracellular matrix contains chondroitin 4-sulfate and reacts histochemically to tests for glycosaminoglycans similarly to vertebrate cartilage (Mathews *et al.*, 1962; Cowden, 1967). Hydroxyproline and hydroxylysine, in amounts typical of vertebrate cartilage, and in an X-ray diffraction pattern typical of collagen, have been found (Person and Philpott, 1969b). Both lipid and glycogen are found as storage products, and the presence of cytochrome oxidase indicates potential aerobic metabolism (Person and Fine, 1959; Philpott and Person, 1966, 1970).

As the chondrocytes divide, a phragmasome-like structure is deposited between the daughter cells. This material provides the basis for the new extracellular matrix and is reminiscent of the division of supporting tissue in plants (Person and Philpott, 1969b). As the cells mature, the whole chondrocyte becomes "chondrified." Its cytoplasm fills in with the matrix products and the cell dies. This process is reminiscent of lignification in plants (Person and Philpott, 1963) and may be a partial explanation for the entirely appositional pattern of growth shown by this cartilage, a pattern in contrast to the interstitial growth of *Octopus* cartilage (Cowden, 1967).

Like all invertebrate cartilages, the cartilage of *Limulus* is unmineralized. The presence of lipids, which in vertebrate skeletal tissues play a role in mineralization, is of interest (see Part 5).

3. *Loligo pealii* (Squid), *Octopus vulgaris* (Octopus)

These decapods contain an extensive cranial cartilage that protects the brain, as well as a scleral cartilage protecting the eye. Squids also possess epidermal cartilaginous scales (Person, 1969). Unlike the cartilage in both *Busycon* and *Limulus*, these cranial cartilages have extensive amounts of extracellular matrix and contain very sparsely scattered cells, with cell densities more typical of vertebrate bone than of cartilages. The matrix may be quite fibrous and often has a canalicular (vascular?) system coursing through it. The predominant glycosaminoglycan is chondroitin 4-sulfate (like *Limulus*), and collagen is present (Person 1960; Person and Philpott, 1963; Philpott and Person, 1970). Histochemical evidence indicates less glycosaminoglycan and a different spectrum of matrix proteins than in *Limulus* cartilage (Cowden, 1967), a finding that emphasizes the need for more biochemical characterization of the matricial products.

Cytochrome oxidase is not found in *Loligo* cranial cartilage, and consequently aerobic metabolism is presumed not to occur (Person and Philpott, 1963).

4. *Eudistylia polymorpha* (Marine Polychaete), *Sabella melanostigma* (Feather-duster Worm)

These polychaetes have extensive, branched, and segmented cartilages, which are present at the base of the tentacular complex, within the tentacles, and in the pinnae that branch from the tentacles. These cartilages give considerable support and flexibility to this food-gathering organ. The cartilaginous matrix is sparse and vascularized, causing it superficially to resemble vertebrate osteoid. The glycosaminoglycan is a highly sulfated chondroitin 6-sulfate (Person and Mathews, 1967). Ultrastructural studies performed on the cartilage of *Sabella* by Cowden and Fitzharris (1975) show heterogeneity of matrix structure.

5. Mineralization of Invertebrate Cartilages

No instance of *in vivo* mineralization of invertebrate cartilage has been recorded, nor has osseous tissue or bone been found in any invertebrate (the vascular cartilage of polychaetes and that of decapods being morphologically the closest to vertebrate osteoid). [For descriptions of the mineralized tissues and organic matrices that mineralize in invertebrates, the reader is referred to Watabe (1965); Travis *et al.* (1967); Travis (1968), and Schrarer (1970).] Many of the individuals containing unmineralized cartilage possess calcified exoskeletons, raising the obvious question of why the cartilages do not mineralize as well. Do they lack the ability, or are they inhibited?

Recently, Eilberg *et al.* (1974, 1975) showed that preparations of cartilage from *Loligo*, *Limulus*, and *Busycon* can mineralize *in vitro* if incubated in a solution metastable to hydroxyapatite, but only at temperatures (37°C) well above their normal environmental temperatures (ca. 20°C). The mineralization of the invertebrate cartilage is in the form of hydroxyapatite, which is the normal form of calcium phosphate in vertebrate cartilage. The calcium that accumulates does so by initial formation of perichondrial and extracellular matrix granules, and finally, by progressive intracellular mineralization of the chondrocytes (Libben *et al.*, 1976). This intracellular mineralization is normally associated with cell death in vertebrate chondrocytes (Hall, 1972a).

The rate of mineralization shows a positive correlation with the concentration of phosphatidylserine within the cartilage (Libben *et al.*, 1976; Rabinowitz *et al.*, 1976). This is consistent with the involvement of lipids in the mineralization of vertebrate cartilage and bone (Irving and Wuthier, 1968; Irving, 1973). Mineralization of cartilage in the invertebrates appears to be effectively blocked by environmental temperature, perhaps coupled with the absence of sufficient concentrations of calcium and phosphorus in the sea water. How the mineralization of invertebrate skeletal tissues relates to the question of the evolution of vertebrate mineralizing skeletal tissues has been discussed by Hall (1975a).

IV. ENAMEL

Enamel and dentine will be discussed only briefly, for although these tissues are mineralized, they are not strictly tissues of the skeleton. In addition, Scott and Symons (1974) have provided an excellent review of the structure and function of both of these tissues.

Enamel is the highly calcified, hard, prismatic, avascular, outer layer of mammalian teeth and those of certain other vertebrates. Enamel is much more resistant to wear than dentine. Some 96% of the bulk of enamel is inorganic, with only 0.5% organic material being present (Glimcher *et al.*, 1965). Enamel is produced by the ameloblasts of the *ectodermally* derived inner dental epithelium (the process of amelogenesis) in two phases. The first phase is the deposition of organic matrix, and the mineralization of that matrix is the second. In addition, the inner dental epithelium is responsible for specifying the shape of the tooth, e.g., whether incisiform or molariform, and through an inductive interaction with the underlying ectomesenchymal cells, is also responsible for the initiation of dentinogenesis (see Chapter 4, Section II,G).

V. DENTINE

Dentine is the tubular, calcified, dental tissue that comprises the bulk of the tooth. Seventy-five percent of its mass is inorganic, and 20% is organic (in contrast to the much lower organic content of enamel). Dentine is produced by the odontoblasts of the dental ectomesenchyme (i.e., it is of neural crest origin) under an inductive influence from the adjacent differentiating ameloblasts. As with enamel, dentine is produced in two sequential phases. The synthesis and deposition of predentine—an organic matrix composed of glycosaminoglycans and collagen of type $[\alpha 1(\mathrm{I})]_2\alpha 2$—are followed by mineralization of that matrix by hydroxyapatite.

VI. INTERMEDIATE TISSUES

It may have been noted that the type of collagen produced by the odontoblasts as they deposit predentine is the same as that produced by the osteoblasts as they deposit osteoid, viz., $[\alpha 1(\mathrm{I})]_2\alpha 2$. This observation suggests a possible relationship between the two types of cells. In fact, the cells that produce the *four* mineralized tissues have been grouped together and called *scleroblasts* by Moss (1964a). By grouping them in this way, he implied that these cells represent a family or race of cells with affinities among individual cells. The synthesis of the same type of collagen by cells producing two quite different mineralized tissues is one such affinity.

Normally, there is no difficulty in assigning tissues and cells to one of the four categories of mineralized tissues. This process represents one way of recognizing the stability of the differentiating or differentiated phenotypes of osteoblast (bone), chondroblast (cartilage), ameloblast (enamel), and odontoblast (dentine). However, there is a group of mineralized tissues that I have chosen to refer to as "intermediate tissues." On the basis of one or more of the criteria outlined in Sections I to IV, these tissues or cells show features of two of the four classes of mineralized tissues. Such tissues represent for the developmental biologist what the mutant is for the geneticist, or the disease state for the pathologist—a way of analyzing differentiation, not only within the intermediate tissues, but in the more readily identifiable classes of skeletal tissues as well.

A. Intermediate Tissues Involving Dental Tissues

In the same study (1859) in which he reported the existence of acellular bone in the teleosts, Kölliker described four types of teleost mineralized tissues. These were (a) cellular bone; (b) osteodentine, bone with dentinal tubules but containing osteocytes; (c) acellular bone, his osteoid-lacking cells or tubules; and (d) osteoid, with occasional tubules. The teleost fishes seem quite capable of modulating cellular activity between odontogenesis and osteogenesis. In *Amia calva,* a holostean fish possessing a cellular dentine, the processes involved in allowing persistence of the odontoblasts within this tissue evidently are identical to those allowing persistence of osteocytes within the bone (Moss, 1964b). Vasodentine (a vascularized dentine) is found on the lingual side of the incisors of the rodent, *Perognathus longimembris* (Moss-Salentijn and Moss, 1975).

Cementum is a supporting tissue that aids in anchoring the tooth into the socket. The guinea pig molar contains a coronal cementum that possesses some features of cementum, some of cartilage, and some that are unique (Listgarten and Shapiro, 1974).

Enameloid (mesoenamel, vitrodentine, durodentine) is a mineralized tissue that coats the teeth of both teleosts and larval urodeles, although the teeth of adult urodeles are capped by true enamel. The enameloid matrix, which is penetrated by fibers and by cellular processes from the dentine, is also aprismatic. In the ballan wrasse *(Labrus bergylta)* and in the eel *(Anguilla anguilla)*, *odontoblasts* produce the collagen while *ameloblasts* produce the other proteins of the enameloid matrix. As mineralization occurs, the collagen is removed—a feature unique to enameloid (Shellis and Miles, 1974). Enameloid, then, is a tissuc produced by the combined and orchestrated action of both odontoblasts and ameloblasts. [Baratz (1974) and Moss (1977) may be consulted for some possible mechanisms.]

In the urodele amphibians, *Ambystoma mexicanum* and *Triturus vulgaris*, the ameloblasts of the larval inner dental epithelium produce enameloid, while those of the adult produce enamel (Meredith Smith and Miles, 1971). This finding presumably reflects the differential activation of regulatory genes during ontogeny and during pre- and postmetamorphosis. Similar regulation has been proposed for the changes in the production of mineralized tissues that have occurred during the evolution of the vertebrates (Hall, 1975a; Maderson, 1975).

B. Tissues Intermediate between Bone and Cartilage

Examples of tissues intermediate between cartilage and bone are also known. Many of the tumors, sarcomas, and neoplasms of the skeleton would fall into this category (Willis, 1962; Jaffe, 1968; Ashley, 1970). The tissue in the antlers of deer has also been compared to neoplastic bone (see Section VI, C).

In the development of the epiphyseal tubercle of the tibia by chondrogenic replacement of the patellar ligament, an unusual fibrocartilage is produced (Badi, 1972b). The ligamentous tissue undergoes metaplasia to a fibrocartilage, which then calcifies and is replaced by bone. However, the calcification occurs without the chondrocytes undergoing hypertrophy or accumulating stores of glycogen, processes which are normally associated with maturational changes in cartilage that precede its calcification (see Chapter 7, Section II,F,3). The articular tubercle of the temporal component of the human temporomandibular joint undergoes transformation into *chondroid bone*. This process involves the formation of a predominantly collagenous extracellular matrix having the basophilia characteristic of bone, but containing nests of cells in lacunae resembling chondrocytes (Wright and Moffett, 1974). Hall (1972b) has presented evidence for a similar transformation of secondary cartilage to chondroid bone in paralyzed embryonic chicks, and has analyzed the available literature on such transformations. The changes in the matrix are the result of the entrapped chondroblasts and chondrocytes assuming osteoblastic activity and modifying the extracellular matrix toward an osseous type. This type of changed activity represents a true metaplasia and is also seen in the repair of fractured long bones in lower vertebrates such as lizards and frogs (Pritchard and Ruzicka, 1950) and at the distal tip of penile bones (Beresford, 1970, 1975a; Beresford and Clayton, 1977).

Evidence for the metaplasia is primarily cytological. It would be very nice to have an analysis of the changes in collagen type produced by these cells, for some transformations are not as readily interpreted as those mentioned above and reviewed by Hall (1972b). Lufti (1971) describes "cartilage bone" lining marrow spaces of the tibiae of 1- to 2-day-old chicks as the first step in the process of endochondral ossification; however, his figures are reminiscent of extension of staining from the surface periosteum. Although Dodds and Cameron (1939) described osteoblasts located in the opened lacunae of cartilage in rachitic rat metaphyses, these could be interpreted as locally transformed chondrocytes or as transformed migratory marrow cells, either of which could have deposited the surface osteoid. A similar phenomenon has been seen in articular cartilage exposed to excess vitamin A in organ culture (Barratt, 1973), a phenomenon which was regarded not as "true osteogenesis," but as "selective collagenous regeneration by the liberated chondrocytes." These semantic refinements illustrate the difficulty one encounters in interpreting the cellular activity underlying such changes. The resolution of these difficulties will be one major step toward understanding the interrelations between skeletal cells and how they modulate their activity (see Chapter 6).

When skeletal tissues are maintained *in vitro*, the substrate itself can initiate modulation of cell activity. When rabbit articular chondrocytes are cultured on decalcified bone, they produce a chondroid matrix having features of both cartilage and bone (Green and Ferguson, 1975); however, when cultured on gelfoam sponge, these same chondrocytes produce a chondromyxoid tissue (Green, 1971). The culture of embryonic chick tibiae and mandibles in complement-sufficient antisera results in the production of extensive amounts of a chondro-osteoid tissue and secondary cartilage from periosteal cells, as well as the release of chondrocytes and their modulation to fibroblasts and osteoblasts (Fell *et al.*, 1968). Similarly, the microenvironment *in vivo* can initiate modulation of skeletal cell activity. When either fresh or lyophilized bone is placed in diffusion chambers with either marrow or muscle, and is implanted into host mice, the cells of the host are induced to form mature bone, while inside the chamber, chondro-osteoid or very cellular bone is produced (Koskinen *et al.*, 1972; Upton, 1972; Simmons *et al.*, 1973; Buring, 1974). The differences in tissue types produced are attributed to the reduced oxygen tension within the chamber.

Sufficient examples are available to indicate that, at a variety of sites within both embryonic and adult skeletons, skeletal cells can produce tissues whose characteristic features in at least one sense are intermediate between those of cartilage and bone. Another example (which will be discussed in more detail) is the tissue that constitutes the body of the antlers of deer and other ungulates.

C. Histogenesis of Antlers

The nature of the cellular processes leading to the ossification of antlers has been a source of debate for at least the past fifty years. Whether a cartilaginous phase characterizes antler formation; if so, whether the ossification process is typical endochondral ossification or whether a metaplasia of cartilage to bone occurs; whether the tissue is only "cartilage-like" and not true cartilage, or whether both cartilage and chondroid are absent and ossification is entirely intramembranous are some of the points at issue. The tissues of the developing antler provide a very useful model for testing the veracity of various distinctions: between endochondral and intramembranous ossification, between vascular bone and avascular cartilage, and between the appositional growth of bone and the interstitial and appositional growth of cartilage.

To the layman, the horns and antlers of male ungulates are all subsumed beneath the heading of cephalic appendages. In actual fact, there are fundamental differences between these two sets of structures, the major of which have been listed in Table I. There are five families of ungulates that have either horns or antlers. These are:

1. Rhinocerotidae (rhinos), which have one or two midline nasal horns;
2. Bovidae (cattle, sheep, goats, antelopes), which have paired, symmetrical horns;
3. Antilocapridae (pronghorn antelope), which have paired, symmetrical, deciduous horns;
4. Cervidae (moose, deer, caribou, elk), which have paired antlers in the males (female reindeer and caribou are also antlered); and
5. Giraffidae (giraffe, okapi), which have permanent, paired "knobs."

I shall not consider the horned ungulates further. Although they may have a bony core, horns are really epidermal, and not skeletal, structures. Antlers, on

TABLE I

The Major Features Distinguishing Horns from Antlers

Horns	Antlers
Living	Dead
Cornified (±osseous core)	Ossified
Unbranched	Branched
Permanent (except pronghorn antelope)	Deciduous (usually annually)
Cannot regenerate	Regenerate
Basal growth	Apical growth

the other hand, may be massive skeletal structures, best known from the "Irish elk" (*Megaceros*) and from hunter's trophies. (The factors involved in locating the antlers on the skull are discussed in Chapter 4, while antler growth and its control are considered in Chapter 9.) The histogenesis of antlers can be treated best species by species. The process of antlerogenesis may vary from one species to another; unfortunately, no comparative study on antler histogenesis is available. For good reviews of the early literature, the reader is directed to Modell and Noback (1931) and Goss (1969a, 1970).

1. *Odocoileus virginianus* (Virginia or White-Tail Deer)

Modell and Noback (1931) examined the growing antlers of the white-tail deer, the American elk (wapiti) (*Cervus canadensis*), and the European fallow deer (*Cervus dama*). They described various zones of cells: a zone of mesenchymal cells at the antler tip underlying the corium, followed basally by a column of preosseous tissue, then by a zone in which degeneration of preosseous tissue was seen, and most basally, by an ossifying zone. Figure 1 shows the location of these zones within the antler. The use of the term "preosseous zone" was *not* to denote a zone of bone-forming tissue, but rather, to denote a zone of tissue that *preceded* the laying down of bone. This tissue was described as

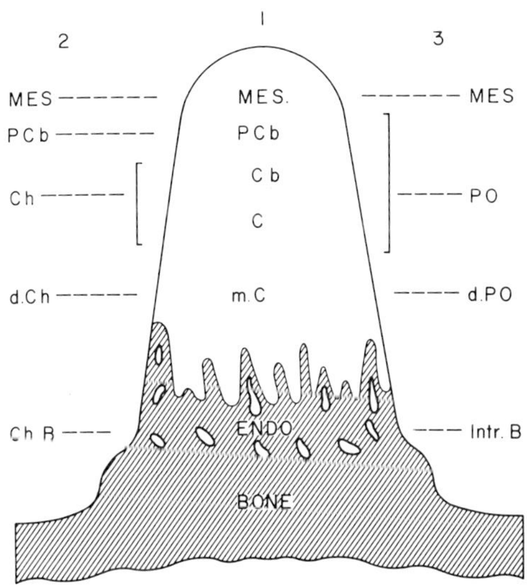

Fig. 1. A diagrammatic representation of the major cellular zones in a typical antler according to (1) Banks (1974), (2) Wislocki *et al.* (1947), and (3) Modell and Noback (1931) and Modell (1969). Abbreviations for the zones: C, chondrocyte zone; Cb., chondroblast zone; Ch, chondroid zone; Ch. B., zone of chondroidal bone formation; d. Ch., zone of degenerating chondroid; d. PO, degenerating preosseous tissue; ENDO., endochondral bone formation; Intr. B, intramembranous bone formation; m.C, mineralizing chondrocyte zone; MES., mesenchymal tissue; PCb, prechondroblastic zone; PO, preosseous zone.

"somewhat resembling hyaline cartilage in slides stained with hematoxylin and eosin" and as a supporting tissue "playing the role of cartilage," but not itself cartilaginous.

Further toward the base of the antler these cartilagelike cells degenerate, leaving behind intact longitudinal columns of matrix ("fibrils," to use Modell and Noback's term). These fibrils provided the scaffolding upon which the osteoblasts settled before beginning the deposition of osteoid and bone. Because this scaffold was not regarded as cartilage, the ossification process was regarded as intramembranous, even though the bone was not being deposited in a membrane. Rather, it was laid down onto a preexisting skeletal extracellular matrix.

Wislocki *et al.* (1947) examined the antlers of the white tail deer using histochemical procedures. They reported that the capsules of the cells in the preosseous zone were strongly metachromatic and had pronounced basophilia, indicative of the high acid mucopolysaccharide content of cartilage. Histologically, the tissue was not assessed as equivalent to hyaline or to elastic cartilage, but rather, as similar to fibrocartilage or to chondroid. Wislocki's group also confirmed that neither was the process of ossification typical of that in long bones, nor of intramembranous ossification (because the presosseous zone was cartilagelike). They coined the term "chondroidal bone formation" for this process of ossification and regarded it as intermediate between intramembranous and endochondral bone formation. However, this process is intermediate only in the sense that the tissue being replaced is not the cartilage typically seen in long bones. This particular type of cartilage is very similar to the secondary cartilage of such membrane bones as the mammalian dentary or the avian quadratojugal (Chapter 6, Section IV), and may reflect a pattern of chondrogenesis exhibited by the cranial neural crest cells of the skull [Chapter 4, Section (11)].

Kuhlman *et al.* (1963) described ossification as occurring on preosseous fibrocellular columns. Although Modell (1969), describing the bone of the antlers of the elk (*Cervus canadensis*) as equivalent to malignant bone sarcoma in its histology and rapid development, denied the existence of a cartilaginous phase, he described ossification on a fibrous framework, much as Modell and Noback did in their 1931 paper.

Recent studies (employing light and electron microscopic, as well as biochemical techniques) on the histogenesis of the antlers of the white tail deer have confirmed the cartilaginous nature of the preosseous tissue of Modell. Banks (1974) described the replacement of cartilage by bone as a process of modified endochondral ossification, a process which is responsible for increasing the length of the antler. The cartilage is arranged in columnar fashion, as in the growth plate of a long bone. Similar to the secondary cartilages, there is scant extracellular matrix, and the cartilaginous columns are separated by highly vascularized spaces. This highly vascularized nature of the cartilage was obviously one of the stumbling blocks to its correct identification. The cartilaginous columns mineralize and are replaced by columns of mineralized bone, occasionally containing islands of chondrocytes embedded within the osteoid. Figure 1 shows the arrangement of the major zones as described by Banks (1974), along with the interpretations of both Wislocki and Modell. Banks provides a critical review of the literature pertaining to the existence and fate of cartilage in the antler and has taken the best micrographs of its structure.

At the circumference, intramembranous ossification adds to the width of the growing antler (Banks, 1974). This pattern, coupling endochondral ossification, which increases the length of the antler, with intramembranous ossification, which adds to its width, parallels the endochondral and subperiosteal ossification taking place in the growing long bones.

2. *Odocoileus hemionus* (Rocky Mountain Mule Deer)

Sayegh *et al*. (1974) and Newbrey and Banks (1975) have examined antler histogenesis at both light and electron microscopic levels. In both studies, cartilage was positively identified as the primary skeletal tissue of the antler. Intramitochondrial accumulation of mineral was observed in those chondrocytes immediately distal to the zone of matrix calcification. Evidently, calcification is initiated within chondrocytes as an intracellular event, as it is in mineralization of the cartilage of long bones.

Histological and biochemical analyses of the beams of mature antlers during the rapid growth phase showed the presence of chondroitin 4- and 6-sulfates and hyaluronic acid in the terminal cap of reserve cells, as well as the synthesis and accumulation of glycosaminoglycans in the subjacent cartilaginous zone (Frasier *et al*., 1975).

In June, at peak antler growth, the remodeling of ribs, metacarpals, metatarsals, and tibial haversian systems is also at a peak (Banks *et al*., 1968a,b; Hillman *et al*., 1973). Even when deer are kept on a high calcium diet, cortical bone is mobilized from the ribs—the bone density, as well as the levels of calcium, phosphorus, and magnesium, all show decreases in ribs biopsied when the mineralization of the antlers is taking place. This "cyclic physiological osteoporosis" was thought to provide calcium and phsophorus for the mineralization of the antlers; indeed, mineralization of antlers may be a good model for osteoporosis, the metabolic bone disease.

3. *Cervus nippon* (Sika Deer)

Goss (1969a, 1970) describes a proliferative zone at the tip of the antler that differentiates into cartilage. He describes a process of metaplasia of this cartilage into bone as the antler matures. Further studies on this species are required.

The histogenesis of antlers, then, illustrates the ability of cells to produce a tissue having features of both cartilage and bone. It also indicates that the process of ossification need not be the strict endochondral or intramembranous sequence described in most textbooks.

These few examples from the range of tissues intermediate between two generally recognized classes of skeletal tissues illustrate some of the questions that will be asked in the next few chapters. How and when did the skeletal tissues arise during vertebrate evolution? What is the relationship between the various types of scleroblasts? What determines the formation of cartilage in one location within the body, the formation of bone in another, and in still another, chondroid bone? What regulates the type of synthetic activity engaged in by scleroblasts, and whether or how that activity can change with time or with variations in scleroblast microenvironment?

2

The Evolution of Skeletal Tissues

Fossilized skeletal tissues now can be processed for both light and electron microscopy (Moss, 1961b; Ørvig, 1965; Halstead, 1969a) and for chemical analysis (Ho, 1967; Mathews, 1967a; Biltz and Pellegrino, 1969; Kobayashi, 1971; Pawlicki, 1977). Hence, considerable information now is available on the structure of these tissues; as well, it is possible now to interpret developmental processes over the course of time and the functional significance of the skeletal tissues during evolution. An examination of the skeletal tissues found in the ancestral vertebrates, even in the earliest chordates found in the fossil record, indicates a surprisingly high degree of homology with the skeletal tissues in present-day vertebrates, in terms of the types of tissue present, as well as in their complexity and degree of specialization (see Hall, 1975a, for a recent discussion).

All four types of mineralized tissue can be found among representatives of the Ordovician Agnatha in the form of fossilized cephalic shields (Table II). These are the oldest recorded chordates, although whether they possessed a notochord (or any other uncalcified tissue such as uncalcified cartilage) is a matter of inference. In the Agnatha, these dermal skeletons consisted of cellular bone, dentine, enamel, and calcified cartilage among the Osteostraci, and acellular bone (aspidin), dentine, enamel, and calcified cartilage in the Heterostraci. Serving as a reservoir for calcium and phophorus, these tissues also acted as a barrier against loss of calcium and phosphorus to the aquatic environment, and as a protective armor against Eurypterids (Denison, 1963; Romer, 1963). Mechanical support was probably of lesser importance for these chordates than it was for later terrestrial groups.

TABLE II

A Summary of the Time of First Appearance of Mineralized Tissues and of the Classes of Vertebrates within the Fossil Record

Geological period	Class	Mineralized tissue	
Quaternary			
Cretaceous			
Jurassic	Mammals		
Triassic	Birds[a]	Enamel, dentine,	
Permian		cartilage, membrane	
Pennsylvanian	Reptiles	bone, aspidin	
Mississippian			
	Amphibians		Endochondral
Devonian	Chondrichthyes[d]		bone
Silurian	Placoderms		
	Osteichthyes[b]		
Ordovician	Agnatha[c]		
Cambrian			
Pre-Cambrian			

[a] Enamel and dentine were secondarily lost in the birds.
[b] Cartilage was secondarily lost in the Osteichthyes.
[c] Aspidin is only found in the Agnatha, acellular bone arose again in the teleost fishes.
[d] Endochondral bone was secondarily lost in the Chondrichthyes.

With further evolution of the vertebrates, adaptations of skeletal tissues arose in response to local environmental demands, utilizing the ability of scleroblasts to modulate their own synthetic activity. However, there is no evidence for a generalized trend toward progressive specialization during evolution. For example, the degree of skeletal mineralization in Ostracoderms (Schaeffer, 1961), Catacea (Felts and Spurrell, 1966), and in the Stegocephalia (de Ricqlès, 1975), reflects adaptations for equilibrium, buoyancy, conservation of temperature and energy, and not progressive evolution within the groups. In the non-weight-bearing skeletons of aquatic vertebrates, there appears to have been no selective advantage to one bony structure over another (Moss, 1963). For a detailed discussion of the significance of histological variation in fossil vertebrates, see Enlow and Brown (1956, 1957, 1958), Enlow (1969), and de Ricqlès (1968, 1969, 1972, 1973, 1974a–c).

We can conclude from the histological studies of fossilized skeletal tissues that these tissues in the earliest Chordates were no less specialized than those of today's vertebrates, such as ourselves; rather than involving major changes in cell or tissue type, evolution has provided the adaptive modulation of a highly plastic series of scleroblasts as dictated by local conditions.

Given this ability to modulate, and given the existence of intermediate tissues in recent vertebrates, one might expect to find evidence of many such intermediate tissues in fossil vertebrates as tissue organizations were tried out in response to changing environmental conditions. Ørvig (1951), who has directed considerable attention to three types of dentine, as well as to tissues intermediate between dentine and bone, and between bone and calcified cartilage in the early vertebrates, is a good source for the speculative thought on this subject. When aspidin, the acellular bone, was identified, considerable debate arose over whether it was bone or dentine. From the presence of intermediate tissues in the early vertebrates, there were then indications that the osteoblasts, odontoblasts, and chondroblasts could modulate their synthetic activity much the same as they do today. Acquisition of such adaptable cells so early in their evolution was no doubt a major reason for the later success of the vertebrates. Halstead (1969a–c, 1973) and Hall (1975a) present more detailed discussions of the significance of these intermediate tissues for the evolution of the vertebrates.

It has been indicated (e.g., in Table II) that aspidin and cellular bone arose contemporaneously during the Ordovician period. This was not always recognized. Indeed, the source of hot debate in the past has been the relationships between the two tissues. Did they evolve independently (Ørvig, 1965; Moss, 1968a)? Did aspidin evolve from cellular bone (Ørvig, 1957, 1968)? Or, did

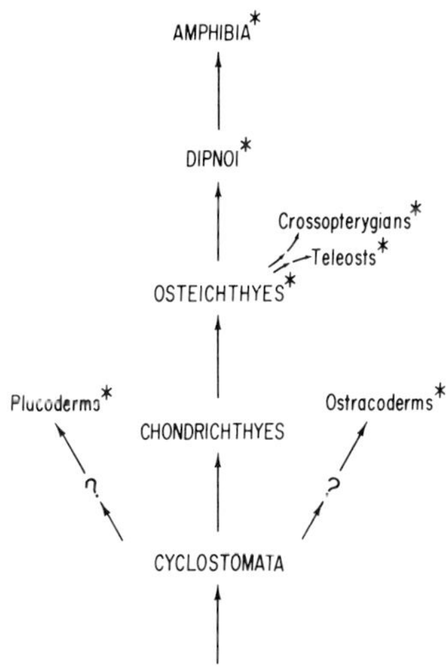

Fig. 2. A phylogeny of the lower vertebrates based on the assumption that cartilage arose before bone and that the Chondrichthyes (cartilaginous fishes) were a primitive group. Groups possessing bone are indicated with an asterisk. Modified from Hall (1975a) with the permission of the publisher.

cellular bone evolve from aspidin (Denison, 1963)? As indicated in Chapter 1, Section II,B, the development of acellular bone from cellular tissues during the ontogeny of the modern teleost provides the necessary clue to understanding the evolutionary problem. For discussions on this point, see Tarlo (1964), Ørvig (1965), Halstead (1969b), Hall (1975a), and Schaeffer (1977).

Similarly, bone and calcified cartilage arose contemporaneously during the Ordovician (Table II) so that the conundrum of whether cartilage arose first during evolution, as it does during ontogeny, has been finally put to rest. However, our inability to identify *un*calcified cartilage in the fossil record still provides an obstacle to settling the time of origin of endochondral ossification. I have discussed this question elsewhere (Hall, 1975a) and have related it to phylogenies of the vertebrates and to modulation of scleroblast activity. Because the embryonic skeleton is initially cartilaginous in many areas, because cartilage is the forerunner of bone in endochondral ossification, and because the Chondrichthyes (cartilaginous fishes) were assumed to be primitive—the logical extension to evolution was that cartilage arose before bone. It became necessary to construct evolutionary trees

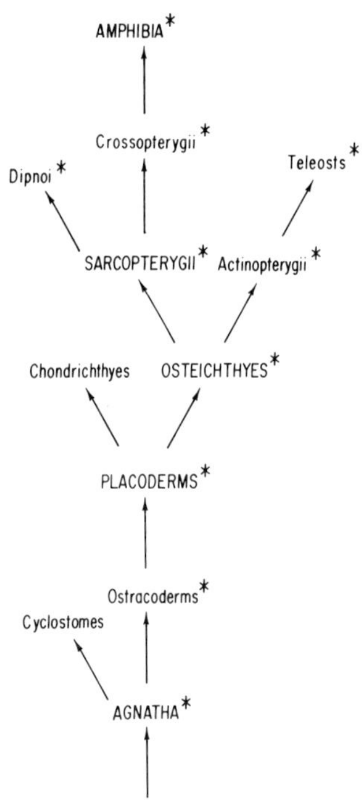

Fig. 3. A phylogeny of the lower vertebrates constructed on the basis of bone as arising early in vertebrate evolution (see Table II). Groups possessing bone are indicated with an asterisk. Reproduced from Hall (1975a) with the permission of the publisher.

such as the one shown in Figure 2. With the recognition that bone and cartilage arose contemporaneously, a quite different phylogeny of the vertebrates could be established (Fig. 3).

This general view of the modulation of scleroblast activity during evolution in response to epigenetic factors (growth rates, muscle attachments, vascularity, habitat, feeding cycles, etc.) is in accord with current notions on the genetic changes necessary for directing evolution. These changes have been shown to be regulatory mutations rather than changes in structural genes [King and Wilson (1975), although see Løvtrup (1974, 1977) for a contrary view]. Basic structural elements are conserved with surprising fidelity. Mathews (1975) has exhaustively summarized the available data on the macromolecular structure of connective and skeletal tissues during evolution, and more recently, Fietzek *et al.* (1977) have provided data on the conservation of the cross-linking region of the collagen molecule during evolution.

3

The Origin of Skeletal Cell Types

I. STEM CELLS

E. N. Willmer of Cambridge University has provided (1970) one of the most thought-provoking analyses of the origin of cell lineages during both development and evolution. The pedigree of stem cell types envisaged by Willmer as leading to the skeletal, connective tissue, and to myogenic and hemopoietic cells is summarized in Figure 4. There is an initial broad subdivision of zygotic blastomeres into ameboblasts (forerunners of the hemopoietic and osteoclast series) and mechanoblasts (forerunners of the skeletal, connective tissue, and myogenic series). The identification of these two classes of cells and of a third, the epitheliocytes, is based on the *in vitro* properties of cells, in particular, of the cells that grow out from explant or organ cultures. These cells differ from one another in their morphology, physiology, and metabolic requirements (Table III), as well as by being stable and by reproducing true to type. They represent three major embryonic cell races found in a wide variety of both vertebrate and invertebrate organs.

Willmer's scheme postulates a dichotomous branching of stem cells, with determination progressively restricted during each step. At each stage during development, the stem cell population undergoes a subdivision into two or more stem cell pools, each of which is able to differentiate into a particular pathway. Within the mechanocyte series, there arise two stem cells, namely, the myxoblast (mucus-secreting), from which the scleroblasts, fibroblasts, and reticulocytes arise, and the myoblast, from which the cardiac, striated, and smooth myoblasts arise (Fig. 4). The myxoblast and the myoblast each have a more restricted potential than did their parent mechanocyte. A note of caution must be added:

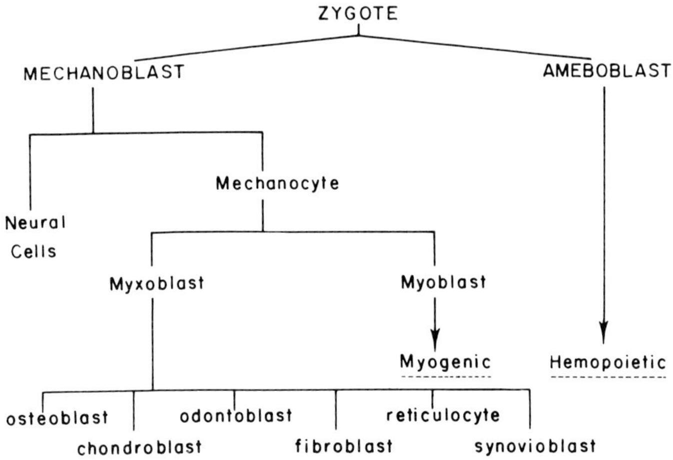

Fig. 4. The origin of skeletal, dental, and connective tissue cell lines by progressive segregation of precursor cells during embryonic development. In addition to ameloblasts, osteo- and chondroclasts arise from the Ameboblast series. Consult text for further details. The terminology of the precursor cells is taken from Willmer (1970).

While the lines joining cell types in the pedigree of Figure 4 are shown as solid, simple, one-way paths, in actual fact, few data are available on the actual sequence or number of steps involved in proceeding from one cell to another within the same stem-cell pool(s). "Cells, by reversing their development, appear to be able, to some extent, to start again on another line" (Willmer, 1970). We know, for example, that the various skeletal cells have more in common with one another than with the myoblasts (Hall, 1970). Unstimulated mechanocytes from bone can form either bone or cartilage *in vitro* (osteoblasts and chondroblasts are modulations of the one mechanocyte), whereas mechanocytes obtained from muscle either cannot form bone or cartilage, or else require stimulation in order to do so (see Section III). Whether this justifies establishing two stem-cell pools (the myxo- and the myoblasts) or whether skeletal, connective, and myogenic cells can form from the one stem cell is unknown (see Chapter 6, Section V,A). It is similarly not known whether all of the embryonic myxoblasts initially possess six potential fates (to form bone, cartilage, dentine, connective tissue, reticular tissue, or synovium), with the particular cell type produced depending on epigenetic factors acting on the myxoblast at particular locations, or whether the myxoblasts represent six pools of stem cells, each possessing one of the above fates, or three stem cells, each with two possible fates, etc. As well, this may vary from location to location and from time to time during development.

Conrad *et al.* (1977) have shown that fibroblasts from embryonic chick cornea, heart, and skin differ, both morphologically and in responses to enzymes

TABLE III

A Summary of the Features Distinguishing Epitheliocytes, Mechanocytes, and Amebocytes from One Another[a]

	Epitheliocytes	Mechanocytes	Amebocytes
1.	Grow as a sheet	Grow as a net	Isolated
2.	Close cell-to-cell contact	Contact inhibition	No contact
3.	±[b] Desmosomes	± Desmosomes	No desmosomes
4.	Cells move as a sheet	Cells move with polarity	Ameboid movement
5.	Grow at surfaces	Some ability to penetrate the substrate	Penetrate substrate
6.	± Phagocytic	Not phagocytic	Phagocytic
7.	Do not require embryo extract to divide	Require embryo extract	Fatty degeneration in presence of embryo extract
8.	—	Do not support Rous chick sarcoma virus	Support Rous chick sarcoma virus
9.	Rough and smooth E.R.[c]	Rough E.R.	Smooth E.R.
10.	Intra- and extracellular mucosubstances	Extracellular muco-substances	Intracellular muco-substances
11.	Produce keratin	Produce collagen	Produce albumin and fibrinogen
12.	± Acid phosphatase	No acid phosphatase	Acid phosphatase in lysosomes
13.	± Alkaline phosphatase	Alkaline phosphatase	No alkaline phos-phatase
14.	—	Cause epithelial differentiation	Do not cause epithelial differentiation

[a] Adapted from Willmer (1970).
[b] ±, presence or absence of.
[c] E.R., endoplasmic reticulum.

or to chelating agents. Some alternatives for chondrogenic precursors within the avian limb bud are discussed in Chapter 6, Section VI,B.

A classical study in skeletal biology is the pioneering work of Fell (1932, also see review in 1969) on the nature of the cell populations within the periosteum of the long bones of the embryonic chick. The periostea are bilayered, consisting of an outer fibrous and an inner cambial layer. When intact periostea were stripped from the tibiae of 6-day-old embryonic chicks and organ-cultured on embryo extract–plasma clots for ten days, bone formation occurred. However, only fibrous tissue formed when the outer fibrous layer of the periosteum was cultured alone. Fell concluded that the outer fibrous and inner cambial (cellular) layers of the periosteum consisted of two separate populations of cells, each having different potentials. She considered the fibrous layer to contain fibroblast progenitor cells (myxoblasts), which were not able to form bone, while the inner layer consisted of osteoblast progenitor cells, which were able to form bone.

At other skeletal sites, the progenitor cells are not so restricted. Indeed, several lines of evidence support the concept of progenitor cells as having dual potential. The fact that osteoblasts and chondroblasts are modulations of a single mechanocyte has already been mentioned and is discussed in depth by Hall (1970, 1972b) and in Chapter 6. Willmer (1970) has presented some of the evidence indicating that fibroblasts, chondroblasts, osteoblasts, synovioblasts, and odontoblasts represent "a little subfamily of cells." The ability of scleroblasts to modulate their activity and to undergo metaplasia (Haines and Mohiuddin, 1968; Hall, 1970, 1972b), as seen in the existence of tissues intermediate between various mineralized tissues (Chapter 1), indicates that separate classes of myxoblasts are an unlikely possibility and that progenitor cells derived from them are likely to possess at least *two* potential fates or determinations. Examples would be progenitor cells capable of producing cartilage or bone; bone or fibrous tissue; fibroblasts or reticulocytes. (Such modulations of stem cell behavior that occur *after* the stem cell line has been established will be discussed in Chapter 6.) The fact that differentiated products of myxoblasts, e.g., fibroblasts, can show a wide range of synthetic activities lends further support to the possibility of alternative pathways being open to stem cells.

II. MODULATION OF SYNTHETIC ACTIVITY

A. Fibroblast Transformations

There is a sizeable literature on the modulation of cell activity between chondroblasts and fibroblasts. With cultured chondrocytes, this transformation has often been regarded as a dedifferentiation, although now there is considerable evidence indicating that the synthetic activity of the cells changes from typically chondrocyte to that typical of fibroblasts (Mayne *et al.*, 1973, von der Mark *et al.*, 1977a). If bovine articular chondrocytes are treated either with liver lysosome or lysosomal enzyme, or if they are subjected to prolonged culture, they acquire a fibroblastic morphology and switch to the synthesis of collagen of type $[\alpha 1(I)]2\alpha 2$ (Deshmukh, 1974). Under diseased conditions, chondrocytes modulate to fibroblasts and modify their patterns of protein synthesis, e.g., in chondromalacia of the human patella (Zimny and Redler, 1969, 1972). Perhaps even more remarkable is the transformation, under the influence of cAMP and/or testosterone, of Chinese hamster ovarian cells to fibroblasts with concomitant initiation of collagen synthesis (Hsie *et al.*, 1971).

B. Synthetic Activity

Many studies have shown synthesis, secretion, or degradation of both glycosaminoglycans and collagen to be altered when cells are perturbed *in vivo* or *in vitro* (Hsie *et al.*, 1971; Goggins *et al.*, 1972; Kulonen and Pikkarairen, 1973). Among others, Matalon and Dorfman (1966), Green and Goldberg (1968), Conrad (1970), and Levene and Bates (1970) have shown that populations of fibroblasts from a variety of sources synthesize both collagen and either hyaluronic acid or chondroitin sulfate when maintained *in vitro*. Matalon and Dorfman's study on fibroblasts from patients with Hurler's syndrome heralded the subsequent use of the biochemical analysis of cell lines in the assay of synthetic shifts in metabolic bone diseases. In fact, most cell types synthesize glycosaminoglycans when maintained *in vitro* (Daniel *et al.*, 1973), and quite dissimilar fibroblasts (e.g., those from skin and gingiva) and chondrocytes produce similar glycosaminoglycans *in vitro* (Goggins and Billups, 1972; Mayne *et al.*, 1973). Clonal cultures of cells from epithelia, melanoma, liver, HeLa cells, pituitary, kidney, and neuroblastoma all produce collagen *in vitro* (Green and Goldberg, 1968; Langness and Udenfriend, 1974).

The same fibroblast can also vary its synthesis depending on the culture conditions. While fibroblasts from the embryonic chick amnion incorporate glucosamine into hyaluronic acid when cultured at low cell densities, they incorporate it one-to-one into hyaluronic acid and chondroitin sulfate at high cell densities (Mayne *et al.*, 1971).

These studies illustrate the effect of *in vitro* perturbations on *populations* of cells, and demonstrate the existence of similar synthetic activity in apparently dissimilar cell types. To determine whether *single* cells are capable of synthesizing more than one extracellular matrix product, and if so, whether it takes place simultaneously or sequentially, is a more difficult problem. More knowledge on this question is required before searches for bipotential precursor cells are undertaken.

Perhaps the most compelling evidence for synthetic ability in single cells is that provided by Gay *et al.* (1976a) and by Müller and Kühn (1977). These authors took advantage of the distinctive primary structure of collagen of types I and II and developed an immunofluorescent technique, a method which enabled them to visualize tagged antibodies against either of the collagen types. With this technique, type I and type II collagen were both demonstrated to be present in human dermal fibroblasts. Type I collagen is normally regarded as a product of fibroblasts, bone, ligament, cornea, and "dedifferentiated" chondrocytes;

type II is thought to be a product of cartilage and notochord. These differentiated fibroblasts were shown to be synthesizing both collagen types.

C. Degradative Activity

Also pertinent are the studies of Richard Ten Cate and his colleagues in Toronto, who have used electron microscopy to show that a single fibroblast can synthesize and degrade collagen simultaneously. This conclusion is based on the identification of intracellular, membrane-bound, banded collagen fibrils within the fibroblasts of mouse and guinea pig periodontal ligament (Ten Cate, 1972) and in fibroblasts of healing skin wounds in the mouse (Ten Cate and Freeman, 1974) at the same time as these cells are synthesizing collagen. These cells contain alkaline phosphatase, with acid phosphatase also present within the membrane-bound fibrils (Deporter and Ten Cate, 1973; Ten Cate and Syrbu, 1974), a finding which indicates degradative activity. Ten Cate and Deporter (1975) and Ten Cate *et al*. (1976) have reviewed the available data.

These data on synthetic and degradative activity of fibroblasts are certainly consistent with a search for precursor cells with more than one potential. So, too, are the recent findings on the presence of myosin within nonmyogenic cells.

D. Distribution of Actin and Myosin

Until quite recently, it was thought that actin and myosin were unique biochemical and morphological markers for myoblasts and muscle. The production of these "luxury molecules" was diagnostic for contractile function—a function that was seen as the prerogative of cells within the myogenic cell compartment.

Reports of identification of actin and/or myosin in nonmyogenic cells have begun to appear in the literature, accompanied by a controversy over methodology and interpretation. The controversy centers on whether actin and myosin were really present, and if they were, whether they functioned in the same way as actin and myosin functioned in myoblasts. If they were serving a contractile function, was contractility a more general property of cells than previously had been thought? Did actin and myosin play a role in the intracellular transport of materials or in such major cellular events as cytokinesis? And what of the notions of early separation of myogenic from other cell lines during development?

As an indication of the diversity of cell types shown to possess actin and/or myosin, these molecules have been shown to be present in, among others, platelets, leukocytes, and fibroblasts (Adelstein and Conti, 1971; Adelstein *et al*., 1971, 1972; Shibata *et al*., 1972); in the myofibroblasts that surround implants (Ryan *et al*., 1973); in the brush border of intestinal cells (Tilney and

Mooseker, 1971), and in many tissues of the early embryonic chick, including the primitive streak; and in several sources of skeletal mesenchyme, viz., the somites, the lateral plate mesoderm, and the wing buds (Orkin *et al.*, 1973).

The myofibroblasts that form the capsule surrounding implanted blood clots are especially interesting, for these are cells having characteristics of both fibroblasts and smooth muscle cells (Ryan *et al.,* 1973). This identification is based on both ultrastructural and functional evidence. Like fibroblasts, the cells contain intracellular accumulations of collagen and bundles of microfilaments, but they bind anti-smooth muscle serum and respond to pharmacological stimulation like smooth muscle cells. These cells provide an excellent example of the ability of differentiated cells to express functions normally thought to be properties of separate populations of cells, reaffirming the need for caution when identifying a cell type solely on the basis of the performance of one differentiated function.

Two reports of myosin and/or actin in skeletal and dental tissues have come to my attention. Histochemical visualization of the A-band was used to demonstrate the presence of myosin in fibroblasts of the dental pulp and in the processes of odontoblasts (Leonard and Sharawy, 1974). Mouse calvaria, when cultured for three days in the presence of parathyroid extract (1 IU/ml), show binding of actin-like filaments to heavy meromyosin. These 50–70 Å-diameter filaments are found beneath the plasma membrane and in the cell processes of osteoblasts, osteocytes, and osteoclasts (King and Holtrop, 1974, 1975). The structural genes specifying actin may differ from tissue to tissue (Storti and Rich, 1976), so that specificity may be present at the genomic level. Nevertheless, the ubiquitous distribution of actin and myosin raises questions of the possible contractility of nonmyogenic cells, and of the role this might play in both developmental and physiological processes.

III. PRECURSOR CELLS

A. Fibroblast Colony-Forming Cells

Aside from Willmer, Alexander Friedenstein and his colleagues at the Gamelaya Institute, Academy of Medical Sciences, Moscow, have contributed extensively to the resolution of the origin and diversity of mechanocyte precursors. Friedenstein and Lalykina's 1973 monograph provides the most detailed analysis of the work of this group; however, this monograph is in Russian. Friedenstein (1973, 1976) provides the most complete review in English. All of their work has been conducted on precursors of mechanocytes in adult mammals, whereas Willmer has concentrated on the origin of the stem cells during embryonic development. One particular class of mechanocytes, viz., the stromal reticular cells of hemopoietic and lymphoid organs, has been very closely examined. The approach utilizes the technique of selective cloning under monolayer conditions to produce clones of fibroblasts from large, mononuclear stem cells, i.e., fibroblast colony-forming cells (FCFC). Cell cultures from bone marrow, spleen, thymus, and lymph nodes under these conditions yield fibroblast colony-forming cells in quantities specific to each tissue (Friedenstein *et al.*, 1976). While all of these cells form colonies of fibroblasts having similar properties, the fibroblast colony-forming cells have other properties that vary depending on their tissue of origin. For example, the fibroblast colony-forming cells from bone marrow can undergo osteogenesis, while those from spleen cannot. Such differing potentials (determinations?) of apparently similar mechanocytes form the basis for the thesis that mechanocytes or their precursors differ in different hemopoietic organs.

Friedenstein's group has also produced evidence that mechanocytes or their precursors circulate within the blood vessels of the adult. Fibroblast colony-forming cells can be isolated from blood, albeit in very low concentrations (e.g., one fibroblast colony-forming cell/100,000 leukocytes in guinea pig blood; Friedenstein, 1976). This finding has considerable relevance to the problem of ectopic osteogenesis (see Chapter 4, Section VI).

B. Osteogenic Precursor Cells

The studies carried out over the past two decades by Friedenstein and his colleagues on the origin of *osteogenic* precursor cells in the postnatal mammal will serve as an excellent example of how to search for races of cells and for determined precursor stem cells; accordingly, these experiments will be discussed in some depth. Three basic approaches have been taken.

The first approach was to place transitional epithelium (a known inducer of ectopic osteogenesis) within a diffusion chamber, to implant the chamber *in vivo*, and to show that normally nonosteogenic cells of the host could be induced to undergo ectopic osteogenesis (Friedenstein, 1962; Petrakova and Friedenstein, 1965; Friedenstein *et al.*, 1967; Abdin and Friedenstein, 1972). Lalykina and Friedenstein (1969) and Friedenstein and Lalykina (1970, 1972) then showed that lymphocytes from thymus, spleen, peritoneal fluid, and blood, but not from lymphatic nodes, could respond to transitional epithelium *within* diffusion chambers by producing ectopic bone. These organs, then, are considered to possess a population(s) of cells capable of responding to transitional epithelium by producing bone. As shown above, they also possess populations of fibroblast colony-forming cells. Are these two populations one and the same? Yes; see below.

Their second approach was to transplant bone marrow carrying a chromosomal marker under the kidney capsule of hosts with a different chromosomal marker. Under such transplantation conditions, osteogenesis was evoked within the population of marrow cells, so identified on the basis of their distinctive chromosomal marker. Therefore, existing within bone marrow is a line of cells that can be induced to form bone ectopically. Do these arise from the stem cells of the hematopoietic tissue or are they a separate stem cell line? Reverse transplantation of repopulated marrow back to donor bone marrow indicated that the ectopic bone was not resorbed by donor cells, i.e., it was immunologically the same as the donor tissues and therefore derived from donor osteogenic cells in the marrow. Host marrow had replaced the graft cell population and *was* resorbed upon transplantation back to the original donor. They concluded that a line of osteogenic stem cells existed in marrow, a cell line which was separate from the stem cell line producing the hematopoietic tissue. This was confirmed by irradiation experiments (Friedenstein and Kuralesova, 1971; Kuralesova, 1971) in which marrow was transplanted to lethally irradiated individuals in whom osteogenic precursors are maintained but hematopoietic precursors destroyed. In such transplants, ectopic bone forms, but hemopoietic tissue does not (confirmed by Amsel *et al.*, 1969; Amsel and Dell, 1971, 1972; Elves and Pratt, 1975). Storage at 0°C for 1–2 weeks inactivates the potential for ectopic osteogenesis (Bierley *et al.*, 1975), but does not destroy the population of determined precursor cells (Nelson, 1975).

Their third approach was to reimplant *in vivo* the cultures of embryonic spleen, liver, and marrow established by clonal culture of fibroblast colony-forming cells. When bone marrow is placed in diffusion chambers, it routinely forms reticular tissue unless the cell density is very high, in which case bone forms (Friedenstein *et al.*, 1966). The progenitor cells from marrow have been cloned by Friedenstein *et al.* (1970). When fibroblasts obtained from these monolayer cultures are implanted into diffusion chambers, they form bone routinely. Although they form only reticular tissue under similar implantation conditions, fibroblasts obtained from the monolayered cloning of spleen cells will form bone if placed in contact with transitional epithelium of the urinary bladder (Friedenstein *et al.*, 1968, 1970). The fact that the fibroblast colony-forming cells formed bone ectopically was confirmed when these cells were eliminated from suspensions of spleen or thymic cells (by taking advantage of their stronger adhesion to the substrate) and as a result, the ability of the population to undergo osteogenesis was also eliminated. This, and the formation of bone from clonally derived cells (Friedenstein, 1973; Friedenstein and Lalykina, 1973), provide strong evidence for the dual potential of these stem cells.

The osteogenic cell population within bone marrow *can* undergo osteogenesis in the absence of an inductive stimulus. Ectopically produced bone becomes populated with bone marrow. The progenitor cells from the marrow of the ectopic bone *do* require induction to produce their own bone. Even though this difference in osteogenic potential exists, the fibroblast colony-forming abilities of the two populations are similar (Friedenstein, 1976). A greater *amount* of bone is produced by cells obtained from the marrow of nonectopic bone if an inducer is present, although *initiation* of osteogenesis is not dependent on the inducer.

Therefore, either normal bone marrow contains some inducible cells, or the activity of the determined osteoprogenitor cells is increased by an inducer. Within the spleen, the fibroblast colony-forming cells cannot initiate osteogenesis without an inductive stimulus.

The general conclusions from all of these studies as they apply to osteogenesis is that the adult mammal possesses two categories of osteogenic precursor cells, namely, a population of determined osteogenic precursor cells within bone and bone marrow, and a population of precursor cells requiring induction to form bone (inducible osteogenic precursor cells) located within the blood, spleen, thymus, and peritoneal fluids. The studies of Owen (1970, 1971, and see Chapter 6) support this conclusion.

Bone formation can be induced by transitional epithelium at sites locally irradiated with 2000–5000 rads confirming that the inducible osteogenic precursor cells can migrate (Friedenstein, 1973), although they need not (Hall, 1975c).

The determined osteogenic precursor cells of marrow have been shown to be separate from the hematopoietic precursor cells of the bone marrow. There is evidence in the literature (Hall, 1970, 1975c) that lymphocytes, monocytes, and macrophages (descendants of the ameboblast stem cells) may also form bone. In the light of Willmer's and Friedenstein's studies, these cells must be regarded as a separate line from the determined cells of the marrow and from the stromal mechanocytes, which also undergo osteogenesis. The approaches taken by these authors have been the most profitable so far in the identification and localization of determined stem cell races within the skeletal and connective tissues. Some information on the *number* of stem cells within the embryo may be found in Chapter 9, Section II,A.

IV. SPECIFICITY OF EMBRYONIC INDUCTION

Another line of evidence for the separation of mechanocytes into separate races is the differential effect that mesenchyme of different origins has on epithelial differentiation. In order for its growth to continue, it is essential that epithelium interact with mesenchyme. The differentiation displayed by the epithelium depends upon the type of mesenchyme with which it comes into contact. A classical study is that of McLoughlin (1961), who combined epidermis from the hindlimb bud of 5-day-old chick embryos with isolated mesenchyme from various sites, and then allowed the mesenchymal–epithelial combinations to differentiate *in vitro*.

The epithelium of the limb bud keratinizes when cultured alone, when in contact with limb bud mesenchyme, and when it is in contact with fibroblasts obtained from embryonic heart. Limb bud epidermis in contact with mesenchyme from the embryonic gizzard does not keratinize; rather, it becomes and remains mucus secreting and ciliated. In contact with mesenchyme of the embryonic proventriculus, mucus secretion is initiated but not maintained, and after seven days *in vitro* the limb bud epidermis reverts to keratinization. Myoblasts from the embryonic heart cultured in contact with the limb bud epidermis elicit neither keratinization nor mucous secretion, but rather, direct the differentiation of a squamous epithelium. The behavior of mesenchymal cells can be altered by cultivation on basement lamellae isolated from epithelia (Overton, 1977). Presumptive evidence for similar interactions in limb morphogenesis will be considered in Chapter 4, Section III,A,8.

The mechanocytes and/or their extracellular products from these various mesenchymes are quite obviously separate races with respect to their ability to direct differentiation of limb bud epithelium along several different and quite separate pathways. Of course, these mesenchymes themselves were already quite highly differentiated at the ages used in this experiment. The reciprocal of this interaction, i.e., the effect of the epithelium on the mesenchyme both during and after the stage at which the mesenchymal cells become determined, provides an essential clue to the mechanism that controls sites within the embryo where various skeletal elements will form.

4

Location of the Skeleton within the Embryo

I. INTRODUCTION

Sufficient evidence has now accumulated to substantiate the thesis that the location of skeletal elements within the embryo is determined by interactions between the skeleton-forming mesenchyme and the epithelia with which that mesenchyme interacts. The available evidence will be considered by discussing the localization of several major skeletal elements. These elements fall into two broad embryological categories, based on the two general origins of skeletal tissue: either from mesenchyme derived from the presumptive embryonic mesoderm or from mesenchyme derived from the embryonic neural (ganglionic) crest. The latter mesenchyme is known as ectomesenchyme, in order to signify that its origin is ectodermal.

II. THE SKELETON OF THE NEURAL CREST

A. Introduction

The neural crest represents the ridge of cells seen at the junction of the presumptive epidermal and neural ectoderm during early neurulation. As the neural plate transforms into neural folds, the cells of the neural crest, now in the apices of the neural folds, begin to leave the crest and to migrate away from it. This remarkable behavior is a prelude to the migration of cells from the neural crest to many areas of the body where they differentiate into a diversity of cell and tissue types. Spinal ganglia, the sympathetic nervous system, the adrenal medulla (chromaffin cells); the dental papillae, odontoblasts, and the dentine of teeth; melanocytes, connective tissue cells, craniofacial skeletal elements, and antlers all arise from descendants of these wandering cells.

I shall present first the evidence which indicates that ectomesenchyme from the neural crest forms skeletal and connective tissues. Notable among these are the tissues of the hyoid and mandibular arches (the visceral skeleton), the cranium, and the anterior trabecular cartilages.

B. Skeletogenic Capacity of the Neural Crest

In light of the renewed interest in the neural crest, especially the possibility that many craniofacial and dental anomalies and malformations may originate in the defective migration, differentiation, or growth of neural crest cells (Fitch, 1957; Ossenberg, 1974; Bergsma, 1975; Johnston, 1975, 1977; Poswillo, 1975, 1976; Morriss and Thorogood, 1978), the neural crest deserves a book to itself. Unfortunately, it seldoms receives more than a passing reference, even in texts on the skeleton, and will receive less space here than it deserves. The available literature up to about 1948, most of which was obtained from experimentation on amphibians, has been summarized and reviewed in depth by Hörstadius (1950) and to a lesser extent by both de Beer (1958) and Holtfreter (1968); thus, I shall concentrate on more recent selected research efforts.

The evidence that neural crest derivatives produce cartilage, bone, and dentine is derived from: (1) selective ablation of neural crest zones in the neurula or early embryo (so-called deficiency experiments), or explantation to other sites of either neural crest, other tissues in place of the neural crest, or exchange of neural crests between different species; (2) use of nuclear, cellular, or isotopic markers in conjunction with (1); and (3) *in vitro* cultivation of neural crest cells.

1. Ablation and Transplantation Experiments

Hörstadius (1950) has reviewed the early experiments and has summarized the use of nuclear and cellular markers. Of these studies, the experiments of Ross Harrison are especially elegant. They took advantage of the differential growth rates of *Ambystoma punctatum* and *A. tigrinum* larval organs, and exchanged branchial neural crests between neurulae of the two species. Oversized or undersized visceral skeletons confirmed their origin from the grafted ectomesenchyme [see Hörstadius (1950) for a discussion].

Hörstadius devoted only one paragraph to the question of the origin of membrane bones from the neural crest. Even as recently as 1970, Weston was not able to marshall much evidence for the neural crest origin of cranial bones. However, in more recent transplantation experiments with amphibians, it has been confirmed that cartilage, bone, and dentine are *all* derived from the ectomesenchyme of the neural crest (Cassin and Capuron, 1972, 1977; Cassin, 1977), a conclusion echoed by studies with avian embryos (Section 2).

Some caution must be exercised in interpreting the deletion and transplantation experiments, because cells from the neural crest of one side migrate across the midline to the other side (especially in the trunk). As a result, unilateral experiments may not yield true results (Hörstadius, 1950). Furthermore, cells from the neural tube or from the lateral mesoderm may, in the absence of the neural crest cells, regulate their own cellular differentiation and form tissues normally produced only by neural crest cells (Weston, 1970).

The neural crest of the embryonic chick was first described by His (1868, cited by Hörstadius, 1950). Yntema and Hammond (1954) showed that visceral ganglia were of neural crest origin. Hammond and Yntema (1953, 1964), who performed deletion experiments on embryos having 6 to 12 pairs of somites, showed that removal of the postorbital neural crest resulted in the failure of the hyoid skeleton to form, whereas deletion of preotic neural crest resulted in deficiencies in Meckel's cartilage of the mandibular arch. Their embryos were examined too prematurely to detect whether deficiencies in cranial bones had occurred.

2. Marker Experiments

M. C. Johnston (1964, 1966, 1975) and Johnston and Listgarten (1972) pioneered the use of isotopically-labeled transplants to follow the fate of the neural crest cells. Neural crests from embryos incubated for 30 hours were replaced with similar-sized pieces previously labeled with tritiated thymidine. The cartilage and connective tissue that developed in the head and visceral arches were subsequently found to be radioactively labeled.

Weston (1963, 1970) and Weston and Butler (1966) have used autoradiographic analysis after isotope labeling to follow the pathways of migrating neural crest cells, and to detect whether determination occurs either before the cells leave the crest, while they are migrating, or after they have reached their final site. Deficiencies did not occur when neural crests from older embryos were transplanted into younger embryos, indicating that final determination occurs after migration has begun. Thymidine labeling has also been used to identify the unique migration patterns of cells from each area of the avian cranial neural crest, and to demonstrate that the pathways are not fixed before migration starts—the pathways of migration are normal even if the regions are exchanged (Noden, 1975). As we shall see below (Section 3), interaction with epithelia during and after migration is of prime importance for the subsequent differentiation of cells from the neural crest and for the continued growth of the epithelia.

The technique of transplanting neural crests with a nuclear marker has been used most elegantly and profitably by Nicole LeDouarin and her colleagues in their studies on the exchange of neural crests. The systems used were embryos of the Japanese quail (*Coturnix coturnix japonica*), whose cells, which contain nuclei with large accumulations of heterochromatin, were exchanged with cells from embryos of the common fowl (*Gallus domesticus*), whose cells contain nuclei with uniformly distributed heterochromatin. The technique is reviewed in detail by LeDouarin (1973, 1974), and has been used to show that, among others, the following cell types are of neural crest origin: adrenal medulla (LeDouarin and Teillet, 1973a); parasympathetic nervous system and ganglia (LeDouarin and Teillet, 1973b, 1974); calcitonin-producing cells of the ultimobranchial body (LeDouarin *et al.*, 1974); and craniofacial and visceral skeletal and connective tissues (LeDouarin, 1974, 1975; Le Lièvre, 1974; Le Lièvre and LeDouarin, 1975). (Detailed maps of the origin of the various skeletal elements can be examined in the two papers by Le Lièvre.) The ability of the neural crest to form cranial skeletal elements extends down to the level of the fifth pair of somites. Below that level, cartilage of neural crest origin does not form. Within the avian head, only the occipital bones and the bones of the otic capsule are of entirely mesodermal origin, although some bones, such as the frontal, are formed from cells of both mesodermal and ectomesenchymal origin.

Recently, Pictet *et al.* (1976) described a technique for removing the neural crest from mammalian (mouse) embryos. This method necessitated the removal of the entire ectoderm, and organ culture of the remaining cell layers. Although primarily used to show that insulin-producing cells are of neural crest origin, this method also might be used profitably for the study of skeletal elements and their origins.

3. *In Vitro* Culture

The most direct way of determining whether, or at what age and under what conditions, neural crest cells can form skeletal or connective tissues is to culture the neural crest, either in isolation from, or in combination with, other tissues. In practice, it is difficult to culture ectomesenchyme once the cells have left the neural tube (they rapidly spread among the mesenchyme), although Cohen and Konigsberg (1975) have devised a culture technique that allows cells of the neural crest to grow out from the neural tube *in vitro* under conditions in which clonal cultures can then be set up.

The nature of the extracellular environment into which cells from the neural crest migrate *in vivo* may have an important influence on their subsequent differentiation. In *Pleurodeles waltlii* (a urodele amphibian), this matrix initially consists of collagen and hyaluronic acid. The latter is removed by hyaluronidase, and chondroitin 4- and 6-sulfates then begin to accumulate (Corsin, 1974, 1977). In the embryonic chick, the cranial neural crest cells migrate into a cell-free, hyaluronic acid-rich space between the epidermal ectoderm and mesoderm (Pratt *et al.*, 1975; Fisher and Solursh, 1977). Some of this material may be synthesized by the neural crest cells themselves, because such cells from both chick and

Japanese quail synthesize hyaluronic acid and chondroitin sulfate when cultured alone (Greenberg and Pratt, 1977; Manasek and Cohen, 1977). The migratory behavior of neural crest cells does appear to depend on the nature of this extracellular material, because the migratory behavior of isolated neural crests can be modified by varying the extracellular material of the substrate on which they are cultured (Maxwell, 1976). When whole embryos are cultured in the presence of exogenous vitamin A, the migration of neural crest cells into the first visceral arch is inhibited, apparently because of the interference with either the extracelluar matrix or the ability of the neural crest cells to interact with that matrix (Hassell *et al.*, 1977). Further data on the significance of the extracellular matrix as a modifier of cell behavior are discussed in Chapter 5, Section IV.

4. Ectomesenchymal–Epithelial Interactions

Given that the nature of the environment encountered during cell migration is both diverse and changing over time, one can ask whether cranial neural crest cells are competent for differentiation into skeletal tissues prior to neurulation (when they are still in the neural plate or neural folds), or whether these cells acquire such competence by association with mesoderm, epithelia, or extracellular matrix products during or after their migration. The association of ectodermal or endodermal thickening with the presence of adjacent ectomesenchyme cells has been commented upon several times in the past (Jacobson and Fell, 1941; de Beer, 1947; Hörstadius, 1950; Kingsbury *et al.*, 1953; Holtfreter, 1968; Tonegawa, 1973) and is suggestive of tissue–tissue interactions. Coculture of neural crest cells and pharyngeal endoderm provides the definitive experimental evidence that, in the amphibians, competence to form visceral cartilage is acquired as the ectomesenchyme interacts with the pharyngeal endoderm.

Drews *et al.* (1972) cultured neural crest alone and found that it did not chondrify. However, 60% of the neural crests formed cartilage when cultured in the presence of endoderm from the developing pharynx. Epperlein (1974) and Epperlein and Lehmann (1975) confirmed that *contact* with pharyngeal endoderm was necessary, that neural crest cells did not preferentially migrate toward the endoderm, and that mesenchyme, as well as cartilage, was produced from such neural crests. In another urodele, *P. waltlii*, contact with dorsal mesoderm as well as with pharyngeal endoderm will allow the neural crest to chondrify *in vitro* (Corsin, 1975a). Corsin also tested (1975b) the effect of both hyaluronic acid (30 μg/ml) and testicular hyaluronidase (2 μg/ml) on isolated neural crests to see whether these extracellular matrix products would initiate chondrogenesis, but they did not.

As the cells from the neural crest pass the embryonic pharynx, contact with endoderm provides an interaction necessary for these cells to be subsequently capable of forming visceral cartilage. In addition to this interaction, which confirms chondrogenic potential, there is an earlier restriction in determination within regions of the neural crest itself. From the transplantation studies of Le Lièvre (Section 2), we saw that, within the cranial neural crest, one could distinguish regions competent for the formation of particular skeletal elements. There is an even greater restriction of competence between cranial and trunk neural crest, because trunk neural crest does not possess the ability to form skeletal elements at all. This notion goes back to the extirpation experiments of Detwiler and van Dyke (1934), who showed that in *Ambystoma* vertebral neural arches formed in the absence of trunk neural crest. Hörstadius (1950) has discussed the evidence indicating that within the cranial neural crest of *Ambystoma* there is a qualitative difference between the ectomesenchyme of the mandibular arch and that of the gill arches. On the basis of an extensive series of rotation experiments, he distinguished eight regions of the neural crest. These regions were (from anterior to posterior): i,ii, The most anterior neural crest, which could not form cartilage; iii, a region that could form trabecular cartilage but not visceral arch cartilage; iv, a region that could form mandibular arch skeleton but not trabecular or visceral skeleton; v–vii, regions that formed hyoid and gill skeleton; and viii, trunk neural crest, which could not form skeletal tissues. Figure 5 shows the general location of these zones. Le Lièvre (1974) and Le Lièvre and LeDouarin (1975) present evidence for similar regionalization within the avian neural crest.

Thus, although contact with the endoderm of the developing pharynx is necessary before cartilage will differentiate, some aspects of the determination to form

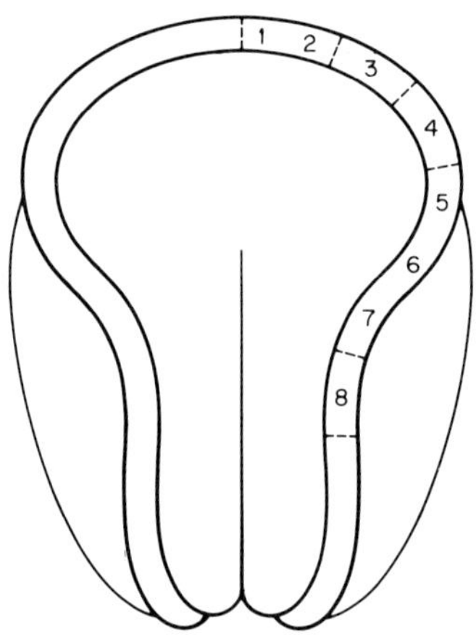

Fig. 5. A dorsal view of the neurula of a urodele amphibian to show the eight regions of the neural crest. Regions 1 to 7 are cranial neural crest. Region 8 is the trunk neural crest. Cartilage does not form from regions 1 or 2. Trabecular cartilage forms from region 3. The tissues of the mandibular arch form from neural crest in region 4. Regions 5 to 7 form the hyoid and branchial skeletons. No skeletal tissues form from the trunk neural crest (region 8). Modified from Hörstadius (1950).

cartilage (and we have no idea what these aspects are) are laid down within the neural tube.

The action of the pharyngeal endoderm is specific. It will allow chondrogenesis in neither somitic nor otic mesoderm.

The interaction of ectomesenchyme with epithelia provides a stimulus for the continued mitotic activity of that epithelium (Hörstadius, 1950; Holtfreter, 1968; Johnston, 1975), perhaps performing the function in the head that mesenchyme performs in the trunk (McLoughlin, 1961). The ectomesenchyme also induces the epidermal ectoderm of amphibian embryos to initiate the development of the balancers, gills, teeth, and the dorsal fin (Hörstadius, 1950) and may play a permissive role in the differentiation of the otic and auditory capsules, even though these tissues are of mesodermal origin.

There is evidence to suggest that the pattern of tooth replacement in the rainbow trout (*Salmo gairdneri*) is related to the order of migration of neural crest cells and to the local competence of the oral epithelium (Berkovitz and Moore, 1974), offering a possible morphogenetic role for the neural crest–epithelium interaction. Whether the pattern of skeletal elements in the head has a similar basis warrants some study (see Section E,2).

In summary, the concept is well founded that much of the cranial and visceral skeletal and connective tissues of amphibians and birds is of neural crest origin. Interaction between the ectomesenchyme of the neural crest and the pharyngeal endoderm is a necessary step in the chondrogenesis of amphibian cranial neural crest cells and is one example of a specific epithelial–ectomesenchymal interaction. Pharyngeal endoderm will allow ectomesenchyme, but not somitic or otic mesenchyme, to form cartilage.

Can we find in the interaction between ectomesenchyme or mesenchyme and epithelia a mechanism for the localization of specific skeletal elements? I believe that we can, and now shall proceed to look at several skeletal elements, both for purposes of introduction, for they will reappear in other contexts, and to discuss the reasons for their formation at various sites in the embryo or, in some cases, in the adult. As examples of skeletal elements derived from ectomesenchyme, Meckel's cartilage and the investing membrane bones of the mandible and the face will be discussed, as well as the scleral cartilage, the dermal bones of the cranium, the scleral bones, antlers, and teeth. I shall discuss the skeleton of the limbs, the vertebral skeleton, and the auditory (otic) cartilages as examples of skeletal elements formed from mesenchyme. Ectopic cartilage and bone formation outside the boundaries of the skeleton also will be discussed.

C. The Mandibular Skeleton

The classical study on the development of the mandibular skeleton is the organ culture investigation of the mandibular arch of the embryonic chick performed by

Jacobson and Fell (1941). They described three centers within the mandible: one osteogenic, one chondrogenic (these two they thought arose within the mandible), and one myogenic. Each was associated with a transitory thickening of the buccal epithelium, and each, when maintained in isolation *in vitro*, formed only one differentiated end product—either bone, cartilage, or muscle.

The mandibular skeleton, along with three membrane bones of the skull (the quadratojugal, squamosal, and pterygoid), is derived from the neural crest at the level of the mesencephalon. Neural crest cells first leave the mesencephalon when the embryo has developed to the 5-somite stage (H.H. stage 8.5).* The last cells leave at H.H. stage 10, at which point the embryo has ten pairs of somites. Neural crest cells first arrive in the mandibular arch at H.H. stage 15 (Le Lièvre and LeDouarin, 1975; C. Le Lièvre, personal communication). The muscle of the arch is of local mesodermal origin.

Initially, the mandibular skeleton of the avian embryo consists of Meckel's cartilage, a central rod of cartilage that begins to chondrify at 5 days of incubation (H.H. stage 26). In the retroarticular process of Meckel's cartilage, an endochondral bone, the articular, begins to develop at 14 days of incubation (H.H. stage 40). The remainder of the Meckel's cartilage persists and is invested with 6 membrane bones, whose ossification commences at 7 days of incubation (H.H. stage 31). It has recently been shown by Tyler and Hall (1977) that, while neither Meckel's cartilage nor the articular are dependent for their development on the presence of mandibular epithelium, the membrane bones of the mandible *are* dependent in the presence of that epithelium (Table IV). Whole mandibular arches, or mandibular ectomesenchyme and mesenchyme deprived of its epithelium by treatment with trypsin and pancreatin, either were maintained in organ culture or were grafted to the chorioallantoic membranes of host embryonic chicks. (For a discussion of the technique of chorioallantoic grafting and its usefulness in such studies, see Hall, 1978a.) Recall that neural crest cells reach the presumptive mandibular arch at H.H. stage 15. Removal of mandibular epithelium at H.H. stage 16 (52–53 hours of incubation) did not prevent differentiation of Meckel's cartilage or of the endochondral articular. In order for cartilage to form, either the ectomesenchyme does not require interaction with epithelium (it certainly does not require interaction with mandibular epithelium beyond H.H. stage 16), or, in line with the results from the culture of amphibian neural crest discussed in Section B,3, there is a transitory interaction with the pharyngeal endoderm during migration at around H.H. stages 14–15. Before

*Throughout the text, the stages of development of the embryonic chick are given according to Hamburger and Hamilton (1951) and are listed as H.H. stages. I have provided an Appendix that gives the times of incubation that correspond to these H.H. stages. Also listed in this Appendix are the major events in skeletal development that occur at each stage.

TABLE IV

Results of Organ Culture of Mandibular Ectomesenchyme after Enzymatic Removal of the Mandibular Epithelium[a]

H.H. stage at which mandibular epithelium was removed	Percent of cultures producing membrane bone[b]
16–18	0 (0/17)
20	0 (0/15)
21–22	0 (0/31)
23	0 (0/26)
24–25	100 (71/71)

[a] Isolated mandibular ectomesenchyme only forms membrane bones if epithelium has been present until H.H. stage 23. [Based on Tyler and Hall (1977) and Hall (1978d).]

[b] Cartilage forms in all of these cultures. Its formation is independent of epithelial influences. The timing of membrane bone formation is similar when mandibular ectomesenchyme is grafted to the chlorioallantoic membrane rather than organ-cultured.

membrane bones could begin to ossify on day 7, mandibular epithelium had to be present until H.H. stage 23 (4 days of incubation, Table IV).

Because cartilage always developed in all-age cultures of mandibular ectomesenchyme, we could not find a non-cartilage-forming area of the mandibular arch such as was described by Jacobson and Fell (1941). The possibility therefore arose that Meckel's cartilage itself, and not mandibular epithelium, was the agent influencing the mandibular ectomesenchyme to ossify. However, we often found that bone some distance from cartilage in our cultures or grafts, and other membrane bones that develop in isolation from cartilage *in vivo*, require interaction with adjacent epithelia before they will differentiate *in vitro* (see Section E,1). Normal differentiation of the mandibular *epithelium* requires the presence of mandibular ectomesenchyme or mesenchyme, although mesenchyme of the chorioallantoic membrane provides partial support for continued epithelial differentiation (Tyler and Hall, 1977) Mandibular ectomesenchyme will also respond to either wing or hindlimb bud epithelium by initiating intramembranous and subperichondrial osteogenesis. The timing of these heterotypic interactions is currently under investigation (Hall, 1978c).

In the amphibian neurula, the location of the mandibular cartilaginous skeleton is determined by specific interactions with the pharyngeal endoderm (Section B,3). We have no information about the factors responsible for the localization of the membrane bones. In the avian embryo, the localization of the membrane bones of the mandible is determined by interactions with the mandibular epithelium. The localization of cartilage of the mandible is not influenced by the mandibular epithelium, at least not from H.H. stage 16 on. For both the bone and cartilage of the mandible, the ectomesenchymal–epithelial interaction ends several days before the *in vivo* or *in vitro* differentiation of bone or cartilage.

There are a number of fragmentary pieces of information on the need for epithelial involvement in the differentiation of the mammalian mandibular ectomesenchyme. Frommer and Margolies (1971), who conducted a histological study of the development of the mouse mandible, maintained that the close spatial relationship between Meckel's cartilage and the initiation of ossification in adjacent membrane bones indicated the induction of membrane bone formation by Meckel's cartilage. Several studies have reported on the properties of isolated mammalian mandibular mesenchyme *in vitro*. Kollar and Baird (1969) observed osteogenesis in cultured isolated mandibular mesenchyme from mice embryos at 13 to 16 days of gestation; however, these are rather late stages and probably are past the point of determination. E. J. Kollar (personal communication) has found that after a period *in vitro*, cells from Meckel's cartilage re-form cartilage when placed in contact with mandibular arch epithelium, but when in contact with foreign (oral, epidermal, foot, or enamel organ) epithelium, these cells modulate to bone formation. This finding points to a directive role of the epithelium on mandibular differentiation. Svajger and Levak-Svajger (1971) used isolated mandibular mesenchyme from younger embryos (12 days of gestation) and observed chondrogenesis *in vitro*. No mention was made of bone formation. And, in their study of mandibles from fetal rats grafted to the avian chorioallantoic membrane, Tenenbaum *et al.* (1976) saw a thickening of the oral epithelium associated with osteogenesis in the adjacent ectomesenchyme. The possibility of interaction between the epithelium and the mandibular ectomesenchyme in mammals is certainly strongly suggested by these experiments; however, it has not been proved, as has been done for the chick.

D. Scleral Cartilage

The vertebrate optic cup is enveloped by mesenchyme, in which the scleral cartilage develops (snakes and mammals are the exceptions, as vertebrates in which the sclera remains collagenous). Morphologically, the scleral cartilage in the chick consists of a thin, curved plate of cartilage whose characteristic morphology can be molded by the mechanics of the growing eye (Coulombre, 1965) or modified by tension artificially applied *in vitro* (Weiss and Amprino, 1940). Despite, or perhaps because of, this adaptability, the basic morphological form of the cartilage is intrinsically determined. Dissociated and reassociated scleral chondrocytes form plates of cartilage *in vitro* (Weiss and Moscona, 1958), as does scleral ectomesenchyme when it is recombined with pigmented neural epithelium and grafted to the chorioallantoic membrane (Newsome, 1972). The localization of the scleral cartilage corresponds to the area of the pigmented epithelium underlying the developing optic cup. Does the pigmented epithelium induce the ectomesenchyme to chondrify? Before answering this question, I should like to comment on the origin of the ectomesenchyme.

The periocular ectomesenchyme is derived from the neural crest and migrates to the optic cup by H.H. stage 14 (Stewart and McCallion, 1975). Chondrogenesis does not commence until day seven or eight of incubation, although the ectomesenchyme has been determined for chondrogenesis prior to this point. Type II (cartilage-type) collagen can be detected by immunofluorescence from H.H. stage 30 on (von der Mark *et al.*, 1977b). Weiss and Amprino (1940) indicated that determination occurred by the fourth day of incubation, because periocular ectomesenchyme maintained *in vitro* would chondrify only if had been taken from older embryos. However, isolated periocular mesenchyme from some embryos as young as H.H. stage 19 (3–3½ days) will form cartilage when grafted to the chorioallantoic membrane. These experiments of Newsome's (1972) are summarized in Table V. The determination of the periocular mesenchyme for chondrogenesis occurs after H.H. stage 18. Because neural crest cells reach the optic cup at H.H. stage 14, determination must occur both after migration and within the optic cup. Several lines of evidence indicate that the agent responsible for their determination is the pigmented neural epithelium.

If all but the pigmented epithelium and the adjacent ectomesenchyme is removed from embryos of H.H. stages 12 or 13, scleral cartilage subsequently develops (Reinbold, 1968). Newsome (1972), grafting pigmented epithelium from embryos of H.H. stage 25 along with periocular mesenchyme from embryos of H.H. stages 17 and 18 onto the chorioallantoic membrane, obtained cartilage formation in 70% of the 20 grafts. He concluded that the determination

TABLE V

Percent of Grafts of Periocular Ectomesenchyme that Form Scleral Cartilage as a Function of the Age of the Donor Embryo[a]

Donor age (H.H. stago)	Percent of grafts with cartilage[b]
17	0
18	0
19	14
20	38
23	82

[a] Adapted from data in Newsome (1972).

[b] Periocular ectomesenchyme was grafted to the chorioallantoic membrane. Determination of ectomesenchyme for chondrogenesis occurs between H.H. stages 18 and 19.

that occurs *in ovo* at H.H. stage 19 was dependent upon interaction with the pigmented epithelium. Both epithelium alone and mesenchyme alone failed to chondrify. Stewart and McCallion (1975) provide details of similar experiments (see Fig. 6) in which they obtained chondrogenesis, provided that the ectomesenchymal donor embryo was at least H.H. stage 14.

Whether pigmented epithelium can induce nonscleral mesenchyme to chondrify has been both supported and refuted by recent experiments. Wedlock and McCallion (1969) grafted H.H. stage 26 pigmented epithelium, neural retina, or H.H. stage 18 spinal cord onto the chorioallantoic membrane; ectopic chondrogenesis was obtained only when the pigmented epithelium or spinal cord were used. They claimed that the cartilage observed was derived from cells of the host chorioallantoic membrane: "It is certain that the cartilage observed in the present study was formed from the mesenchyme of the chorioallantois;" however, they had no biological or isotopic marker to confirm this interpretation. Newsome (1972) grafted H.H. stage 25 pigmented epithelium to the chorioallantoic membrane, but did not observe ectopic chondrogenesis. When periocular ectomesenchyme was deliberately or accidentally included with the graft, chondrogenesis was observed, raising the possibility that donor scleral ectomesenchyme was included among the grafts of Wedlock and McCallion. Stewart and McCallion (1975) grafted H.H. stage 16 Japanese quail pigmented epithelium or spinal cord onto the chick chorioallantoic membrane without observing chondrogenesis.

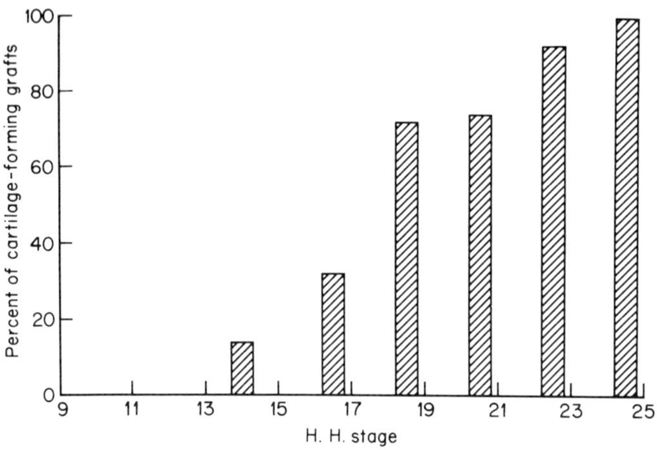

Fig. 6. These data show the percentage of grafts forming scleral cartilage when periocular ectomesenchyme and pigmented neural retina from embryos of H.H. stages 11 to 24.5 are grafted to the chorioallantoic membranes of host embryonic chicks. Cartilage is obtained only if the donor embryos are H.H. stage 14 or older, indicating that neural crest cells first reach the optic cup at H.H. stage 14. Based on data in Stewart and McCallion (1975).

But, when pigmented epithelium was grafted into cephalic mesenchyme, ectopic cartilage was induced (Reinbold, 1968; Stewart and McCallion, 1975). There is, then, some specificity in the ability of scleral ectomesenchyme to respond to pigmented epithelium by initiating chondrogenesis (see Chapter 3, Section IV).

The inductive action of the pigmented epithelium on the scleral ectomesenchyme is evidently mediated by products of the extracellular matrix of the epithelium. Pigmented epithelium produces a collagenous, acellular, extracellular matrix at the basal surface of the cells, a matrix closely associated with the basement membrane (Newsome and Kenyon, 1973). Cloned pigmented epithelial cells deposit similar material onto filters *in vitro*. If these cells are removed from the filters, and periocular ectomesenchyme or neural crest is added, chondrogenesis occurs (in 39% of 155 cases; Newsome, 1976). Purified type I or II collagen from rat tail tendon, chick skin, or cartilage will not substitute for the epithelial extracellular material.

The localization of the scleral cartilage as a ring encircling the developing optic cup is a result of the specific interaction between extracellular products of the pigmented retinal epithelium and competent ectomesenchymal cells of the periocular mesenchyme.

E. Membrane Bones of the Facial and Cranial Skeleton

1. The Quadratojugal

The membrane bones that develop in close association with the perichondrial surface of Meckel's cartilage have been discussed. It has been shown that, in order for these membrane bones to form, mandibular ectomesenchyme requires the presence of, and the interaction with, mandibular epithelium until H.H. stage 23. Similar experiments have been performed on isolated ectomesenchyme from the suborbital region of the face of the embryonic chick. This is an area of ectomesenchyme in which the quadratojugal–jugal bar differentiates at seven days of incubation. These bones are not associated with the primary cartilaginous skeleton, as are the extraperichondrial membrane bones of the mandible.

Like the mandibular membrane bones, the quadratojugal is derived from neural crest cells at the level of the mesencephalon (Le Lièvre, 1974; Le Lièvre and LeDouarin, 1975), and again like the mandible, the quadratojugal ectomesenchyme requires the presence of an epithelium until at least H.H. stage 23 in order for intramembranous ossification to be initiated at H.H. stage 31 (Table VI). Similar results are obtained when the isolated ectomesenchyme is grafted to the chorioallantoic membrane (Tyler, 1978). The possibility of influences from adjacent primary cartilage is excluded as a factor in the initiation of these osteogenic cells. Thus, all of the membrane bones of the avian head that have been tested require an epithelial interaction before they can form.

TABLE VI

Results of Organ Culture of Whole Trypsin-Pancreatin-Treated Suborbital (Quadratojugal) Regions of the Embryonic Chick and of Isolated Suborbital Ectomesenchyme[a]

H.H. stage of tissue cultured	Percent of cultures producing membrane bone after 7 days *in vitro*
Intact, enzyme-treated, suborbital tissues	
17–18	100 (8/8)
23	100 (8/8)
26	100 (6/6)
29	100 (7/7)
Isolated suborbital ectomesenchyme	
17–18	0 (0/8)
21	0 (0/6)
23	0 (0/6)
26	100 (5/5)
29	100 (7/7)
32	100 (8/8)

[a] The enzymatic treatment does not prevent the intact tissues from forming membrane bone. The isolated ectomesenchyme forms membrane bone only if epithelium has been present until H.H. stage 23. B. K. Hall (unpublished results).

2. The Cranium

Many other membrane bones do not develop in proximity to the primary cartilaginous skeleton. It has been maintained by Urist (1962, 1970) that these bones are induced to form by interactions between potentially osteogenic mesenchyme and fibrous connective tissues, although there is no experimental evidence to support this notion. The membrane bones surrounding the brain, i.e., those of the cranium, develop from ectomesenchyme whose skeletogenic potential is acquired or activated by interactions with the developing brain and notochord. These membrane bones act as site-specific inducers of dermal skull bone development, as has been elegantly shown by Jean Schowing (1961, 1968a–c) and by Benoit and Schowing (1970). Their basic technique has been to remove selected areas of the developing brain or notochord at 35–40 hours of incubation and to observe any deletions in the dermocranium at 15 days of incubation. The map of inductive influences that has been obtained is shown in Figure 7. For example, the differentiation of ectomesenchyme into frontal bones requires influences from both the prosencephalon and the mesencephalon. If both are excised, the skull develops, but without frontal bones. It is normal in all other respects.

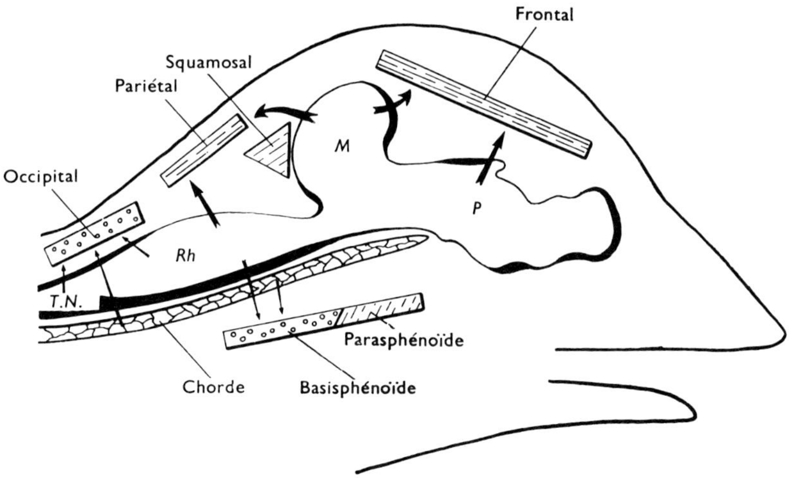

Fig. 7. Site-specific osteogenesis within the developing avian skull is initiated by interaction between cranial ectomesenchyme and specific areas of the developing brain, neural tube, and/or notochord as shown by the arrows in this figure. Chorde, notochord; M., mesencephalon; P., prosencephalon; Rh, rhombencephalon; T.N., neural tube. Reproduced from Schowing (1968c) with the permission of the author and publisher.

Explantation of the ectomesenchyme of presumptive cranial bones to the chorioallantoic membrane indicates that the interactions occur between 48 and 60 hours of incubation. Reversal of the developing brain, i.e., placing the rhombencephalon into the position normally occupied by the prosencephalon, will not allow differentiation of the frontal bones, a finding which confirms the site specificity of the interactions. The brain induces the cranial vault, and the notochord induces the cartilaginous floor of the skull (Fig. 7). Spinal cord will not substitute for brain (Schowing, 1974), although quail brain will substitute for chick, producing a normal, albeit small, skull (Schowing and Robadey, 1971; Robadey and Schowing, 1972). Diethelm and Schowing (1973, 1974) removed the embryonic brain, treated it with actinomycin D, and then replaced it. These embryos subsequently developed cyclopia, and the skull bone failed to form. Schowing neatly summarized the results of the investigations: "In cranial development, the contents induce the container."

Although these interactions are best understood in the developing avian embryo, de Beer (1937) has presented some evidence for various species. There is even some information for mammals, e.g., in skull regeneration in the dog, new bone is formed only when there is contact of ectomesenchyme with the underlying dura mater (Kantorova, 1972). While mechanical stresses associated with skull expansion play a role in patterning the shape of the skull and in directing its growth, they are not required for the initiation of cranial osteogenesis (Murray, 1936).

3. Scleral Ossicles

The scleral ossicles are a ring of trapezoidal, overlapping plates, usually consisting of membrane bone, that surround the margin of the cornea in many submammalian vertebrates. They have been studied primarily in birds, and especially in the embryonic chick. In the Elasmobranchi, they are cartilaginous rather than osseous, while in teleosts these ossicles arise by endochondral replacement of the scleral cartilages rather than by separate intramembranous ossification adjacent to the scleral cartilage as is the case in the birds. However, no comprehensive comparative developmental studies have been performed.

Although the scleral ossicles develop in the periocular ectomesenchyme, they are preceded by a series of epithelial thickenings (scleral or conjunctival papillae), whose number and position correspond exactly to the number and future position of the scleral ossicles. These papillae arise during the eighth day of incubation in the embryonic chick. One to two are present at seven days, 11 by seven and a half days, and 14 are present by eight and a half days of incubation. The majority of individuals have 14 scleral ossicles in the eye, although other avian species may have as many as 18. Coulombre and Coulombre (1973) found the distribution of ossicle number in 612 embryonic chicks that had been incubated for 17 days to be: less than 1% of individuals with 13 ossicles; 80% with 14 ossicles; 19% with 15 ossicles, and 1% with 16 ossicles. Production of a fixed number of ossicles evidently is strictly regulated. There is a tendency toward production of an extra ossicle, rather than toward the elimination of an ossicle. This conservation of developmental information will be seen again when limblessness in vertebrates is discussed (Section III).

The histological development of the scleral papillae, from their appearance during the eighth day to their disappearance during the twelfth day, has been studied by Murray (1941, 1943), O'Rahilly (1962), Coulombre *et al*. (1962), van de Kamp (1968), and Ambrosi *et al*. (1973). Initially, there is ingrowth of an epithelial placode into the subjacent ectomesenchyme. This is followed by withdrawal and outpushing of the epithelium, processes which leave epithelially derived collagen and granular material within the ectomesenchyme. Finally, extensive cell death occurs within the epithelium, and during the twelfth day, the papillae disappear. Intramembranous ossification then begins in the adjacent ectomesenchyme, initially at separate centers isolated from one another. During days 13 and 14, adjacent centers gradually come to overlap, and an imbricated ring of bony ossicles is produced.

The correspondence between the numbers and distributions of the papillae and the subjacent ossicles, even when ossicle number deviates from the normal, and the timing of papillae removal and onset of osteogenesis suggests a possible inductive interaction between the epithelial papillae and the adjacent periocular ectomesenchyme. The accumulation of a critical mass of ectomesenchymal cells in the position of the future ossicle (the formation of the condensation) certainly depends upon the presence of the papilla. Between seven and ten days of incubation, the cell density within the ectomesenchymal primordium (underlying a papilla) increases from 4×10^5 cells/mm^3 to 14×10^5 cells/mm^3, while the cell density between primordia (in ectomesenchyme lying between adjacent papillae) increases to only 9×10^5 cells/mm^3 (Hale, 1956a). The implication, therefore, is that proximity of ectomesenchyme to a papilla results in increased cellularity, because the papilla either stimulates cell division in the adjacent ectomesenchyme (as Hale thought was the case) or influences migration of ectomesenchyme to a position beneath its center (a postulate being investigated by David Fyfe in my laboratory). Although not tested, by removing papillae and examining changes in cell density in the subjacent ectomesenchyme, it looks as if the stimulus for the origin of the ectomesenchymal condensation is the presence of the adjacent scleral papilla. The papilla itself arises because of the aggregation of epithelial cells toward a center (Putchkov, 1964).

Whether osteogenesis within the ectomesenchyme requires the presence of the scleral papilla has been investigated. This was done by removing a single papilla (number 12) on either day seven, eight, or nine of incubation and then observing whether ossicle number 12 appears later in embryonic development. If removed early (at seven days of incubation), the older embryos either lacked ossicle 12 (70% of cases) or had a much reduced ossicle number 12. If the papilla was removed later (nine days of incubation), 90% of the later embryos developed a normal ossicle number 12, and no embryo lacked the ossicle (Coulombre *et al.*, 1962). Therefore, the presence of the papilla influences the development of an intramembranous ossicle in the immediately subjacent ectomesenchyme and does so between seven and nine days of incubation. Grafting periocular ectomesenchyme denuded of epithelium to the chorioallantoic membrane of a host embryo provides further information on the timing of ectomesenchymal determination for osteogenesis (Coulombre *et al.*, 1962). As the data in Table VII indicate, determination for osteogenesis occurs during the eighth day of incubation.

Hydrocortisone has also been found to selectively reduce the number of papillae, either completely or partially, depending on the time and number of treatments given during days six to eight. The development of the adjacent ossicles is prevented or slowed by this treatment (Johnson, 1973).

A further line of evidence pertaining to the induction of scleral ossicles comes from the study of the scaleless chick mutant (Palmoski and Goetinck, 1969, 1970; Abbott, 1975). In the scaleless, the epithelial structures are altered or

TABLE VII

Percent of Grafts of Periocular Ectomesenchyme that Form Scleral Ossicles as a Function of the Age of the Donor Embryo[a]

Age of donor embryo		
Days	H.H. stage	Percent of grafts with bone[b]
7	31	0
8	34	7
9	35	50
10	36	100

[a] Determination for osteogenesis occurs during the eighth day of incubation. Adapted from data in Coulombre *et al.* (1962).

[b] All grafts examined 7–10 days after initial grafting to the chorioallantoic membrane.

missing. The affected structures include the feathers, scales, spurs, footpads, and scleral papillae. Only three papillae develop in the scaleless mutant, and only one of these persists beyond the ninth day of incubation. Consequently, scaleless embryos have only one ossicle in each eye. Reciprocal grafts between scaleless and wild-type embryos have established that the genetic defect is initially epidermal. The inductive interaction between ectomesenchyme and epithelium prior to the onset of osteogenesis is given additional support by the loss of papillae and the resultant loss of ossicles; also confirmed is how precise ossicle distribution is controlled relative to the location of the preceding papillae.

In addition, the number of ossicles that form, their spatial distribution in the cornea, and their pattern of overlap depend in part on the growth rate of the optic cup and the eye between four and eight days of incubation, and in part on the growth of the ossicles themselves between days 12 and 14 (Coulombre and Coulombre, 1975). The slower the growth of the eye, the fewer the ossicles that form—evidently, a problem in the minimal spacing between papillae. The morphogenesis of the ossicles is also apparently influenced by cell death, which destroys cells within the osteogenic centers (Hale, 1956b).

F. Antlerogenesis

The position of the antlers on the heads of deer, moose, etc., is taken for granted. However, examining the initiation of the formation of either the first or subsequent set of antlers shows that precise interactions involving specific attributes of the frontal bone, its bony pedicle, which presages the antler bud, and the overlying epidermis, all participate in localizing the position of the antlers on the head.

1. The Shedding Cycle

Mature antlers are solid, osseous appendages lacking vascularization or innervation. They are joined to the frontal bone of the skull by a bony pedicle and are covered with an epidermal covering of very short straight hairs (the velvet). The histogenesis of these tissues was described in Chapter 1.

Early in the spring, after the mating season and over the course of just a few days, osteoclasts erode the base of the antler at its junction with the pedicle (Goss, 1963). The source of the osteoclasts, whether local or from migratory cells, and the stimulus for their accumulation and action at this junction remain unknown (testosterone may be involved in the latter process; see Chapter 9). The antler is shed, leaving a wound on the pedicle, with bone, vasculature, and nerve endings all exposed. Skin overgrows the wounded pedicle and forms a flat antler bud which is highly vascularized and heavily innervated. The rapid growth of this bud over the course of the spring and summer produces the antler of the new season. The wounded pedicle serves as a vital stimulus for the renewal of antler growth, as has been convincingly demonstrated by Lincoln and Fletcher (1976) for congenitally polled ("hummel") red deer (*Cervus elaphus*). These antlerless animals possess frontal pedicles; if the apices of these pedicles are amputated, growth of an antler is initiated.

2. Initiation of Antler Formation

If the bony protuberance on the frontal is removed from a fawn that has not produced its first set of antlers, no pedicle forms and antler development is not initiated (Goss *et al.*, 1964; Goss, 1970). If the frontal bone is transplanted under epidermis other than that overlying the frontal bone (e.g., elsewhere on the head, or on the metacarpal), pedicle formation *is* initiated (Hartwig, 1968, 1972; Goss, 1970; Hartwig and Schrudde, 1974). Thus, it is the frontal bone, and not the overlying epidermis, that possesses the ability to initiate pedicle development.

If the pedicle itself is removed, formation of an antler bud and hence, an antler, is prevented (Goss, 1961; Goss *et al.*, 1964); however, if only the bone of the pedicle is removed, the wound heals, and an antler begins to develop (Goss, 1964, 1970). Thus, while the frontal bone initiates pedicle formation, the bone of the pedicle does not initiate antler formation. Rather, it is the epidermal–dermal covering of the pedicle that initiates development of the antler bud. The normal response of mammalian dermal tissue to wounding is the formation of scar tissue. But it is the absence of this response on the pedicle that allows the antler to regenerate. Transplantation of head or ear skin to the pedicle is followed by formation of an antler covered with velvet (Goss, 1964; Hartwig and Schrudde, 1974). That is, the epidermal requirement of the pedicle is not specific in terms of initiation of antler development. Epidermis other than pedicle epidermis can be induced to respond to a pedicle by producing velvet. Specificity of the dermal component has not been thoroughly tested. Thus, in antlerogenesis, the position of the pedicle depends upon properties of the frontal bone, while formation of the antler bud on the pedicle depends upon interactions between the pedicle and the epidermis and perhaps also with the dermis.

G. Tooth Development

The mammalian tooth develops because of reciprocal interactions between the buccal epithelium and its derivatives—the enamel organ and the ameloblast—and similar interactions between the buccal ectomesenchyme and its derivatives, namely, the ectomesenchymal (dental) papilla and the odontoblasts (Fig. 8). The histogenesis of these tissues has been described by Peyer (1968), Dahlberg (1971), and Scott and Symons (1974), and the interactions between epithelium and ectomesenchyme by Hörstadius (1950), Slavkin and Bavetta (1972), Kollar (1975), and Butler and Joysey (1978).

While neither the epithelium nor the ectomesenchyme will continue to differentiate in the absence of the other tissue, both continue to differentiate when cultured transfilter to one another (Koch, 1967, 1972; Fromme *et al.*, 1971). Cell contact between the interacting tissues is required (Thesleff *et al.*, 1977; Katchburian and Burgess, 1977).

Buccal epithelium can be induced to differentiate structures other than enamel organ and ameloblasts if cultured with heterotypic mesenchyme. For example, in the presence of snout mesenchyme, the buccal epithelium produces hair follicles (Kollar and Baird, 1970). Only recently has it been shown that dental mesen-

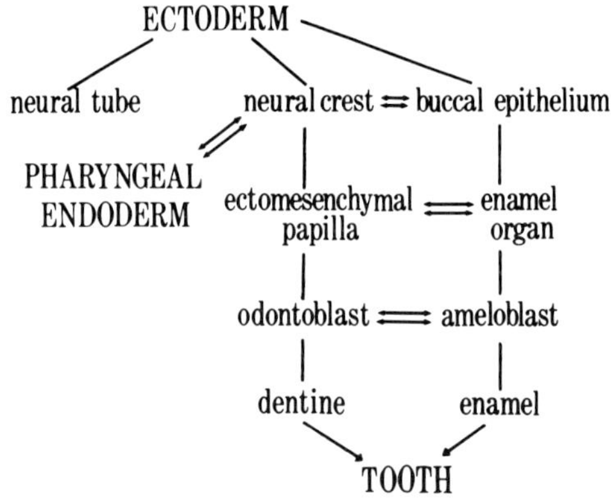

Fig. 8. The major ectodermal–ectomesenchymal interactions in the formation of the mammalian tooth. Double arrows indicate inductive interactions. Single arrows indicate the developmental sequence. Reproduced from Hall (1975a) with permission of the publisher.

chyme can exhibit a nondental differentiation. Silbermann *et al.* (1977) recombined ectomesenchyme from the mandibular first molar of the 17-day-old embryonic mouse with ectoderm from the chick limb bud (age unspecified). They then grafted the recombinant to the chorioallantoic membrane. The dental mesenchyme differentiated into cartilage having the morphology of a tooth. The cartilage, which had scant amounts of extracellular matrix (like the secondary cartilages) containing considerable fibrous tissue, was invested with a perichondrium. The epithelium of the limb bud evidently elicited this unusual differentiation; however, the ectomesenchyme expressed its morphogenetic properties, and organized the cartilage into a toothlike morphology.

Previous studies had shown that the pattern of the tooth, e.g., whether an incisor or a molar, was determined by the ectomesenchymal dental papilla. This was elegantly demonstrated in the mouse by Kollar and Baird (1969). Incisor epithelium recombined *in vitro* with molar ectomesenchyme allowed differentiation of a molariform tooth, whereas molar epithelium recombined with incisor ectomesenchyme allowed differentiation of an incisiform tooth. Therefore, the shapes of the incisors and molars of the mouse are under ectomesenchymal control and are independent of extrinsic factors. A similar conclusion applies to tooth shape in the frog *Rana pipiens* (Howes, 1977a). The structures of tooth attachment, such as the osteodentine, cementum, and alveolar bone, are dependent upon extrinsic factors for their normal morphogenesis (Ten Cate *et al.*, 1974; Freeman *et al.*, 1975; Ten Cate, 1975; Howes, 1977b).

The factors that determine the location of the tooth primordia along the jaws have been more difficult to uncover. In the rodent, teeth are absent from the diastema of the jaw because of deficiencies in the ectomesenchyme of that region. If epithelium from the toothed area of the buccal cavity is combined with ectomesenchyme from the diastema, no tooth develops; however, if diastema epithelium is combined with tooth ectomesenchyme, ameloblasts develop within the epithelium and a tooth forms (Kollar, 1972). If this result is typical of dental ectomesenchyme in general, it indicates that the position of the teeth depends upon discontinuous distribution of competent ectomesenchyme along a continuous, inductively active buccal epithelium. Because the ectomesenchyme migrates in from the neural crest, presumably we must look for factors which stop either the migration of the neural crest cells or their localization beneath the epithelium. The position of taste buds is determined by influences from local nerve endings. The discontinuous distribution of nerve endings along the dental lamina, the localized activation of the primordia of enamel organs, and their subsequent influence on migrating neural crest cells constitute one possible mechanism (E. J. Kollar, personal communication; Pearson, 1977). Such fields of activity might then lead to fields of inhibition around each primordium, as Berkovitz and Moore (1974) have shown to occur in the replaceable teeth of the rainbow trout. Once localized, the ectomesenchyme would begin to interact with the buccal epithelium and set in motion the chain of interactions depicted in Figure 8.

III. THE LIMB SKELETON

Like the ectomesenchymal skeleton, the mesodermally derived skeleton is positioned within the embryo as a result of epithelial–mesenchymal interactions. The best studied of these are the interactions between the presumptive limb mesenchyme and the apical ectodermal ridge (AER) and those occurring between the somitic presumptive vertebral mesenchyme and the notochord and spinal cord. The evidence obtained from experiments on the developing limb buds of the embryonic chick will be discussed in some detail. Somitic chondrogenesis has recently been reviewed (Hall, 1977a) and thus will be treated in less detail.

Because the limb bud will be used to illustrate aspects of cell condensation (Chapter 5), chondrogenesis (Chapters 6 and 7), polarity, cell death, and morphogenesis (Chapter 8), as well as growth (Chapter 9), I have prepared a summary table of the major morphogenetic and cytodifferentiational events of limb bud development up to H.H. stage 30 (7 days of incubation) (Table VIII). This table should serve to supply the reader with the details of avian limb bud development and to provide a quick reference to the events occurring at particular developmental stages.

I propose, then, that the evidence for interactions between the ectodermal ridge and the underlying mesoderm be considered in three categories: (1) the directive role of the ectoderm upon the mesoderm; (2) the induction of the apical ectodermal ridge by the mesoderm, and (3) the maintenance of the ectodermal ridge by the underlying mesoderm.

A. The Directive Role of the Ectoderm

The early limb bud of the avian embryo consists of a core of mesoderm (the mesoblast) that underlies a thin cap of cuboidal ectoderm. Late in H.H. stage 17, the ectodermal cells at the apex of the limb bud become columnar, and by H.H. stage 19, the more rapid postaxial developmental of the limb bud has produced an asymmetrical bud with a nipplelike ridge of ectoderm at the distal tip. As development proceeds, this asymmetry becomes more pronounced.

For almost a hundred years, it has been known that a portion of the ectodermal covering of the developing limb bud thickens into a ridge [Kölliker (1879) and Balfour (1885) cited by Kelley (1973)]; however, it was not until 1948 that experimental evidence was obtained to indicate that this ectodermal ridge directed developmental events in the underlying mesoderm. In that year, Saunders surgically removed the ridge [which he named the apical ectodermal ridge (AER)] from the wing buds of embryonic chicks *in ovo* and allowed the ridgeless wing mesoderms to continue development in their normal body sites. Subsequent outgrowth ceased, and the wings that did develop lacked some or all of the more distal skeletal elements, such as radii, ulnae, or digits. The proximal limb elements (humerus, pectoral girdle) were normal in both morphology and size. The overall shortening of the limb resulted from the absence of the distal cartilages and not from an overall decrease in total limb length.

Saunders noted that the older the embryos at the time of extirpation of the AER, the less marked the abnormalities in wing development; in particular, he observed that increasingly distal cartilages were present when the ridge was removed from progressively older embryos. Saunders interpreted these results to mean that the AER controlled the development of the mesodermal skeleton by specifying a proximodistal sequence of limb elements and limb bud outgrowth. Early in development, the mesoderm adjacent to the ectoderm would be specified for proximal skeletal elements such as the humerus. Progressively later in development, the mesoderm near the ridge (the distal mesoderm) would be determined for increasingly distal elements. Indeed, the cartilages of the limb bud do differentiate according to such a sequence.

It followed from this interpretation of limb development that the form of the mature limb skeleton could be mapped out in the limb bud at successive stages of development, i.e., that fate maps could be constructed. Saunders (1948) mapped out the presumptive areas of the wing bud using carbon marking and showed that the wing bud mesoderm was derived from the somatic mesoderm of the lateral plate rather than from somitic mesoderm (but see Chapter 6, Section VI,B,5 on the latter point). Some controversy still remains as to the precise demarcation of presumptive areas in the early wing bud; see Saunders *et al.* (1962), Stark and Searls (1973), Summerbell (1974b, 1976), Summerbell *et al.* (1973), and Amprino (1976, 1977a).

If the apical ectodermal ridge does have a controlling role in limb development, then variations in the size, shape, and location of the ridge in the early limb bud (the ridge is active from H.H. stage 17 to 19; see Table VIII), should be reflected in deviations from the normal later in development. I shall consider three possibilities, and shall discuss whether such limb buds provide evidence in support of the directive role of the ectodermal ridge. First, I shall consider the limb buds that lack an apical ectodermal ridge. In nature, these limb buds are found in limbless vertebrates and in skeletal mutants, and as Saunders showed, they can be produced surgically. Second, I shall discuss limb buds having either an additional or enlarged apical ectodermal ridge. Again, mutants of these types exist (Abbott, 1967, 1975), and they can be produced experimentally. Third, I shall consider limb buds having a narrow or a subdivided apical ectodermal ridge.

TABLE VIII

A Summary of the Major Events in the Development of the Wing Bud in the Embryonic Chick

H.H. stage[a]	Length of H.H. stage (hrs)	Developmental event	Reference
7	3	Presumptive limb mesoderm can chondrify ectopically.	13, 15
10	7	Earliest point at which flank mesoderm can become limb.	3
10–11	10	Anterior–posterior axis already fixed. Dorso-ventral axis fixed at this point.	2
12	3	Earliest point at which wing mesoderm will induce formation of an AER[b] from flank epidermis.	17
15	2–3	Earliest age for autonomous differentiation.	13
17	14	Limb buds present (wing bud adjacent to somites 15–19; leg bud adjacent to somites 26–32). AER present. Wing mesoderm loses ability to induce AER in the flank and flank ecto-derm loses ability to respond to limb mesoderm.	10, 16
18	6	Limb mesoderm forms cartilage *in vitro.* Humerus specified.	20, 25
20	4	Vascular pattern initiated. Ulna and radius specified.	1, 25
21	8	Specification of wrist starts (lasts until H.H. 24).	25
22	6	S uniformly distributed over mesoderm. Sinusoids demark myo- from chondrogenic areas. Proximal, central mesenchyme con-densing. PNZ[c] irreversibly determined. Mitotic index starts to decline, especially centrally.	9, 12, 19, 24, 27
23	12	Myosin first appears. First innervation of hind-limb bud.	6, 11
24	6	Y-shaped condensation and opaque patch appear. Vascular pattern well established. Cell packing in prechondrogenic mesenchyme in-creased 60% over that at H.H. 22. Proximo-distal gradient in cell division present. First necrotic cells in PNZ. Ability to form cartilage stabilized between H.H. stages 24 and 25.	8, 23, 26
25	6	Extracellular matrix of cartilage appears. Proximal mesoderm will no longer support an AER (i.e., morphogenetic properties lost), nor will it produce distal structures. Innervation reached stylopod.	5, 6, 18

continued

TABLE VIII (*Continued*)

H.H. stage[a]	Length of H.H. stage (hrs)	Developmental event	Reference
26	12	Cell density in chondrogenic mesenchyme increased 10% over that at H.H. 24. Ability of cells to segregate appears. Dissociated chondrogenic mesenchyme produces condensations. Digit III and metacarpals specified (H.H. 25–26). Muscle appears.	4, 7, 21, 22, 25, 26
27	12	Cartilaginous models of limb skeleton present. Distal mesoderm will no longer support an AER. Proximal phalynx specified. Innervation reached zeugopod (H.H. 27–29).	5, 6, 25
28	6	Opaque patch disappeared. Distal phalynx specified. Innervation reached autopod.	6, 25, 26
29	12	AER loses inductive ability.	14
30	6	Interdigital cell death (H.H. 30 in hindlimb; H.H. 31 in wing). Definitive nerve pattern.	16

[a] For hours of incubation corresponding to these H.H. stages, see Appendix 1.
[b] AER, apical ectodermal ridge.
[c] PNZ, posterior necrotic zone.

The phases of limb development may be summarized as: H.H. stages 14 to 25–26: morphogenetic phase; H.H. stage 25ff: cytodifferentiation phase; H.H. stages 12 to 22: limb bud regulative; H.H. stages 22 to 24.5: limb bud stabilizing; H.H. stages 25ff: limb bud mosaic; H.H. stages 17 to 29: apical ectodermal ridge active.

References: (1) Caplan and Koutroupas (1973); (2) Chaube (1959); (3) Dhouailly and Kieny (1972); (4) Ede and Flint (1972); (5) Finch and Zwilling (1971); (6) Fouvet (1973); (7) Hilfer *et al.* (1973); (8) Hornbruch and Wolpert (1970); (9) Janners and Searls (1970); (10) Kieny (1960); (11) Medoff and Zwilling (1972); (12) Mottet and Hammar (1972); (13) Pinot (1970); (4) Rubin and Saunders (1972); (15) Rudnick (1945); (16) Saunders and Fallon (1967); (17) Saunders and Reuss (1974); (18) Saunders *et al.* (1959); (19) Searls (1965a); (20) Searls (1968); (21) Searls (1971); (22) Searls (1972); (23) Searls and Janners (1969); (24) Searls *et al.* (1972); (25) Summerbell (1974b); (26) Thorogood and Hinchliffe (1975); (27) Zwilling (1966).

1. Absence of the Apical Ectodermal Ridge

a. Experimental removal of the AER. Saunders' observation concerning the consequence of AER removal and its essential role has been confirmed by other workers (Hampé, 1959; Barasa, 1960; Summerbell, 1974b). Summerbell established the timing of the specification of elements of the wing on the basis of their deletion after timed removal of the AER (see Table IX and Fig. 9). He proposed that this obvious proximodistal sequence was both continuous and graded rather than stepwise. In the latter case, large elements (e.g., the forearm)

TABLE IX

The Timing of Specification of the Skeletal Elements in the Wing Bud of the Embryonic Chick[a,b]

H.H. stage	Skeletal element specified
Before stage 18	Shoulder girdle, humerus
Stages 19 to 20	Ulna, radius
Stages 21 to 24	Wrist
Stages 25 to 26	Metacarpals
Stage 27	Proximal phalynx of digit III
Stage 28	Distal phalynx of digit III

[a] Based on the latest H.H. stage at which deletion of the apical ectodermal ridge results in deficiency of that element.
[b] Based on data in Summerbell (1974b).

would be specified at one time; however, as Fig. 9 indicates, the forearm is determined progressively over just a 12-hour period (H.H. stages 18 to 21), while the much shorter (but perhaps more complex) wrist takes 24 hours (H.H. stages 21 to 24) to be specified. The removal of large pieces of presumptive mesoderm leads to deletions of only small pieces of the limb.

Not all workers attribute these deletions to ectodermal control. Amprino (1965, 1975a,b) and Amprino and Amprino-Bonetti (1967a,b) suggest that damage to the mesoderm must be considered as a factor. For example, more deficiencies occur if all of the ectoderm is removed from the limb bud than if only the AER is removed. These authors take this to mean that there is more mesodermal damage after removal of all of the ectoderm, but the nonridge ectoderm may play a role (see Section 7). Their own experiments could be interpreted in this light, for if only the lateral ectoderm is removed and the AER left intact, deficiencies still arise. Nevertheless, it should be noted that some outgrowth of the mesodermal core of the early limb rudiment occurs before an AER is present so that the AER is either not essential for this early outgrowth (Amprino, 1965), or it can function before it has acquired its characteristic morphology (see Section 8).

Does the AER govern outgrowth by maintaining a high rate of mitosis in the subridge mesoderm? Although the labeling index of the subridge mesoderm at H.H. stages 23 to 24 is slightly higher than in the more proximal mesoderm, removal of the AER at H.H. stage 19 has no effect on this labeling index (Janners and Searls, 1971). A wave of cell death does occur in the subridge mesoderm after removal of the AER (Barasa, 1960; Janners and Searls, 1971). Perhaps the AER slows cell death in that area and therefore indirectly maintains the size of the cell pool, which is still labile and which can undergo determination to form more distal structures. Cairns (1975) has obtained evidence to this point. AER's

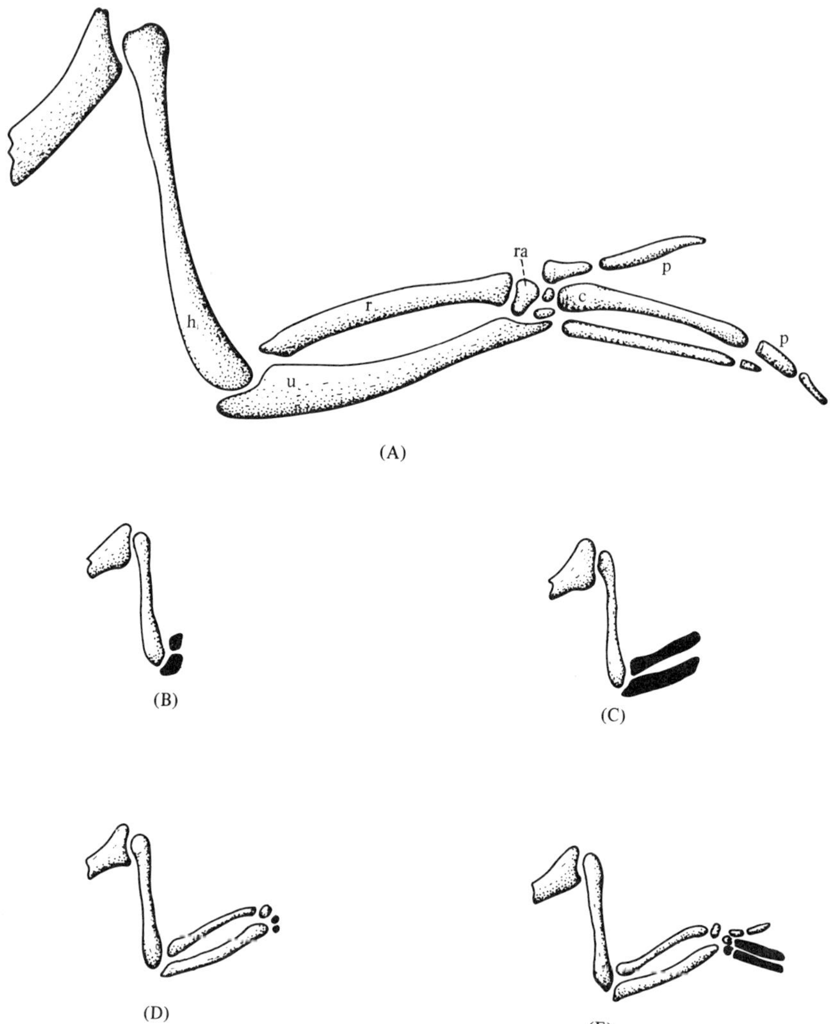

Fig. 9. The effect of timed excision of the apical ectodermal ridge (AER) on the subsequent development of the skeleton of the wing. The later the age at which the AER is removed, the fewer the distal deficiencies in the wing. (A) The normal skeleton of the wing and its appearance after removal of the AER at or after H.H. stage 28. The extent of development of the skeleton (B) after removal of the AER at H.H. stage 18; (C) stages 19 and 20; (D) stages 21–25, or at stage 26 (E) is shown. The terminal deficient skeletal element is shown in black. C, carpometacarpal; h, humerus; r, radius; ra, radiale; u, ulna. Based on data in Summerbell (1974b).

maintained in wire baskets over limb mesoderm denuded of its own AER stop the cell death that normally occurs when the limb mesoderm is cultured alone—presumptive evidence for the transmission of a diffusible stimulus from ectoderm to mesoderm.

On balance, then, the evidence from experimental ablation of the apical ectodermal ridge indicates that this treatment prevents any further development or outgrowth of the mesodermal limb core, supporting the thesis that interaction between the ectoderm and the mesoderm is essential for initiation and maintenance of limb bud development.

b. Wingless mutants. Several other lines of evidence indicate that the AER controls limb development. Zwilling (1949) utilized "wingless" mutants in his initial experiments on the role of the ectodermal ridge. During the normal development of wingless embryos, a wing bud complete with AER begins to form. Rather than being maintained, the AER regresses, so that the limb bud fails to show any further outgrowth or development. Hinchliffe and Ede (1973), after studying another wingless mutant, have proposed that the regression of the AER results from precocious cell death in the mesoderm, an event which prevents production of the factor responsible for maintenance of the ectodermal ridge (see Sections A,2 and C). It seems surprising that a wing should begin to form in a wingless individual; however, this situation turns out to be the norm for such mutants (Abbott, 1967, 1975). In an analogous manner, limbless vertebrates produce abortive limb buds, and hairless mice begin life with a fine coat of hair. Presumably, there is a selective advantage in maintaining the ability to form the organ primordium. This is also seen in the tendency of embryonic chicks to gain, rather than to lose, scleral papillae (Section II E.3).

The tail of the mouse develops under the influence of a ventral ectodermal ridge. In the vestigial tail mutant (*vt*), this ridge is missing (Grüneberg and des Wickramaratne, 1974).

2. Limblessness in Vertebrates

Many species in several major orders of amphibians, (Apoda, Caecilia) and reptiles (Ophidia, Amphisbaenia) lack limbs as adults, and some species show rudimentary limbs of limb girdles. (The larval stage of Anuras is excluded from this limbless group, because the adults possess functional limbs.) These reductions and examples of rudimentation represent degenerative conditions. All of the species are descended from limbed ancestors, and most possess limb buds at some stage of their embryonic development. Limblessness represents arrested limb bud development and subsequent regression, rather than the regression of a fully formed limb. As such, it provides an excellent example of the need for continued interactions between ectoderm and mesoderm for normal and complete limb development.

Limblessness is correlated with many growth patterns: with elongation of the body, with reduction in body diameter and mass, with a sheltering or burrowing mode of life (except for many snakes), and with locomotion by lateral undulation (again with exceptions among the snakes). These points, and other anatomical and ecological modifications of limbless vertebrates, along with a discussion of the evolution of the condition are considered by Gans (1974, 1975).

With respect to the initial development of the limb buds, the most extensively studied limbless vertebrates are the slow worm (*Anguis fragilis*) and its relative, the legless lizard or glass snake (*Ophisaurus apodus*); the snakes *Python reticulatus* and *Tropidonotus tessellata*; two lizards (*Scelotes brevipes* and *S. gronopii*), and two turtles (*Emys orbicularis* and *Testudo graeca*).

In *Anguis fragilis*, four anterior somites send ventral extensions into the somatopleur at the level of the future forelimb bud, while three somites send processes into the position of the future hindlimb bud. An ectodermal ridge forms at the apex of the conical limb bud. This ridge shows both morphological specialization and higher levels of protein and RNA than the adjacent epithelium (Raynaud *et al.*, 1974a, 1975; Raynaud and Adrian, 1975a,b). Rather nice scanning electron micrographs of the developing buds are presented in Raynaud *et al.* (1974b). At this stage of limb bud development, regression sets in. Cellular degeneration, beginning in the tips of the somitic extensions, is followed by degeneration of the ectodermal ridge (substantial amounts of cytolysosomes accumulate), and finally, by degeneration of the somatopleural mesenchyme of the bud (Raynaud *et al.*, 1974a). It has been proposed (Raynaud, 1972) that reduction begins because insufficient somitic processes penetrate the somatopleur. However, as it is, these processes are the first to degenerate; consequently, one might speculate that the somatopleural mesenchyme lacks a factor required to maintain the processes.

In *Ophisauras apodus*, a species from the same family (Anguidae) as *Anguis fragilis*, three somites also send extensions into the future hindlimb bud. An epithelial ridge forms, and the developing limb bud becomes vascularized and innervated. The ridge, the somitic extensions, and the mesenchyme all necrose, reducing the bud to a rod in which a rudimentary skeleton develops (M-Z Rahmani, 1974). In *Python reticulatus*, in the skinks, *Scelotes brevipes* and *S. gronopii*, and in the turtle *Testudo graeca*, a similar pattern, i.e., of somite extensions, formation of an epithelial ridge, and regression, is seen (Raynaud, 1974; Raynaud *et al.*, 1974c; Pieau and Raynaud, 1976; Raynaud and van den Elzen, 1976). This is evidently the characteristic pattern of initial limb development and subsequent regression in all limbless vertebrates.

Vasse (1974, 1977) has provided direct experimental evidence for the necessity of somitic extensions to stimulate proliferation of somatopleural mesenchyme and formation of an initial limb bud in the turtle *Emys orbicularis*. Removal of somites 6–13 at the 20–23 somite stage completely prevented limb bud development. Removal after the 25-somite stage did not affect limb bud development, establishing the latest time of action of the somite on the somatopleur. Mechanical blocks to the path of the somitic extensions also prevented any further limb bud development. Thus, the somitic extensions do indeed play a role in activating the proliferation of somatopleural mesenchyme.

Experimental evidence also supports a role for somites in inducing limb development in the avian embryo. Somitic cells stimulate the limb mesenchyme to proliferate (Pinot, 1970; Kieny, 1971), penetrate the somatopleur, and contribute myoblasts and fibroblasts to the limb bud (see Chapter 6, Section VI,B,5). Position of mesoderm in relation to particular somites, then, is an important factor in positioning the limbs along the embryonic axis. It is a prelude to interaction to the mesoderm with the overlying ectoderm to form the AER (see Section B).

In summary, removal of the AER by surgical procedures, by mutation, or by a developmental process of timed degeneration indicates that limb formation cannot proceed without a functionally active ectodermal ridge interacting with the mesenchyme.

3. Extra or Enlarged Apical Ectodermal Ridges

a. Experimental addition of an AER. Zwilling (1956a) provided the first evidence indicating that when an extra AER was grafted to a limb bud, both the transplant and the host AER would be maintained if kept sufficiently separate from one another. The result was duplication of the limb, with one axis beneath each AER. The earlier the embryo at the time of transplanting the extra AER, the more proximal the duplication. Therefore, the presence of one AER does not inhibit the maintenance of another, and the underlying mesoderm can respond to the second AER by both morphogenetic duplication of the limb axis and cytogenetic duplication of the limb skeleton. Localization of the limb bud does not depend on the ability of the embryo to maintain only one ridge per limb site.

b. Mutants. The converse of the wingless mutant is the polydactylous mutant (limbs having many digits). Embryos carrying these mutations may have limb buds with ridges up to 50% larger than normal, as is the case in dominant polydactyly and in duplicate (Zwilling and Hansborough, 1956), in talpid^2 and in talpid^3 (Goetinck, 1966; Ede and Agerback, 1968), and in preaxial polydactyly in man (Yasuda, 1975a); alternatively, embryos may have a second AER on the limb bud, as in eudiplopodia (Goetinck, 1964; Fraser and Abbott, 1971a,b). Mutants with an enlarged ectodermal ridge are caused by primary defects of the mesoderm, while those with an extra ridge are caused by primary defects of the ectoderm (see Section B).

Zwilling and Hansborough (1956) used the polydactylous mutant, dominant polydactyly, to provide some of the most convincing evidence for the directive role of the ectodermal ridge on limb development. Up to the third day of incubation, normal and polydactylous limb buds are indistinguishable. Early in the third day, the AER of the polydactylous mutant enlarges. If one-third of the AER is removed from the polydactylous limb bud at this stage, a normal limb develops, providing clear evidence that the enlarged AER plays a role in the mutation. If mesoderm from a polydactylous limb bud is grafted beneath ectoderm of a normal limb bud, the AER of the normal bud thickens preaxially and, as in polydactyly, accessory distal limb elements develop. One can manipulate the form of the limb by appropriate combinations of normal and mutant tissue.

Fraser and Abbott (1971a,b) have provided equally compelling evidence for the mutant eudiplopodia, in which distal portions are duplicated on the dorsal face of the limb bud. These arise in response to a second AER, which develops one day later than, and independent of, the normal AER (at H.H. stages 22–23), and provide the first sign of deviation from normal development. As far as is known, this ability of eudiplopodia ectoderm to form a ridge up until H.H. stage 23, when normal limb ectoderm loses the ability at H.H. stage 18, is the primary action of the gene. Ectoderm from these mutants can respond to mesoderm from wild-type, talpid^2, or Japanese quail embryos, by producing a secondary apical ectodermal ridge and distal skeletal duplications in each case.

4. Narrow or Subdivided Apical Ectodermal Ridges

No positive examples of these conditions are known. If apical ectodermal ridges are grafted side by side onto a limb bud, they fuse into one giant ridge; if cultured in contact, the ridges fuse. Unless they are far apart, AER's cannot retain their independence. Cleft hand deformity in man might arise because of subdivision of the AER (Yasuda, 1975b), although the evidence is not compelling.

5. Specificity of the Ectoderm

Limb mesoderm can maintain a second ectodermal ridge (Zwilling, 1956a). Can it also make a ridge from nonlimb mesoderm? Zwilling (1964) tested this possibility. Ectoderm from the flank was grafted to an ectoderm-denuded limb bud after the normal stage of initiation of the AER. The limb failed to develop further. Therefore, either the ability to initiate AER production is time specific or flank ectoderm is nonresponsive. Would younger mesoderm induce flank ectoderm to produce a ridge? Zwilling dissociated limb mesoderm into single cells, reaggregated the cells into a pellet, wrapped them in either limb or flank ectoderm, and grafted them to the chorioallantoic membrane. The mesoderm formed cartilage when associated with limb ectoderm, but not when associated with nonlimb ectoderm. Therefore, the limb ectoderm does have a specific role to play in evoking chondrogenesis within the limb, and this role is both site and time specific. Other experiments involving the response of the flank ectoderm to limb mesoderm will be discussed in Section B, 1, in which the evidence for the time of onset of this specificity will also be considered. It will be observed that, once flank ectoderm has been in contact with presumptive limb mesoderm for a time, it becomes limb ectoderm, and that the ectoderm not having had this contact loses the ability to respond to the mesoderm and can no longer become limb ectoderm. Timing of positioning of limb buds along the axis thereby is specified (see Table XIII).

6. Specificity of the Mesoderm

The mesoderm that normally responds to the presence of the AER is the distal mesoderm of the limb bud, the mesoderm immediately subjacent to the AER. Can more proximal mesoderm of the bud respond to the AER, and can non-limb bud mesoderm respond to the AER? The responsiveness of mesoderm that is not normally presumptive limb mesoderm is considered in Section B,2.

Saunders and his colleagues (1955, 1957, 1959) transplanted proximal prospective thigh mesoderm adjacent to the AER of the wing bud between H.H. stages 18 and 27. Will the proximal mesoderm produce proximal or distal limb structures, and will these structures be typical of foot or of wing? They turn out to be distal, and foot in character (Table X). The mesoderm responds by producing the skeletal elements appropriate to that level of the proximodistal sequence, but retains its limb type specificity by producing a toe rather than a digit typical of a wing. Proximal mesoderm *could* respond, as long as it was taken from embryos younger than H.H. stage 24 and as long as it was in contact with the AER. Placing proximal mesoderm beneath distal mesoderm subjacent to the ridge prevented the former from responding.

Amprino and Bonetti (1964) performed similar experiments using embryos of H.H. stages 25 and 26; however, even though they left 15–20 rows of mesodermal cells at the tip, these workers still obtained distal structures from the trans-

TABLE X
Results of Typical Experiments that Involve Grafting Proximal Hindlimb Bud Mesoderm Subjacent to the Apical Ectodermal Ridge of the Wing Bud[a,b]

Donor age (H.H. stage)	Percent of grafts producing distal foot structures (*n*)
17 to 19	86% (74/86)
20 to 22	62% (24/39)
23 to 25[c]	29% (15/51)
26 to 27[c]	0% (0/33)

[a] Distal foot structures are produced.
[b] Based on data in Saunders *et al.* (1959)
[c] The ability of proximal limb mesoderm to produce distal skeletal structures is lost between H.H. stages 25 and 26.

plant. They concluded that contact of proximal mesoderm with the AER was not essential, but at that late stage the graft may have received the ectodermal message secondarily from the determined distal mesoderm with which it had come into contact. Indeed, the experiment does not necessarily rule out ectodermal involvement at earlier stages.

7. Time Dependency

Rubin and Saunders (1972) examined the time relations of the interaction between ectoderm and mesoderm. Regardless of the respective ages of the two tissues, the ectoderm could elicit complete limbs from limb mesoderm when ectoderm–mesoderm recombinants were grafted as flank grafts. The ages tested were H.H. stages 18–20 mesoderm recombined with ectoderm from embryos of H.H. stages 23–25 (i.e., younger mesoderm with older ectoderm) and H.H. stages 22–25 mesoderm with H.H. stage 19 ectoderm (i.e., older mesoderm with younger ectoderm). Ectoderm from the older embryos was just as capable of eliciting limb development from the mesoderm as the younger ectoderm. Rubin and Saunders proposed that the inductive signals from the ectoderm were constant over time, and that any proximodistal sequence within the limb bud must be a property of the mesoderm.

The AER does finally lose its ability to induce limb outgrowth during H.H. stage 29 (Table XI). This loss of inductive ability was not accelerated by combining ectoderm with mesoderm from older embryos, nor was it slowed by combining ectoderm with mesoderm from younger embryos.

Amprino and his colleagues have produced evidence that they interpret as contrary to the "Saunders–Zwilling" model of ectoderm–mesoderm interaction (Amprino, 1965, 1975a, 1977b). Their view has not gained wide support in the

TABLE XI

Results of Experiments to Test the Time Dependency of the Ability of the Apical Ectodermal Ridge to Induce Limb Bud Mesoderm to Grow and to Produce the Skeleton of the Limb[a,b]

Age of embryo providing limb bud ectoderm (H.H. stage)	Percent of grafts forming wings (*n*)
26 to early 29	78% (36/46)
29	18% (4/22)
30	0% (0/9)

[a] Ectoderm from the limb buds of embryos of various ages was recombined with limb bud mesoderm from embryos of H.H. stages 18 to 20 and maintained as a flank graft. Based on data in Rubin and Saunders (1972).

[b] This ability to induce limb bud mesoderm is lost between H.H. stages 29 and 30.

past. However, a recent analysis by Amprino and Ambrosi (1973) does seem to be gaining some acceptance, and may be synopsized as the "biomechanical role of the ectoderm." Using both wing and leg buds from embryos aged between H.H. stages 18 and 25 (3–5 days of incubation), these authors placed agar plate markers either immediately beneath the ectodermal ridge or within various levels of the apical mesenchyme. During limb growth, they observed a proximodistal sliding of the entire ectodermal cover, moving the AER more and more distally. If the ectoderm was transfixed dorsoventrally, the sliding ceased, growth of the limb bud ceased, and distal deficiencies arose. While these are experimental results normally associated with removal or destruction of the AER, in their study, the AER was not injured; rather, movement of the whole ectoderm was stopped. Amprino and Ambrosi therefore proposed that the ectoderm as a whole plays a role in modeling mesodermal outgrowth. Whether the ultimate control is by ectodermal movement governing mesenchyme growth, or by mesenchyme growth governing ectodermal movement, they do not say. Their work does show, however, that mesodermal outgrowth is greater distally, and that the superficial blood vessels of the limb bud maintain a constant distance from the ectoderm via vascular reconstruction. The possible pressure exerted by this tissue is also viewed as important in mesodermal growth. As Amprino (1974) summarized it: The pressure of the vascularized mesoderm is greater along the proximodistal axis than along other axes, and exerts a tension on the ectoderm that is balanced by division and sliding of ectoderm, i.e., control is initially mesodermal.

The proposal of greater mesodermal growth distally raises the question of its mechanism of growth. Amprino and his colleagues found no evidence for cell migration or cell hypertrophy. Unless the pressure of the superficial blood vessels is sufficient, one is left with the possibility of a proximodistal gradient in cell division within the mesoderm. Discussing the available data, these investigators emphasize that no significant gradient has been found, even though one is needed to simulate limb outgrowth by computer modeling (a topic discussed in Chapter 9). Thus, Amprino *et al.* conclude that some slight gradient must be present and operative.

The notion that sliding of the ectoderm is largely responsible for maintenance of the AER is in line with the absence of mitotic activity within the AER (but, for a contrary view, see Errick and Saunders, 1976; Saunders *et al.*, 1976). Mitosis is slight when the ridge arises (at H.H. stage 17), and by H.H. stage 21, has completely stopped (Amprino, 1974). Cell division does occur in the ectoderm adjacent to the ridge, an event which could bring about a piling-up of cells into a ridge.

8. Apical Ectodermal Ridges within the Vertebrates

The obviously significant function served by the avian AER prompts one to ask whether such structures, and/or ectodermal–mesodermal interactions, are a universal property of vertebrate limb buds. As we shall see below, absence of a ridge does not necessarily indicate absence of ectoderm–mesoderm interaction.

The limbless reptiles (Section 2) represent one group of vertebrates shown to possess an active AER as the limb bud is still developing. Reptiles that retain their limb buds also possess ectodermal ridges (Milaire, 1957), And, although fish have been little studied from this point of view, Bouvet (1974, 1976) has shown that both teleosts and the Elasmobranchi possess ridgelike structures as precursors to appendage development.

The South African clawed toad (*Xenopus laevis*) has been studied and found to have an AER. If this ridge is surgically removed, the digits fail to develop. If it is rotated 180° and replaced, digits form, but with abnormal orientation (Tschumi, 1957). Ultrastructural evidence of specialization in the AER has been provided by Tarin and Sturdee (1971, 1974), and presumptive evidence for AER–mesoderm interaction by Kelley and Bluemink (1974). The latter authors describe the following sequence. Presumptive limb mesenchymal cells arrive at the epithelium at the future limb site and accumulate within the complex extracellular matrix lying directly beneath the epidermis. This matrix is then partially degraded and rearranged, allowing close contact between mesenchymal and epithelial cells. Specialized junctions between adjacent mesenchymal cells then form. Similar disorganization of subepidermal basement membrane and subsequent interaction between ectoderm and transplanted neural fold have been postulated as the basis for the induction of supernumerary limbs in *Bufo* (Balinsky, 1974); Overton (1977) has produced direct evidence for alteration of the morphogenetic behavior of mesenchymal cells when they are cultured on isolated basement lamellae.

Although Lauthier (1974) could find no enzymatic specificity in the ectoderm of the limb buds of *Pleurodeles waltlii*, the penetration of filopodia from the mesenchyme into the basal layer of the ectoderm (Lauthier, 1977) suggests the possibility of interaction. Neither this species nor *Nothophthalamus cristatus* or *N. vulgaris* has a morphologically specialized ectodermal ridge (Sturdee and Connock, 1975).

Milaire (1956, 1963, 1965a) has summarized the presumptive morphological and cytochemical evidence for a functional AER in the mouse limb. Its time of appearance is appropriate for interaction with the underlying mesoderm preceding chondrogenesis. If it is removed from the mouse limb bud earlier than 10½–11 days of gestation, cartilage fails to develop. If removed at or later than 11 days of gestation (the precartilaginous stage of limb development), chondrogenesis is normal (Milaire and Mulnard, 1968). This requirement for ectoderm before chondrogenesis will occur may not be specific to limb ectoderm, for mesoderm from 10-day limb buds will grow and chondrify when in contact with ventral spinal cord. It also appears that nonridge limb ectoderm may direct proximal differentiation, while ridge ectoderm governs distal organization

(Milaire, 1965a); however, experimental verification of this notion has not been obtained. Filopodia from mesenchymal cells contact the basement membrane underlying the ectoderm of the limb bud (Houben, 1976).

Cairns (1965) grafted apical mesoderm from both young and old mouse limb buds beneath the ectoderm of H.H. stages 18–20 chick limb buds. His young mouse mesoderm was equivalent in gross development to that of H.H. stages 18–24 chick limb buds, and allowed the development of mouselike terminal digits on the chick limb buds. In grafts of older mesoderm, growth was slowed, and no digits developed. Maxillary mesoderm also failed to respond. There was no intermingling of chick and mouse cells, confirming the lack of large-scale cell migration during limb bud development. Cairns concluded that mouse mesoderm required interaction with ectoderm, and that the interaction was time dependent. Human limb buds also possess an AER (O'Rahilly *et al.*, 1956). It appears at about five weeks (Streeter's stages 15–16), and by the seventh week (stages 17–18) it is thinning where it is underlain by necrotic mesoderm (see Section C). Gap junctions and bundles of microfilaments characterize the ultrastructure of the inductively active AER on avian, mammalian, and primate limb buds (Kelley and Fallon, 1976; Fallon and Kelley, 1977; Kaprio, 1977). No experimental work has been carried out on human or primate material. Swinyard (1969) and Bergsma and Lenz (1977) offer reviews relevant to human limb development.

In *Xenopus*, the AER can regenerate rapidly and must be re-removed in deficiency experiments (Tschumi, 1957). Given the appropriate circumstances, the avian AER may also regenerate. Searls and Zwilling (1964) cultured ectoderm isolated from H.H. stages 19 and 20 limb buds with agar, somitic mesoderm, or with limb bud mesoderm. In all cases, the AER degenerated within seven to ten hours, but a new ridge formed if limb bud mesoderm was the substrate. This new ridge could support limb outgrowth and skeletal development if the tissues were grafted to the chorioallantoic membrane. Even after contact with agar for ten hours and regression of the AER, limb development could be reinitiated by replacing the agar with limb bud mesoderm and grafting to the chorioallantoic membrane. Searls and Zwilling also extirpated the AER *in ovo* and found that, provided the ectodermal fragments left were close together, a new AER would form.

Errick and Saunders (1974) performed an elegant experiment on re-formation of the AER. After placing H.H. stages 21–23 leg bud mesoderm into inside-out jackets of wing or leg bud ectoderm, they then grafted these composites into the somites. The ectodermal histology was rapidly restored, and, although growth of the limb bud began after 24 hours, it took three days for the AER to reappear. Non-limb bud ectoderm would not substitute for the limb bud ectoderm. This implies, then, that the ectoderm was acting on the mesoderm even though an AER could not be seen. Dissociated ridges grafted with leg mesoderm from H.H. stage 19 embryos also direct limb development (Errick and Saunders, 1976). It is necessary, then, that active ectoderm be sought, and not just the presence or absence of an apical ectodermal ridge.

This cautionary note applies to the work of Amprino and Aiello-Malmberg (1971), in which the developmental role of the ectoderm was questioned. In this experiment, approximately 75% of the distal and central mesoderm was removed from the leg bud of H.H. stages 23–25 embryos. The two peripheral, proximal pieces were joined together, and it was found that, provided the ectoderm had covered the stump, some digits developed from this proximal mesoderm. However, an AER was never seen in these limb buds. They concluded that the mesoderm provided all of the necessary stimulation, but it may well have been that the ectoderm, even though it had not formed a ridge, had nevertheless still acted on the mesoderm to direct limb outgrowth.

It would be valid, then, to generalize that the localization and development of the vertebrate limb require interaction between presumptive limb mesenchyme and a physiologically—and in some cases, cytologically specialized—epithelium. Urodele amphibians may be an exception, although even here caution must be exercised. Even in the avian embryo, where the AER is so prominent, the limb ectoderm has been shown to act upon limb mesoderm when the former is disorganized.

B. Induction by the Mesoderm

It was implied in Section A that the AER was induced within the ectoderm by some influence from the presumptive limb mesoderm. The evidence for this will now be briefly reviewed, along with that on the timing of the interaction (see also Section A,6).

1. Limb Mesoderm in Ectopic Sites

The first experiments, performed on embryonic amphibians, are now classics in experimental embryology (Harrison, 1918). When prospective limb bud mesoderm was grafted under nonlimb ectoderm, as, for example, on the flank, the mesoderm swelled (grew?), an AER was induced within the flank ectoderm, and a supernumerary limb grew out from the side of the host. Limb mesoderm could induce nonlimb ectoderm to form an AER. When prospective limb mesoderm was removed, leaving the ectoderm intact, no limb developed. When mesoderm from outside the prospective limb area (the limb field) was grafted under ectoderm within the limb field, no limb developed. Specificity of mesoderm and evocation of ectodermal ridge by mesodermal action were strongly suggested.

The first definitive evidence for a *mutual* interaction between mesoderm and ectoderm came from the experiments of Zwilling (1956b) on the wingless mutant in the chick. In our previous discussion of this mutant, we indicated that an AER forms but then regresses (Section A,1,b), much as in abortive limb bud development in limbless vertebrates. Zwilling performed recombination experiments involving limb buds from normal and wingless embryos. When mesoderm from normal embryos was grafted beneath limb bud ectoderm of wingless embryos, no limb formed. When mesoderm from a wingless embryo was grafted beneath ectoderm from a normal embryo, over the next three days the existing AER flattened, and the distal structures of the limb failed to form. Information was flowing from mesoderm to ectoderm and *vice versa*. Zwilling postulated the existence of a mesodermal factor that maintains the AER (see Section C).

Kieny (1960), Reuss and Saunders (1965), and Saunders and Reuss (1974) performed similar experiments with limb versus nonlimb sites on normal embryos. Pre-limb bud mesoderm was grafted to the flank. The flank ectoderm produced an AER, and a supernumerary wing developed. The timing of this interaction provides a clue to the normal time course of the interaction of limb mesoderm and ectoderm. Wing mesoderm loses the ability to induce formation of an AER in flank ectoderm, and flank ectoderm loses the ability to respond at H.H. stage 17 (see Table VIII), the stage at which the first sign of a limb bud normally appears. The timing ensures that once one bud has formed, ectopic buds will not be produced. Presumptive limb bud area mesoderm from embryos of H.H. stage 11 or younger will not elicit ridge formation from ectoderm. Although mesoderm from embryos of H.H. stages 12–17 will act, older mesoderm will not, thus localizing the time of interaction somewhere between H.H. stages 12 and 17.

Collagen accumulates beneath flank ectoderm of embryonic chicks between H.H. stages 12 and 17, but does not accumulate beneath prospective limb ectoderm (A. A. Smith *et al.*, 1975). This barrier may represent part of the reason why flank ectoderm loses its ability to support limb development at H.H. stage 17. The inductively active AER of the human limb bud may also be distinguished from adjacent ectoderm on ultrastructural grounds (Kelley and Fallon, 1976).

Dhouailly and Kieny (1972) have demonstrated that flank somatic mesoderm from embryos of H.H. stages 10–17 can participate in limb bud formation if brought into contact with young (H.H. stages 13–17) pre-limb bud mesoderm. The prospective limb mesoderm was grafted into the flank, where it migrated to the ectoderm, induced the formation of an AER, and participated in the formation of a supernumerary limb. Tritiated thymidine labeling, the nuclear histology of Japanese quail cells, and type of epidermal differentiations in heterospecific grafts, have all indicated involvement of flank mesoderm in the supernumerary limb. Limb bud mesoderm from H.H. stages 19 and 20 has also been grafted beneath the tail ectoderm of H.H. stages 21 and 22 embryos, where it induced tail ectoderm to form an AER. When the composite, i.e., tail ectoderm plus limb mesoderm, is grafted to the chorioallantoic membrane, limb outgrowth and digit formation occur (Searls and Zwilling, 1964).

The specification of the localization of the position of the limb bud then seems to rest initially with the presumptive limb mesoderm. This mesoderm, over defined periods of time, induces flank ectoderm to acquire specialized developmental properties that allow it to direct further limb development.

2. Mesoderm Specifies Limb Type

As well as the specification of the *location* of the limb bud along the flank of the embryo, the question of specification of the *type* of limb that will develop has also been studied. It turns out that mesoderm, rather than ectoderm, specifies whether a forelimb or a hindlimb will form. This conclusion is based on reciprocal transplantation studies between limb buds of either the same or different species (Cairns and Saunders, 1954; Zwilling, 1955; Saunders *et al.*, 1957, 1958).

If ectoderm from the leg bud is grafted onto an ectodermless wing bud, the distal structures that develop are of wing type, and are covered with wing feathers. Conversely, when ectoderm from wing bud is grafted onto leg bud mesoderm, the skeletal elements and feathers typical of the leg develop. When mesoderm from the leg bud is grafted onto the wing in place of wing bud mesoderm distal digits typical of the leg will form, whereas mesoderm from the wing bud grafted onto the leg bud produces digits typical of wing on a leg stump. The distal structures are always typical of the origin of the mesoderm and *not* of the origin of the ectoderm. When the exchange is between different species (e.g., between common fowl and duck), the results are even more striking—the species-specific characteristics reinforce the mesodermal specification of the limb type. The one known exception of the eudiplopod mutant, in which control is ectodermal, has already been mentioned (Section A,3).

C. Maintenance of the Ectodermal Ridge by Mesodermal Factors

The majority of the evidence for the presence in the mesoderm of a factor that maintains the existence of the apical ectodermal ridge (the so-called apical ectodermal maintenance factor, AEMF) comes from studies on wingless or polydactylous mutants in the chick. Zwilling (1956b) and Zwilling and Hansborough (1956) provided the initial evidence that the AER in the wingless mutant regressed because of absence of AEMF. When mesoderm from the limb bud of a wingless mutant is grafted into a mesoderm-denuded normal limb, the normal AER regresses and a wingless embryo is produced.

In the wingless, wing buds last only to the third day of incubation. If wingless mesoderm is combined with normal ectoderm and grafted to the dorsal face of a normal wing bud, the AER persists, outgrowth of the mesoderm occurs, and some skeletal structures develop. If an intact early limb bud from a wingless mutant is grafted to the same site on the normal limb bud, its AER persists for longer than normal and there is more than the usual amount of mesodermal outgrowth (Zwilling, 1974). Zwilling thus concluded that AEMF from the normal host limb diffused into the grafts, and maintained and continued limb development. He did not record grafting to an ectopic site (such as somite) as a control.

AEMF is distributed within the postaxial half of the limb bud [except for the posterior necrotic zone (PNZ), which lacks AEMF (Fig. 10)]. Placement of a filter across the limb bud dividing it pre- and postaxially results in regression of the AER (Saunders and Gasseling, 1963). If AEMF is located solely postaxially, then 180° rotation of the distal half of the limb bud should place AEMF in the preaxial half of the limb bud, where it should induce a second AER and a duplicated limb. Saunders and Gasseling (1968) report that, in fact, this is the case (Fig. 10).

The polarity of the AER is maintained by a gradient from the AEMF within the mesoderm. MacCabe and Parker (1975) cultured preaxial wing mesoderm alone, with flank mesoderm, or with anterior limb mesoderm, and found that the AER flattened under all three conditions. When preaxial mesoderm was cultured with postaxial limb mesoderm, the AER was maintained with no evidence of cell death. When cultured with middorsal limb mesoderm (i.e., from along the proposed gradient) there was some loss of AER and some cell death. A cell-free extract of postaxial but not of preaxial limb buds has now been shown to possess AEMF activity (Calandra and MacCabe, 1978).

Mitotic inhibitors administered *in ovo* diminish the rate of accumulation of mesodermal cells, a diminution which in turn leads to premature loss of the AER and to hypophalangy (Kieny, 1975) or to polydactyly (W. J. Scott, *et al.*, 1977).

Recently, Globus and Vethamany-Globus (1976) have provided evidence for the production of AEMF by limb mesoderm cells maintained in monolayer culture. When such cells are covered by an AER, a growth pattern results which produces a limblike bud covered by an AER of normal configuration. Nonridge ectoderm did not promote mesodermal cell accumulation. Whether nonlimb mesodermal cells would have maintained the AER was not tested.

In the mutant diplopodia^4 studied by MacCabe *et al.* (1975), more AEMF is produced preaxially, resulting in preaxial thickening of the ectoderm and in preaxial polydactyly. The same argument has been used for diplopodia^1 and for talpid^2 (MacCabe and Abbott, 1974).

It was stated above that the posterior zone of cell death within the wing (the PNZ) lacks AEMF. This evidence was provided by Gasseling and Saunders (1964), and has been summarized by Saunders and Gasseling (1968). If the PNZ of the wing or the equivalent area of the leg bud is grafted below the AER, the AER regresses and disappears; however, a new one forms immediately preaxial

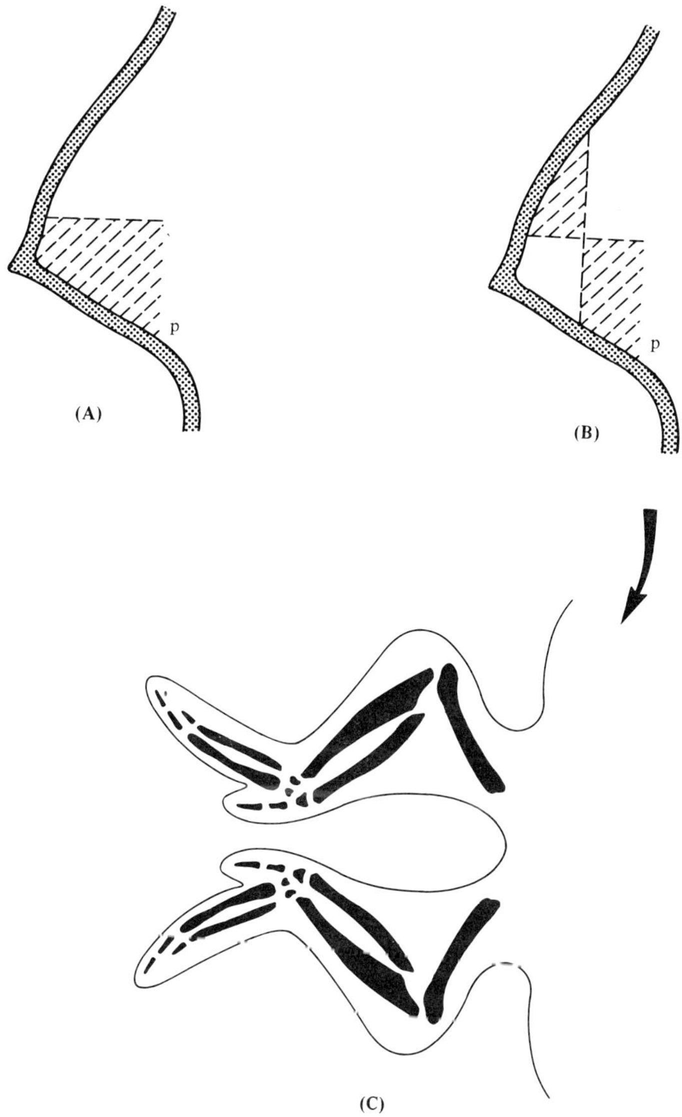

Fig. 10. (A) The location of the apical ectodermal maintenance factor (AEMF) in the wing bud is shown by the dashed area; (p) is the posterior necrotic zone, a zone lacking AEMF. If the distal tip of the wing bud is rotated 180° and replaced, AEMF is placed into the preaxial portion of the limb bud (B) and a mirror duplication of the wing skeleton develops (C). See Fig. 9 for identification of the skeletal elements.

to the graft and a limb grows out from this preaxial location. Paradoxically, then, the PNZ does not produce AEMF, although it does induce adjacent mesodermal cells to produce it. Perhaps this explains the normal postaxial location of the AEMF and the asymmetry of the AER—its position is partially governed by the location of the PNZ. However, except for cell death-related phenomena, removal of the PNZ does not interfere with normal limb development. Therefore, if the PNZ does influence production of AEMF *in vivo*, its action is either very rapid and/or early in development, or it passes long-lasting messages into the mesoderm. Summerbell (1974a) has recently confirmed that grafting a portion of the PNZ subectodermally results in preaxial thickening of the AER. Since the AER is normally thicker at the posterior margin, it is postulated that this zone plays a role during normal limb development *in ovo*.

The properties of the limb mesoderm may be summarized under five headings: (1) It responds to the AER by proliferation and outgrowth; (2) it maintains or supports formation or regeneration of an AER; (3) it transmits AEMF largely in a proximodistal direction; (4) it forms the skeleton in a proximodistal sequence, and (5) it contains limb-type specificity. This reciprocal interaction between mesoderm and ectoderm, then, determines the site, outgrowth, and differentiation of the limb buds within the vertebrates.

IV. THE VERTEBRAL SKELETON

A. Introduction

Chondrification of the sclerotomal portion of the segmentally arranged somitic mesoderm takes place along the embryonic axis; the presumptive vertebrae thus formed result in part from an inherent bias toward chondrogenesis that these cells either possess or acquire and in part from an inductive interaction with the adjacent notochord and spinal cord. This interaction augments, rather than imposes, the chondrogenic bias and localizes the vertebrae within the somitic mesoderm adjacent to the spinal column. Just as the brain induces the ectomesenchyme of the neural crest to produce the skull as its container, so the spinal column induces its skeletal container, the vertebral column. The topographical relationship between the sclerotomal cells as they migrate away from the remainder of the somite (the dermamyotome, which produces connective tissue and muscle) and the spinal cord and notochord as they appear in an embryonic chick incubated for about 50 hours is shown in Fig. 11.

The evidence for the conclusion that notochord and spinal cord determine the localization of the vertebral cartilage could be the subject for a discussion as

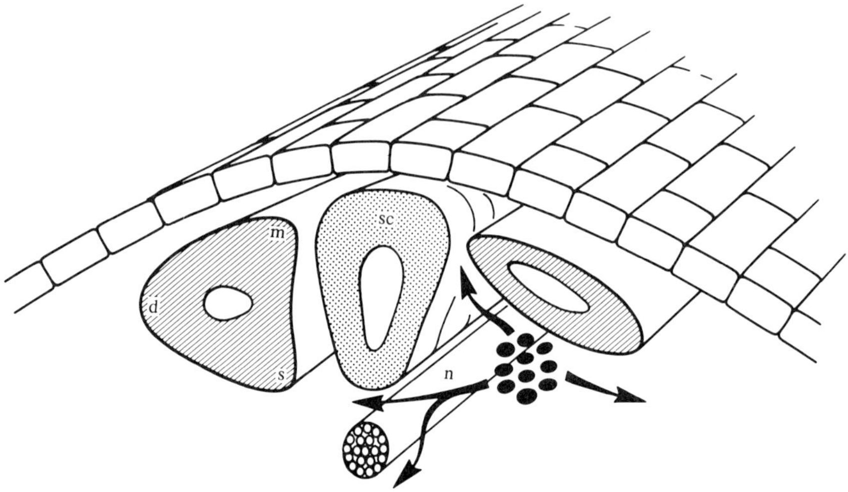

Fig. 11. A transverse, diagrammatic view of the axial structures of the embryonic chick as they appear at ca. H.H. stage 10 [left of the spinal cord (SC) and notochord (n)] and after initiation of migration of sclerotomal cells (black) at ca. H.H. stage 14. The routes of migration are shown by the arrows. Abbreviations: d, dermatome; m, myotome; s, sclerotome. Reproduced from Hall (1977a) with the permission of the publisher.

in depth as that of limb development. However, because chondrogenesis of the somitic mesoderm has just been extensively reviewed (Hall, 1977a), I shall outline only the major pieces of evidence herein. For a detailed discussion, see the above-mentioned review.

B. The Dermamyotome

As shown in Figure 11, the somite consists of three major portions, namely, the sclerotome, dermatome, and myotome (dermamyotome). Vertebral cartilage develops only within the mesenchyme of the sclerotome—not because the dermamyotome cannot chondrify, but because it does not normally come into contact with the inductively active notochord and spinal cord. While removal of the sclerotome in urodele amphibians does not prevent development of vertebral cartilage, it does reduce muscle development, even though myotome is present and intact. Implantation of spinal cord or notochord into the dermamyotome of amphibians or of embryonic chicks induces a secondary set of vertebral cartilages from the nonsclerotomal cells (Holtzer and Detwiler, 1953; Strudel, 1953b). Chondrogenesis takes precedence over myogenesis within myotome placed next to the embryonic axis.

C. Interaction between Notochord–Spinal Cord and Sclerotome

Knowledge of the conditions under which sclerotomal mesenchyme is maintained is crucial for interpreting the combined role of the notochord and spinal cord in eliciting chondrogenesis. For example, cartilage forms from younger-aged somites when they are grafted to the chorioallantoic membrane than when they are maintained *in vitro* (Avery *et al.*, 1956). Somites from H.H. stage 16 embryonic chicks will not chondrify if cultured in the absence of agar or serum; some will chondrify if serum or agar is added to the medium, and all will chondrify if cultured adjacent to spinal cord or notochord (Strudel, 1962, 1963; Lash, 1964; Lash *et al.*, 1964; Zilliken, 1967). Somites from embryos as young as H. H. stage 9 (30 hours of incubation) will chondrify in the absence of spinal cord if given appropriate supplementation to the culture medium (Ellison *et al.*, 1969).

The mass of somites cultured also influences how much cartilage will be produced, or indeed whether chondrogenesis will occur at all [see Hall (1977a) for a discussion]. This chondrogenesis in the absence of spinal cord or notochord may represent a "self-induction" by feedback from extracellular products produced by the somite (Kosher, 1976), mimicking the role of the extracellular matrix of the notochord or spinal cord (see Section D).

Evidently, then, there is a chondrogenic bias within somitic mesoderm that can be elicited by macromolecules of agar, serum, or the extracellular matrix when the mesoderm is cultivated *in vitro* in the absence of spinal cord or of notochord.

In vivo, the extracellular matricial products surrounding the spinal cord and notochord elicit somitic chondrogenesis. The epithelium adjacent to the embryonic axis also may play a role, although if it does, this epithelial influence is subordinate to that of the spinal cord and notochord (Hall, 1977a).

In amphibians, the notochord is essential for morphogenesis, and the spinal cord is essential for differentiation of vertebral cartilage (Hörstadius, 1944; Kitchin, 1949; Holtzer, 1951, 1952a,b; Holtzer and Detwiler, 1953). Ablation experiments show that, in the avian embryo, the notochord is responsible for induction of the centra of the vertebrae, while the spinal cord is responsible for induction of the neural arches (Strudel, 1953a, 1955; Watterson *et al.*, 1954; also see Fig. 12). Only the ventral half of the spinal cord is inductively active (Grobstein and Holtzer, 1955; Flowers and Grobstein, 1967; also see Hall, 1977a).

A very considerable literature on the cultivation of somites in contact with the spinal cord or notochord *in vitro* indicates that somites from young embryos that do not chondrify when cultured alone *do* chondrify when in contact with spinal cord or notochord. In addition, there is extensive evidence that somites from older embryos produce more cartilage, and that they produce it more rapidly

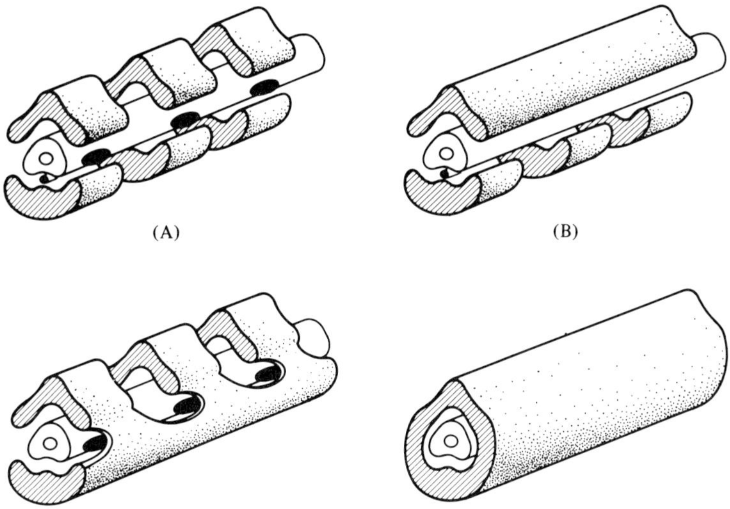

Fig. 12. The effect of extirpation of the spinal ganglia and/or notochord on morphogenesis of the vertebral column in the embryonic chick. (A) A normal embryo to show the segmental arrangement of the spinal ganglia (black) and of the neural arches and centra (stippled, neural arch above and centrum below the tubular spinal cord). (B) Excision of the spinal ganglia does not influence morphogenesis of the centra but does result in the formation of unsegmented neural arches. (C) Excision of the notochord does not influence morphogenesis of the neural arches but does result in the formation of unsegmented centra. (D) Excision of both spinal ganglia and the notochord results in the formation of a totally unsegmented vertebral column. Reproduced from Hall (1977a) with the permission of the publisher.

when cocultured with spinal cord or with notochord (Lash *et al.*, 1957; Holtzer, 1964a,b; Lash, 1967; Holtzer and Mayne, 1973; Gordon and Lash, 1974; also see Hall, 1977a).

D. The Nature of the "Inducer"

Considerable research time has been spent in attempting to isolate an active component (inductive molecule) from spinal cord or from notochord that would permit or enhance the chondrogenesis of somitic mesoderm. No appreciable success has been achieved. Hall (1977a) may be consulted for a discussion.

Over the past seven or eight years, emphasis has shifted to the extracellular matrix (see Løvtrup, 1974, 1977; Manasek, 1975; also the volume edited by Slavkin and Greulich, 1975). Both the notochord and the ventral portion of the spinal cord synthesize and secrete a highly localized extracellular matrix composed of collagen, chondroitin sulfate, and hyaluronic acid. The sclerotomal cells migrate into this matrix as they approach the embryonic axis. Notochord and spinal cord that have been trypsinized to remove their extracellular matrices and then cultured in the absence of enzyme resynthesize their matrices (Cohen and Hay, 1971; Carlson *et al.*, 1974; Carlson and Upson, 1974; Hay and Meier, 1974), confirming that the extracellular matrix is indeed the product of the notochord and spinal cord. Trypsinized notochords such as these will not stimulate somitic mesoderm to chondrify unless the extracellular matrix has been allowed to reform (Kosher and Lash, 1975). But addition of the individual components of the extracellular matrix (chondroitin 4- and 6-sulfate or type II collagen) to somites *in vitro*, while maintaining ongoing chondrogenesis, does not initiate chondrogenesis (Kosher *et al.*, 1973; Kosher and Church, 1975; also see Hall, 1977a).

In conclusion, the location of chondrogenic centers, and therefore the initiations of vertebral development within somitic mesoderm, is dependent upon interactions between that mesoderm and the notochord and/or spinal cord or their extracellular products.

V. THE AUDITORY CARTILAGE

The auditory (otic) cartilage is mesodermal rather than ectomesenchymal in origin (Hörstadius, 1950; Holtfreter, 1968). Its position and cytodifferentiation depend upon an interaction between the mesenchyme and the otic epithelium (otocyst); this has been demonstrated in turtles (Toerien, 1965), in birds (Benoit, 1960a,b, 1963; Benoit and Schowing, 1970; Simons, 1974; also see Milaire, 1974, for a discussion), and in mice and men (Goedbloed, 1964; Pugin, 1972). Chick otic epithelium will not induce ectomesenchyme or somitic mesoderm to chondrify, although the ectomesenchyme of the neural crest may provide a permissive environment for differentiation of the otic mesenchyme (Hörstadius, 1950).

VI. ECTOPIC SKELETOGENESIS

A. Transitional Epithelium of the Urinary Bladder

The transitional epithelium was first discussed in Chapter 3, Section III,B, in the context of the induction of osteogenesis by inducible osteogenic progenitor cells from a variety of tissues in the postnatal mammal. This ability of transitional epithelium to determine the position of ectopic nodules of cartilage and bone extraskeletally provides a further example of an interaction involving epithelium and mesenchymal cells that leads to the initiation of skeletogenesis. In this case, the epithelium not only plays a role in determining the location of the skeletal nodule, but also elicits skeletogenic differentiative activity from mesenchymal cells that normally do not ossify or chondrify.

Charles Huggins and his colleagues carried out the classical studies on the induction of bone under the influence of the transitional epithelium in the 1930's (Huggins, 1931a,b; Huggins and Sammett, 1933; Huggins *et al.*, 1936). These authors implanted transitional epithelium either intramuscularly or subcutaneously, and showed that host connective tissue underlying the epithelium formed ectopic bone. No bone arose when transitional epithelium was transplanted into kidney, liver, or spleen unless fascia or connective tissue was included in the implant. They concluded that only some tissues possessed a population of cells capable of responding to transitional epithelium. The location of ectopic bone thus depends on the location of a population of inducible cells and on their proximity to an inducer, in this case, their proximity to transitional epithelium. The ability to initiate synthesis of alkaline phosphatase is one prerequisite to the ossification of connective tissues (Johnson and McMinn, 1956).

Several modifications of the original system have been attempted. Transitional epithelium of the pelvis will induce ectopic bone when transplanted subcutaneously (Tavassoli and Crosby, 1971). Ectopic bone can also form *in situ* in the bladder, and tumors of the urinary bladder evoke ectopic bone and cartilage formation (Willis, 1962; Damjanov and Urbanke, 1969; Delides, 1972). These growth phenomena are of interest in light of the relationship between epithelial proliferation and inductive ability, for the tumors are also proliferating rapidly. Ioseliani (1972) has correlated heightened mitotic activity of the epithelium (based on autoradiography after tritiated thymidine) with inductive ability. Huggins had observed that only actively growing epithelium was inductively active; the same may hold for the tumors and for the epithelial cell lines that evoke ectopic osteogenesis (see Section B). Formation of this induced bone continues only when the epithelium is present. This is also characteristic of somitic cartilage, which has been induced to form *in vitro* (Hall, 1977a) and may be a general feature of induced skeletal tissues.

Transitional epithelium will also act on osteogenic progenitor cells located *intra*skeletally. Beresford and Hancox (1965, 1967) used pieces of bladder dispersed in polyurethane sponge or in protein-free bone matrix to accelerate the repair of fractured cranial bones in both guinea pigs and in rats. Gilbert and Gorman (1971) used urinary bladder mucosa to accelerate fracture repair in the dog. Whether the epithelium commits more cells to osteogenesis or whether it accelerates the rate of bone formation from already committed cells is not known; however, the evidence offered by Friedenstein's work (Chapter 3) indicates that it may be the latter.

Thus, in at least some examples of ectopic bone formation, the location of the bone is determined by the interaction of competent cells with an epithelium.

For a broader review of these interactions, Bridges (1959) and Ostrowski and Wlodarski (1971) may be consulted.

B. Epithelial Cell Lines

There is now a body of evidence to indicate that a variety of epithelial cell lines are capable of inducing ectopic cartilage or bone formation. This ability was demonstrated first by H. C. Anderson and his colleagues in the U.S. and then was taken up by K. Wlodarski in Poland.

Cultured human amniotic cells (FL strain), when injected intramuscularly into cortisone-treated mice (so treated to slow or prevent the immunological rejection of the cells), produce tumors in which the host "fibroblasts" modulate to chondroblasts. The ectopic cartilage then is replaced by calcified bone, which by ultrastructural criteria, is indistinguishable from normal bone (Anderson, 1966, 1967; Anderson *et al.*, 1964). Injection of thymidine-labeled amniotic cells shows that only approximately 10% of the host fibroblasts exhibit very light labeling three days postinjection. This is interpreted as representing reutilization of label from amniotic cells rather than transformation of amniotic cells to cartilage (Anderson and Coulter, 1967). The host cells that modulate most readily in response to injection of amniotic cells are the fibroblasts of skeletal muscle (Anderson, 1976a). Perhaps this would be a useful model for *Myositis ossificans progressiva*.

While species or generic specificity is not required for induction of this ectopic cartilage (Wlodarski *et al.*, 1971a), both newborn and X-irradiated mice fail to form cartilage (Wlodarski *et al.*, 1970, 1971b). Subcutaneous injection is also ineffective (Hancox and Wlodarski, 1972).

Wlodarski (1969) also showed that a variety of established normal and neoplastic, murine and nonmurine, epithelial cell lines (HeLa, WISH, CLV-J3, CLV-4, CLV-X, CLV-14) would evoke chondro- and osteogenesis. Primary cultures would not. Some of these cell lines are not epithelial in origin but are virus-transformed fibroblasts (the CLV series) that exhibit epithelial growth patterns after prolonged culture. It seems to be the epithelial pattern of growth, and not the epithelial origin, that gives these cells their inductive ability (Ostrowski and Wlodarski, 1971). These active cell lines also are all heteroploid and agglutinate with concanavalin A, whereas noninductive cell lines are diploid and do not agglutinate (Wlodarski *et al.*, 1974). This raises the possibility that the cell membrane, as well as being a useful marker for inductively active cells, may carry information, and also that contact of the epithelial cells with host fibroblasts might be necessary for modulation to occur. Transfilter studies have established that the induction fails to cross the filter barrier, implicating cell-to cell contact. Nor will coculture or cografting of thigh muscle and epithelial cells to the chorioallantoic membrane elicit cartilage or bone formation, although both form upon intracerebral transplantation (Anderson, 1976a).

As with most (all?) ectopic or heterotopic bone, the bone induced by established cell lines is only maintained as long as the epithelial cells survive—in practice, several months. The bone then undergoes resorption (Wlodarski *et al.*, 1970, 1971b).

These studies may be compared to those which use transitional epithelium to evoke ectopic skeletogenesis, and to studies of epithelial–mesenchymal or epithelial–ectomesenchymal interactions during skeletal development. By this comparison, we see that aspects of the interactions relating to cell–cell contact, to mitotic activity of the epithelium, to the pattern of epithelial growth, and to the basis of the response of host fibroblasts, are the potentially profitable areas for further investigation.

There is at least one epithelium that actively stimulates the *resorption* of bone. The gingival epithelium of man possesses a 10,000–70,000-MW factor that stimulates resorption of the adjacent alveolar bone, especially in periodontal disease (Goldhaber *et al.*, 1973).

There also is a body of literature providing circumstantial evidence for the *metaplasia* of epithelial cells to cartilage or to bone, for example, the ectopic cartilage of the thyroid and pancreas (Kay and Harrison, 1969; Finkle and Goldman, 1973). That epithelial cells can produce sulfated glycosaminoglycans and collagen, the characteristic macromolecules of skeletal extracellular matrices, is well established (Bischoff, 1971; Hay and Dodson, 1973; Meier and Hay, 1973, 1974a; Thonard and Wiebkin, 1973), as is the ectodermal origin of many invertebrate cartilages (Person and Philpott, 1969a; Moss, 1972a; Hall, 1975a). The evidence for metaplasia of vertebrate epithelia to cartilage or to bone is much less direct, and has been reviewed by Willis (1962) and by Jaffe (1968).

5

Initiation of Centers of Skeletogenesis

I. INTRODUCTION

Central to the discussion of Chapter 4 was the evidence indicating that, in a considerable variety of skeletal rudiments, the location of a particular skeletal part, either in the embryo or in the adult, was determined by specific interactions between mesenchyme or ectomesenchyme and epithelia. It was shown that one or the other, or both, of the tissues in these interactions might either (a) possess a developmental bias toward skeletogenesis before the interaction occurred, or (b) acquire that bias as a result of the interaction. In either case, the position of the rudiment was dependent upon the epithelial–mesenchymal or epithelial–ectomesenchymal interaction.

Initially, the mesenchyme forms a uniform, homogeneous framework within the embryo (Grüneberg, 1963; Jollie, 1971). Thus far, virtually no mention has been made of the *mechanisms* by which the mesenchymal or ectomesenchymal cells accumulate at their specific locations within the embryo. The presumptive cells of the vertebrae, the mandible, and the teeth all migrate to their final site, where they accumulate. While we do know that they stop their migration upon contact with specific epithelia, what factor(s) govern(s) the migration? Do these cells migrate as single cells or in clusters? Do they divide as they migrate? Does their accumulation around the embryonic axis, in the mandibular arch, or along the oral epithelium result from the first migrant cell passing information along to the next and so on, or must each cell individually come into contact with the epithelium?

In those skeletal rudiments in which migration takes place over short distances or is nonexistent, as in the limb bud, accumulation of cells adjacent to the epithelia is also seen, and was described in Chapter 4. The accumulations of these progenitor cells form aggregations (condensations, blastemata, anlagen) which have long been recognized as the forerunners of the future skeletal rudiments, even though the synthesis of products characteristic of differentiating skeletal cells might not be apparent within them. Grüneberg (1963) has referred to these condensations as the "membraneous skeleton" of the embryo, and has accumulated evidence which highlights their importance. His data indicate that many skeletal mutants have their primary action on these condensations very early in development. He argued that in condensations that have been reduced below a critical size, chondrogenesis or osteogenesis either would not be initiated or would be delayed. This would result in morphological defects in the skeletons of older embryos or in the adult, as Kochhar (1973) has shown to occur in the mouse.

The role of the condensations in the development of skeletal abnormalities is not, however, quite so clear-cut. In the congenitally hydrocephalic (*ch*) mouse, some of the cartilages (e.g., the trachea) are missing because the blastemata are smaller than normal; some membrane bones (e.g., the vomer) are absent, even though the condensations are larger than normal; and both cartilage and membrane bone develop in some condensations in which either cartilage or membrane bone, but not both, normally develop (Grüneberg and des Wickramaratne, 1974). In the chondrodystrophic (*cho*) mouse, in which the long bones are shortened in the late fetus, the initial condensations are normal in size (Seegmiller *et al.*, 1971). As a result, the importance of the mesenchymal environment created within and around the condensations, rather than their size, came to be stressed as a critical factor in the initiation of skeletogenesis (Grüneberg and des Wickramaratne, 1974). In addition to mutant genes, teratological treatment can disturb this environment. Administration of 6-mercaptopurine to pregnant rats produces a massive reduction in the size of the condensations (Merker *et al.*, 1975). Regrettably, very little is known about the environment within these condensations; thus, any attempts to understand the factors that control the determination and differentiation of skeletal cells, or to extrapolate results obtained *in vitro* to the situation *in vivo*, meet with continual difficulty [see Willmer (1970), Hall (1971), and Saxén (1976) for some comments].

There is an extensive literature dealing with the histochemical characteristics of skeletal condensations, particularly for the mammalian limb bud, as a result of the studies of Milaire in Brussels (Milaire, 1965a, 1967, 1974). I do not wish to review the wealth of detailed histochemical information that is available. For comprehensive reviews, the reader may consult Pritchard (1952), Cabrini (1961), Fullmer (1965), and Kobayashi (1971), and for discussions on the usefulness of histochemical investigations for unraveling aspects of determination and differentiation, one may refer to Hall (1968a), Milaire (1974), and Drews and Drews (1972, 1973). The latter studies relate to the distribution of cholinesterase in skeletal condensations and to its possible role in initiating the synthesis of extracellular matrix in the formation of the condensation. This evidence may be quite important in light of the role of neurotransmitters as inductive stimuli, a view proposed by McMahon (1974), Kebabian *et al.* (1975), and Landauer (1976).

Presently, I wish to turn to the question of whether the initiation of a condensation of cells normally occurs before mesenchymal cells can chondrify or ossify. Since the limb buds of the embryonic chick have been studied the most intensively, I shall concentrate my discussion on this cell system.

II. MESENCHYMAL CONDENSATIONS IN THE DEVELOPING LIMB BUD

Histological observations on the developing vertebrate limb bud date back to at least the early part of this century. Fell (1925) and Fell and Canti (1934a,b) have provided detailed descriptions of the cytology and histology of the developing hindlimb of the embryonic chick. Initially, all of the mesenchymal cells of the limb bud have a uniform cytological appearance. A clustering of mesenchymal cells in the center of the limb bud was described preceding the appearance of extracellular matrix at H.H. stage 25. These cells were described as closely packed, rounded cells with large nuclei. Subsequent ultrastructural studies have shown them to have a high nucleo–cytoplasmic ratio, large nucleoli, poorly developed endoplasmic reticulum, and numerous free ribosomes and small mitochondria (Fig. 13). These are the general characteristics of osteo- or chondroprogenitor cells in a variety of situations (Hall and Shorey, 1968).

The reanalysis of the histology of the limb bud that accompanied the burst of experimental studies on limb morphogenesis in the 1950's reaffirmed, by means of both light and electron microscopy, the existence of mesenchymal cell condensations preceding chondrogenesis in the chick, the mouse, and the rat (Saunders, 1948; Godman and Porter, 1960; Jurand, 1965). More recent studies in some—but not in all—laboratories have cast doubt on the existence of such condensations in limb buds, and by extension, on the necessity of their presence before mesenchyme can chondrify or ossify. It is important to note that some of the studies have been carried out on the wing bud and some on the hindlimb bud. Wing bud development lags behind hindlimb development by about 12 hours and since the period under study (H.H. stages 19 to 25) covers only 36 to 48 hours, this lag could be critical when events in one limb are compared with those in another.

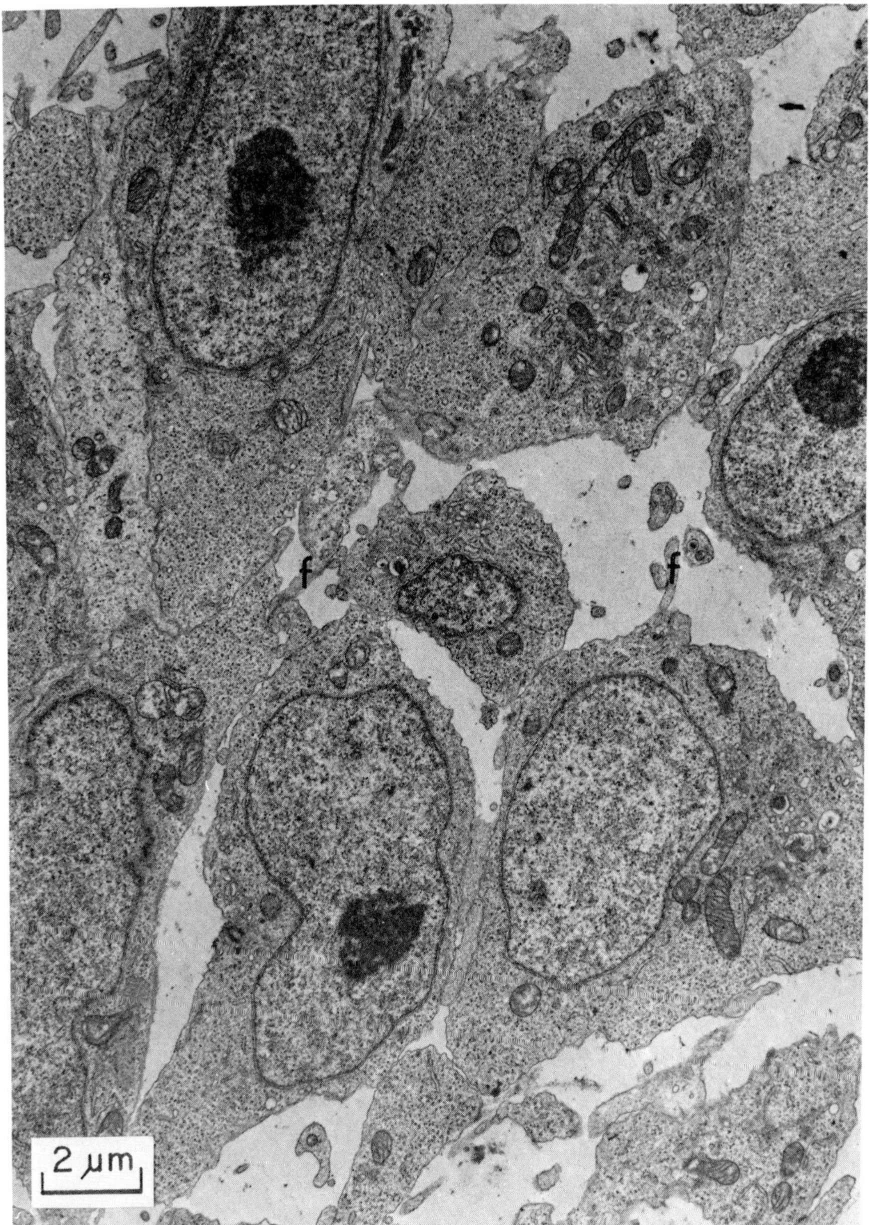

Fig. 13. Mesenchymal cells from the central proximal area of the wing bud of an embryonic chick of H.H. stage 19. Note the broad areas of cell to cell contact, the filopodia (f) and the scant amounts of extracellular matrix. Reproduced from Searls *et al.* (1972) with the permission of the author and publisher.

If condensations are present (and the bulk of the evidence indicates that they are), their mode of formation remains ambiguous. The existence of a condensation requires that there be either an increased number of cells within the condensation and/or a decreased number of cells around it. While there is no evidence for a local increase in mitotic activity at the time of cell condensation in the embryonic chick wing (Janners and Searls, 1970), there is an increase in cell packing (cell density) (Searls, 1973a,b). The possible role of localized increases and decreases in mitotic activity during limb bud development is discussed extensively in Chapter 9, where it is considered in the context of limb growth. Here, it is appropriate only to reiterate the major conclusion of those studies as they relate to the initiation of condensations, namely, that localized changes in mitotic activity do not coincide with initiation of condensations within the limb bud. The increased cell packing within the condensation, then, must be brought about by another mechanism, i.e., either the aggregation or incorporation of cells to a center or the failure of cells to move away from a center.

A consideration of the ultrastructural analysis of Searls *et al*. (1972) will set the stage for a discussion of the changes in cell density during the precartilaginous phases of limb development. These workers examined the wing bud over the period covered by H.H. stages 17 to 27 (2–5.5 days of incubation), and using ultrastructural analysis of transverse sections, identified three stages based on the nature of cell-to-cell contacts and associations.

Between H.H. stages 17 and 23 (a 24-hour period), no regional specialization of cell types could be found either across or along the wing bud, and no alteration in organelle type of configuration was observed (Fig. 13). Cells were in contact over broad areas (up to 4 μ) of their membranes. Between H.H. stages 23 and 25, the number of these broad contacts declined, filopodia began to appear, and many cell-to-cell contacts were now mediated by these projections (Fig. 14). Although the cells in the center of the bud were beginning to differentiate, the extracellular matrix was not being deposited. Searls *et al.* emphasized that the changing nature of the cell contacts from broad areas of cell membranes to restricted areas of projections from the membranes resulted neither from the production of extracellular matrix nor from a decrease in cell density (see mechanisms of origin of condensations, Section III); rather, the altered cell contacts must have reflected changes in the properties of the cell membrane at these stages. The properties of the membrane, relative to the initiation of segregation of limb mesenchymal cells from their precursors, also begin to alter at these same developmental stages (see Chapter 6, Section VI,B,4).

From H.H. stage 25 onward, cell-to-cell contacts were made by filopodial contacts of less than 1 μ (Fig. 15). Searls and associates did not see changes in cell shape or arrangement until H.H. stage 27, when extracellular matrix was being synthesized (Fig. 16).

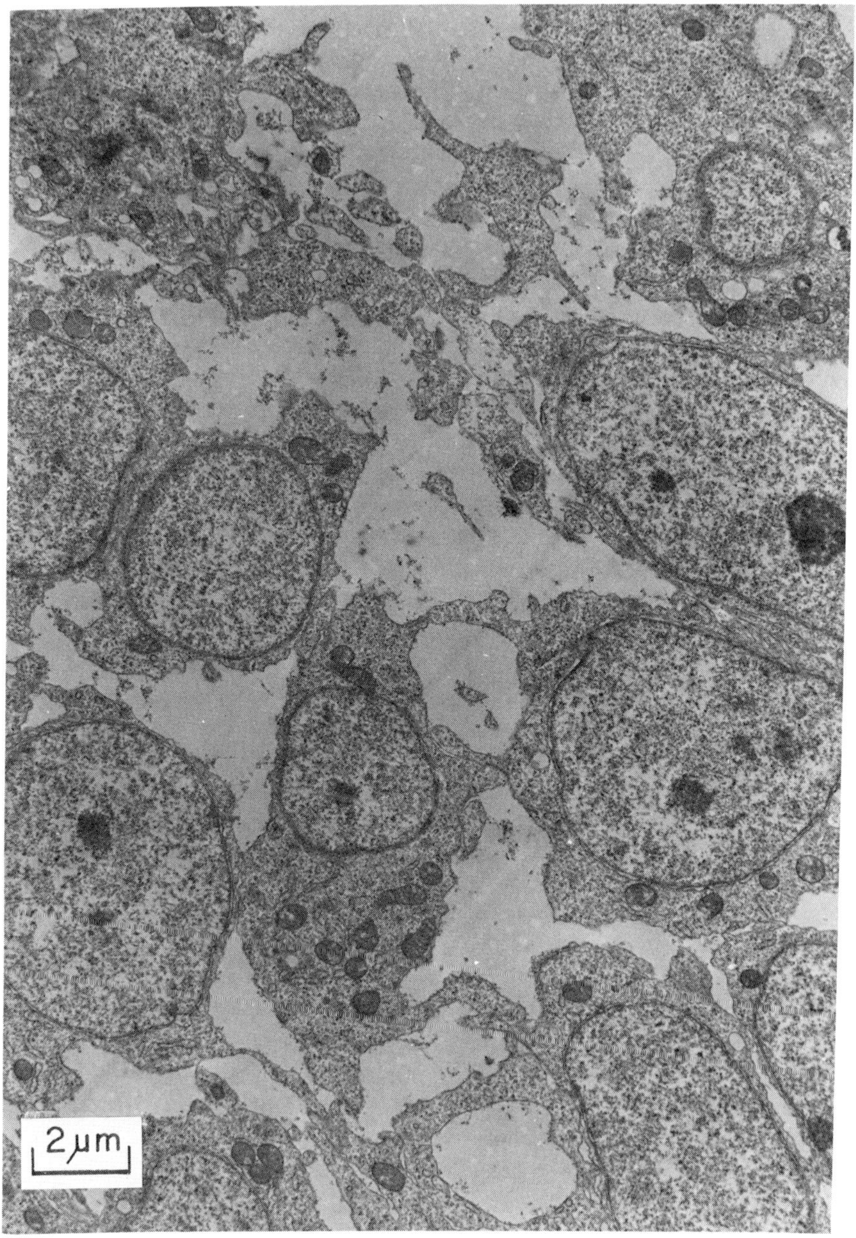

Fig. 14. Mesenchymal cells from the central proximal area of the wing bud of an embryonic chick of H.H. stage 23. More filopodia and narrow cell-to-cell contacts than were present at H.H. stage 19 (Fig. 13) are evident. Reproduced with the permission of the author and publisher from Searls *et al.* (1972).

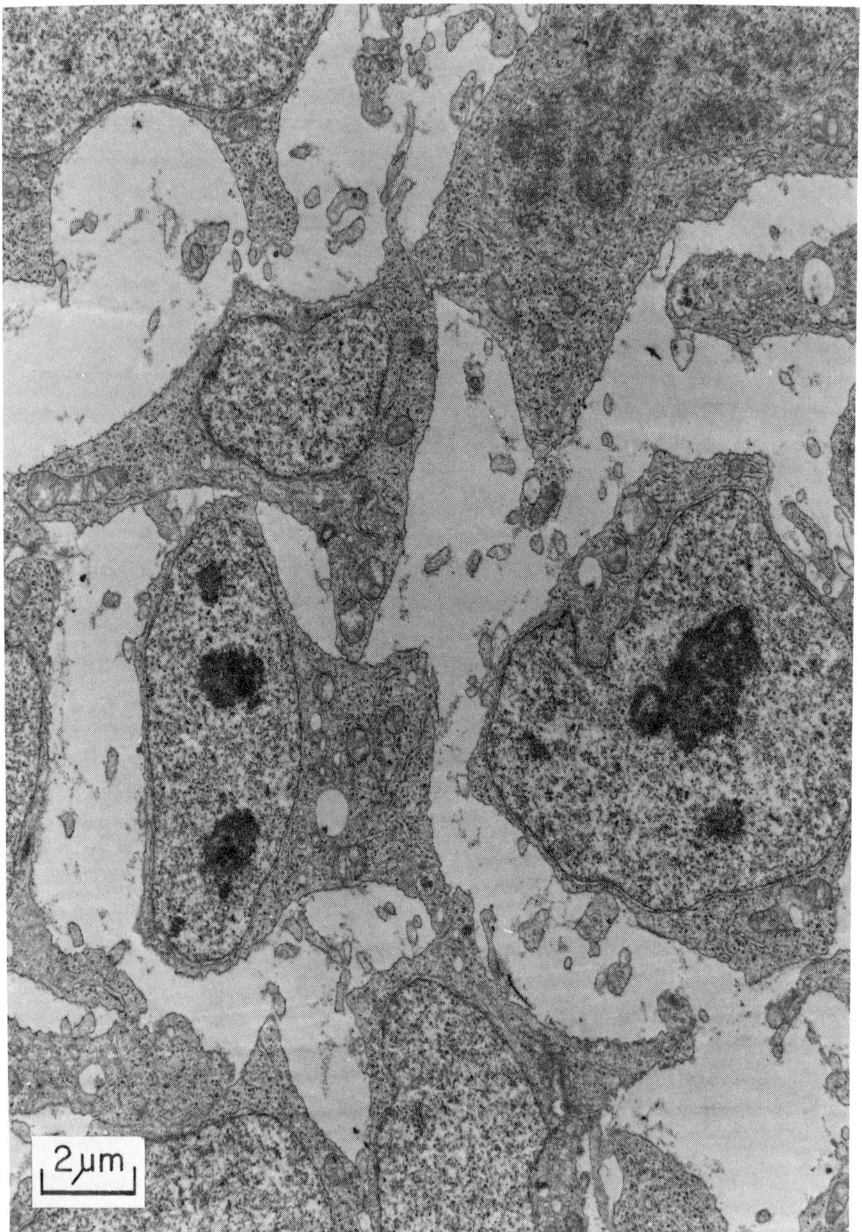

Fig. 15. Mesenchymal cells from the central proximal area of the wing bud of an embryonic chick of H.H. stage 25. There are few contacts between adjacent cells (cf. Fig. 14), and initial accumulation of extracellular matrix and separation of cells are evident. Reproduced from Searls *et al.* (1972) with the permission of the author and publisher.

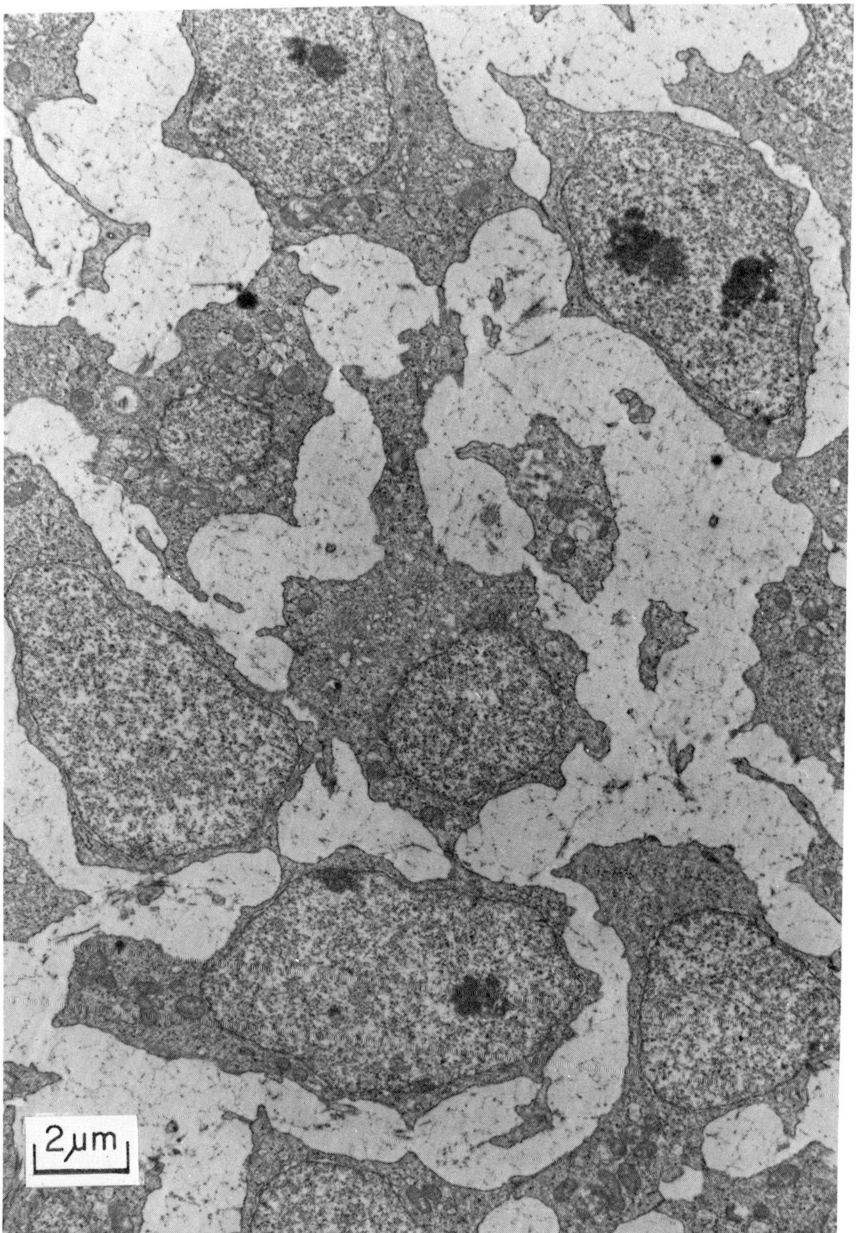

Fig. 16. Prechondrogenic cells from the central proximal area of the wing bud of an embryonic chick of H.H. stage 27. Cell-to-cell contacts are only by filopodia (cf. Figs. 14 and 15). Cell morphology has transformed from mesenchymal (Fig. 15) to prechondroblastic as extracellular matrix accumulates. Reproduced from Searls *et al.* (1972) with the permission of the author and publisher.

In summary, while Searls *et al.* did observe alterations in the nature of cell-to-cell contacts during limb initiation, they did not observe alterations in cell density over H.H. stages 17 to 25, nor did they report any evidence of condensations of mesenchymal cells, the latter result in obvious contrast to the earlier studies using the light microscope. Grafting leg buds from embryos of H.H. stages 18 to 24 onto the chorioallantoic membrane on rafts of Millipore filters showed that chondrification could occur without being preceded by a condensation phase (Searls, 1968). Therefore, these workers maintain that the condensation of mesenchymal cells does not occur in the wing bud *in vivo*, and that it is not a prerequisite for chondrogenesis of the mesenchyme of the leg bud in an ectopic (avascular) site.

In a companion study on myogenesis in the wing bud, Hilfer *et al.* (1973) observed a broadening of cell-to-cell contacts in presumptive muscle-forming peripheral mesenchyme. This contrasted with the narrowing of cell contacts reported in the more central prospective chondrogenic mesenchyme. These features gave the appearance of the presumptive myogenic cells as more compact (condensed) than the presumptive chondrogenic cells.

This impression of condensation of myogenic rather than chondrogenic mesenchyme of limb buds studied at the electron microscopic level was supported by the ultrastructural analysis of Gould *et al.* (1972). They studied the wing bud in transverse section over H.H. stages 19 to 27, as had Searls *et al.* (1972). Although no increase in cell-to-cell *contact* was observed in the prechon-

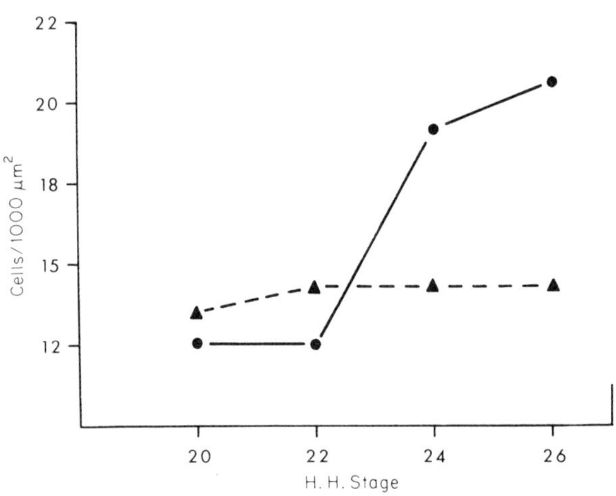

Fig. 17. Cell density (cells/1000 μ^2) increases in the prechondrogenic mesenchyme (●——●) but not in the peripheral mesenchyme (▲--▲) of the hindlimb of the embryonic chick between H.H. stages 20 and 26. Based on data in Thorogood and Hinchliffe (1975).

drogenic mesenchyme, a 30% increase in cell *density* (from 11 to 15 cells/1000 μ^2) was noted. Summerbell and Wolpert (1972) recorded increases in cell density by a factor of 1.5 between H.H. stages 19 and 23, and by a factor of 2.0 between H.H. stages 19 and 25. Myogenic mesenchyme studies by Gould *et al.* showed both increased cell packing and increased cell-to-cell contact, as described by Hilfer *et al.* (1973). If a condensation represents an area of cells both more closely packed and containing more cell contacts than an adjacent area, then the myogenic mesenchyme will show true condensation on ultrastructural analysis, while the chondrogenic area will give this impression under light microscope analysis.

Thorogood and Hinchliffe (1975) examined the condensation process in the *hind*limb bud over H.H. stages 22 to 26. They documented a considerable increase in cell packing within the prechondrogenic mesenchyme over this period (an increase of from 12 to 21 cells /1000μ^2), but observed no change in the myogenic area (Fig. 17). Their ultrastructural evidence indicates increasingly broad cell-to-cell contacts in the chondrogenic, but not in the myogenic, mesenchyme. There is a clear conflict between this study on the hindlimb and those discussed above concerning the wing. Thorogood and Hinchliffe emphasized the difference in timing between wing and hindlimb bud development; they also

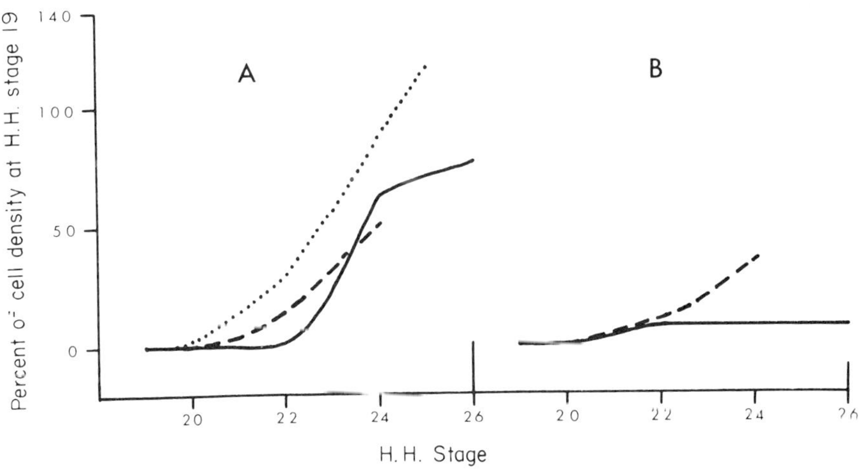

Fig. 18. A comparison of the cell densities (expressed as percent of the cell density at H.H. stage 19) between (A) the central, prechondrogenic and (B) the peripheral premyogenic mesenchymes of the embryonic chick hindlimb bud (———), data from Thorogood and Hinchliffe (1975) and wing bud (--- and. . . . . .), data from Searls (1973a); Summerbell and Wolpert (1972). Note the general agreement that cell density increases after H.H. stage 22 in the prechondrogenic mesenchyme and that cell density is greater in the prechondrogenic than in the premyogenic mesenchyme at all stages.

stressed the transitory nature of the condensations, and the fact that when the fixation procedures of Gould *et al.* (1972) were used, shrinkage occurred and condensations could not be observed with the electron microscope. Obviously, Thorogood and Hinchliffe feel that condensations are present in both sets of limb buds. Further modifications of preparative methods and parallel studies of wing and hindlimb buds using standard techniques are needed before this question can truly be resolved.

An analysis of the density of cells within the prechondrogenic and premyogenic centers of the developing limb bud clearly is of importance in these discussions of the presence or absence of condensations. Some of the available data on cell density have been summarized in Figure 18. Note the consensus regarding the increase in cell density within the prechondrogenic mesenchyme after H.H. stage 22, but not before, and that cell density at any age is greater in the prechondrogenic (central) mesenchyme than in the premyogenic (peripheral) mesenchyme. These data support the existence of condensations of cells in the prechondrogenic areas of the developing limb bud. Disagreement over the nature of the cell-to-cell contacts remains, making the interpretation of ultrastructural studies very difficult. Tissue preparation is evidently critical. The nature of the mechanism that produces the denser packing of cells within the condensations is also hotly debated.

III. MECHANISMS INVOKED FOR INITIATION OF CONDENSATIONS

As shown by Janners and Searls (1970), and as discussed in Chapter 9, there is no localized increase in mitotic activity coincident with the initiation of condensations in the limb bud. Nor does differential division appear to be a factor in the formation of either the dermal lamina or the enamel organ in the mouse (Osman and Ruch, 1975). Differential rates of cell division *are* involved in the formation of condensations in some other skeletal rudiments. Jacobson and Fell (1941) considered the origin of the preskeletal condensations of the mandibular arch and presented cytological evidence that cell division in these condensations was more rapid than in the surrounding mesenchyme. As discussed in Chapter 4, Section II,E,3, the conjunctival papillae stimulate the rate of cell division in presumptive scleral mesenchyme, thereby producing a condensation in which osteogenesis can occur to form the scleral ossicle (Hale, 1956a).

Alternate mechanisms have been proposed to account for the increased cell density within condensations of the limb buds. These explanations are as follows: (1) the absence of movement of cells away from a center and the dispersal and flattening of cells at the periphery (Gould *et al.*, 1972, 1974), and (2) the aggregation of cells toward a center (Ede and Agerback, 1968; Thorogood and Hinchliffe, 1975).

Gould and his colleagues have examined both wing buds of embryonic chicks (H.H. stages 19–28) and the regeneration blastema of amputated limbs of adult *Triturus cristatus*. Figures 1–5 in their 1974 paper provide an excellent visualization of the formation of a condensation during digit development. The form of the condensation, it has been proposed, is generated by a mechanism involving the synthesis, secretion, and accumulation of extracellular matrix, especially at the center of the cell mass. This accumulation of matrix would lead to a separation of the central cells, enlargement of the cell mass, and orientation of the cells perpendicular to the long axis of the condensation. The cells at the periphery of the condensation produce little extracellular matrix and flatten circumferentially, forming a whorl of perichondrial cells.

Whether there is sufficient spatial restriction at the edges of a condensation to allow flattening of the peripheral cells has been questioned by Goel and Jurand (1975). One certainly can see that, once the chondroblasts begin secreting appreciable amounts of extracellular matrix, the presence of that hydrated matrix would aid in separating the cells, i.e., in *expanding* and patterning the condensation. However, during the developmental stages when the condensation is being *initiated* (H.H. stages 20–22, see Fig. 13 and 14), little extracellular matrix is being synthesized. Both the presumptive chondro- and myogenic cells of the limb bud are synthesizing glycosaminoglycans at the same rate (Searls, 1965a,b). The differential synthesis of collagen has not been examined; perhaps it should be studied, in view of the studies of Vinson and Seyer (1974), who investigated the initiation of feather condensations in the embryonic chick. The synthesis of tissue-specific collagen is enhanced during formation of these condensations, and the application of collagenase or of a lathyrogen prevents condensation formation. Initiation of alkaline phosphatase synthesis is also involved in the initiation of these condensations. Alkaline phosphatase is not seen in the scaleless mutant, which lacks condensations (Goetinck and Sekellick, 1970).

Evidence from the *in vitro* studies on the amphibian neural crest carried out by Epperlein and Lehmann (1975) supports the concept of cells failing to move away from a center and thereby forming a condensation. These workers found that, after the initial contact of cultured neural folds with pharyngeal endoderm, neural crest cells began to proliferate and increased their adhesiveness to one another. This prevented their migration away from the inductive center and allowed a condensation of determined chondrogenic cells to accumulate. Holtfreter (1968) previously had proposed that this inductive message was spread by diffusion rather than by proliferation of induced cells.

Donald Ede and his colleagues, formerly at Edinburgh and now at Glasgow, have led a major assault on the mechanism of initiation of cell condensations in the wing bud using both normal and talpid^3 embryos (Ede and Agerback, 1968; Ede, 1971, 1976; Ede and Flint, 1972; Ede *et al.*, 1977). Arguing that the reaggregation of dissociated cells is the *in vitro* equivalent of the *in vivo* condensation of cells, and using the knowledge that condensations were defective in talpid^3 embryos (Ede and Kelly, 1964a,b), Ede and Agerback (1968) dissociated cells of the wing buds of H.H. stage 24 and stage 26 embryos and observed their behavior during reaggregation. Cells from the normal embryos produced fewer and larger aggregates than cells from talpid^3 embryos (for details on talpid, see Chapter 8). They proposed that cells behave during condensation as they do during reaggregation, aggregating to a central focus, as e.g., in aggregating slime molds. *In vitro*, this acts to increase the size of the aggregates by additions to the edges. *In vivo*, a condensation is produced by the "huddling" of peripheral cells towards the center of the aggregate. The nature of the central attraction (a morphogen, cell contacts, metabolic gradient, etc.) is not known. The distance between the foci would determine whether or not adjacent foci would fuse. The more distal in the limb bud the greater the distance between centers, and the greater the separation of the skeletal elements. When the cells are more adhesive (as in talpid^3), cell motility should be reduced, condensations should form less easily and should fuse more readily (producing the polydactyly of the talpid embryo). Aggregative movement of cells from the periphery to the center of the condensation (huddling), then, is the mechanism proposed by them to explain the origin of condensations within the developing limb bud. A similar aggregation of cells toward a center is seen during development of the horny process on the posterior ray of the anal fin of the female medaka, *Oryzias latipes* (Uwa, 1969, 1974; Uwa and Nagata, 1976).

Any factor that reduces the adhesion of cells to one another or to the substrate will facilitate the formation of condensations and the initiation of chondrogenesis (Watanabe and Okada, 1975). The studies of Elmer and his colleagues on the brachypod (bp^h) mouse provide additional confirmation of this model (Elmer and Selleck, 1975; Duke and Elmer, 1977; Elmer, 1977). The rate of decline in the number of single cells in rotation culture is greater in bp^h limb mesenchymal cells than in normal cells, and only half as much chondrogenesis occurs in the bp^h cells (see Chapter 7, Section II,F,3,c).

When H.H. stage 26 limb bud cells are allowed to aggregate and to chondrify, chondrogenesis always is preceded by the formation of a condensation. This initiates with a few cells and grows by the addition of cells at the periphery. Without this condensation, chondrogenesis does not occur (Ede and Flint, 1972), although there are examples of chondrogenesis occurring without prior condensation of cells (Grobstein and Holtzer, 1955; Searls, 1968). In general, dispersal of prechondrogenic cells favors a fibrous rather than a chondrogenic differentiation (Niven, 1933; Weiss and Amprino, 1940; Abbott and Holtzer, 1966; Fell, 1969, 1976).

IV. HYALURONIC ACID AND THE INITIATION OF CONDENSATIONS

Recent studies on the synthesis, distribution, and degradation of hyaluronic acid have demonstrated that this component of the extracellular matrix plays a vital role in the initiation of cellular condensations preceding the differentiation of cartilage. Precisely *how* hyaluronic acid exerts this action has not been determined, but that it has an action is now well illustrated in a number of skeletal systems.

Hyaluronic acid is a high molecular weight polyanion carrying a highly negative charge. In solution, it expands tremendously and forms a net of extended, random-coiled molecules. Hydrated, it typically occupies the volume of a sphere 400 nm in diameter—10^3 times greater than the volume of the nonhydrated chain. Hyaluronic acid occurs in low concentrations in many animal tissues and forms the backbone to which proteochondroitin sulfate is attached, thus forming large aggregates within the extracellular cartilage matrix (Wiebkin and Muir, 1973). Both hyaluronic acid and the aggregate impede transport through cartilage and help to maintain the greatly hydrated nature of its extracellular matrix. Although some of the hyaluronic acid is intracellular (Larsson and Kuettner, 1974), most lies within the extracellular matrix.

Hyaluronic acid first was isolated in 1934; Miner (1950) has provided a comprehensive review of the early work on hyaluronic acid and on hyaluronidase, the enzyme which degrades it. Although most of this work is biochemical, more recently an involvement of hyaluronic acid and of hyaluronidase in cell aggregation, cell migration, and cell differentiation has been postulated. The general conclusion of these later studies is that hyaluronic acid is associated with cell proliferation and cell migration, while hyaluronidase and/or the removal of hyaluronic acid is associated with the cellular differentiation that follows proliferation, migration, and/or aggregation. Maurer and Hudack (1952) were the first to point in this direction when they showed that considerable hyaluronic acid was synthesized in the early stages of callus formation during the repair of fractured long bones. Cell proliferation and migration are the chief events occurring early in fracture repair. Since this study, a developmental role for hyaluronic acid has been found for vertebral cartilage and limb bud development; the migration of neural crest cells; regeneration of amphibian limbs; ectopic chondro- and osteogenesis and in corneal development.

Kvist and Finnegan (1970) isolated hyaluronic acid from the axial tissues (the notochord, spinal cord, and somites) of embryonic chicks and found that peak levels occurred between H.H. stages 21 and 25, and that more hyaluronic acid than chondroitin sulfate was present until H.H. stage 25. For example, at H.H. stage 17, the hyaluronic acid:chondroitin sulfate ratio was 2.5:1, whereas at H.H. stage 28 it was 0.4:1. This pattern of high levels of hyaluronic acid early in development and declining levels later in development will be seen in all of the examples to be discussed below. Kvist and Finnegan suggested that the initial predominance of hyaluronic acid over chondroitin sulfate was associated with the aggregation of sclerotomal cells, and that the decreasing amount of hyaluronic acid as vertebral chondrogenesis ensued suggested that both the high initial concentrations and the declining levels played developmental roles.

Bryan Toole and his colleagues at Harvard University investigated several systems, all of which involved a phase of hyaluronic acid synthesis as the cells accumulated, followed by removal of the hyaluronic acid during cell differentiation.

Hyaluronic acid is the major glycosaminoglycan found in the cornea of the embryonic chick at H.H. stages 24 and 25. At this developmental age, mesenchymal cells invade the cornea and accumulate. As differentiation of the cornea progresses, especially between H.H. stages 35 and 38 (8.5–12 days of incubation), the level of hyaluronic acid declines, whereas the levels of both chondroitin sulfate and hyaluronidase rise (Toole and Trelstad, 1971). These opposing trends are shown graphically in Figures 19 and 20. Toole and Trelstad suggested that hyaluronic acid affected the behavior of the invading mesenchymal cells by providing a substrate over which they could migrate. In addition, this hyaluronic acid had to be removed before cytodifferentiation could ensue.

Toole and Gross (1971) studied another developmental system in which cell accumulation was followed by a phase of cytodifferentiation. This involved the production of the blastema and the subsequent regeneration of an appendage that followed amputation of the limb of the adult newt *Diemictylus viridescens*. It was found that hyaluronic acid accumulated during both dedifferentiation of stump tissues and during blastema formation, and that as regeneration ensued, hyaluronidase removed the hyaluronic acid. This decline in hyaluronic acid paralleled the rise in chondroitin sulfate as the cartilage of the regenerate differentiated (Fig. 21). If the limb was denervated *before* amputation, blastema formation was poor, the level of hyaluronic acid within the blastema was low, and no hyaluronidase appeared (formation of the blastema is dependent upon normal stump innervation). In fact, the absence of hyaluronidase synthesis was associated with the absence of the phase of differentiation of cartilage and not *because* of the dependence of hyaluronidase synthesis on the presence of nerves. This was shown by denervation of the limb *after* amputation and blastema formation. In these animals, cartilage differentiated and was associated with the synthesis of hyaluronidase (G. N. Smith *et al.*, 1973b, 1975b). A similar dependence of regeneration on the presence of hyaluronic acid and its subsequent removal by hyaluronidase is being investigated in sabellid annelids by Fitzharris (1976), of interest because it is an indication that this dependence is not restricted to the vertebrates.

Similar relationships have also been explored in the developing limb buds and in the somitic mesoderm of the embryonic chick by Toole (1972, 1973a,b). The

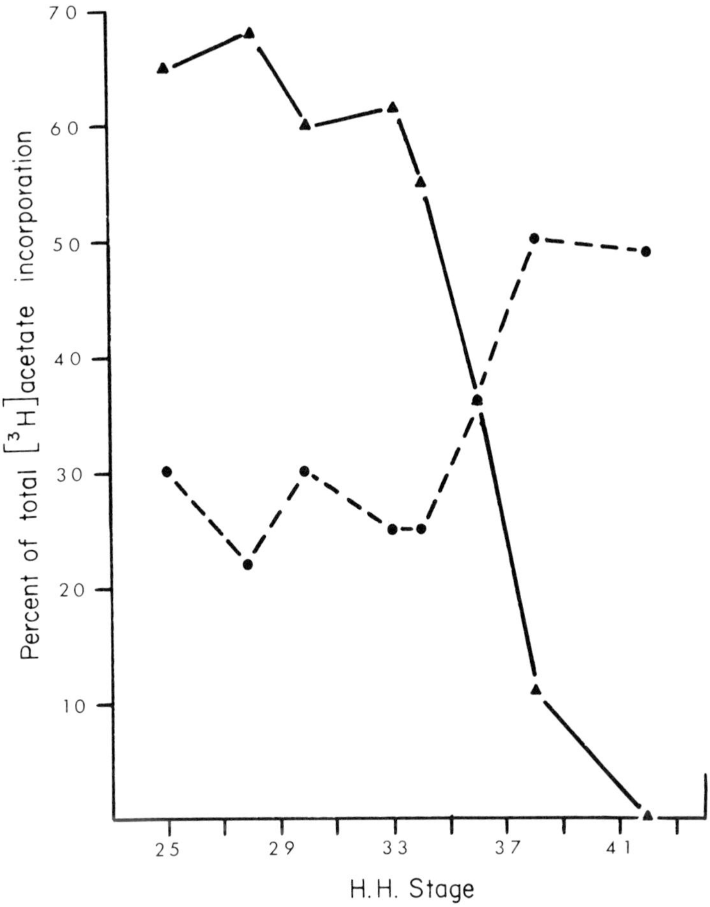

Fig. 19. The inverse relationship betwen hyaluronic acid (▲——▲) and chondroitin sulfate (●--●), both represented as percent of total incorporation of [^{3}H]acetate into glycosaminoglycans, during the development of the cornea in the embryonic chick. Redrawn from Toole and Trelstad (1971).

inverse relationship between levels of hyaluronic acid and chondroitin sulfate and the appearance of hyaluronidase with cytodifferentiation is given in Figure 20.

Some data are available which indicate how the levels of hyaluronic acid and chondroitin sulfate are correlated. At this point, a slight digression to describe these data would be appropriate. If tibial cartilage maintained in organ culture initially is treated for two days with hyaluronidase, and then is cultured for four further days in the absence of hyaluronidase, there is a stimulation of hexosamine synthesis over the level seen in untreated cartilage (Hardingham *et al.*, 1972). Less chondroitin sulfate is synthesized than in untreated cartilage; the chondroitin

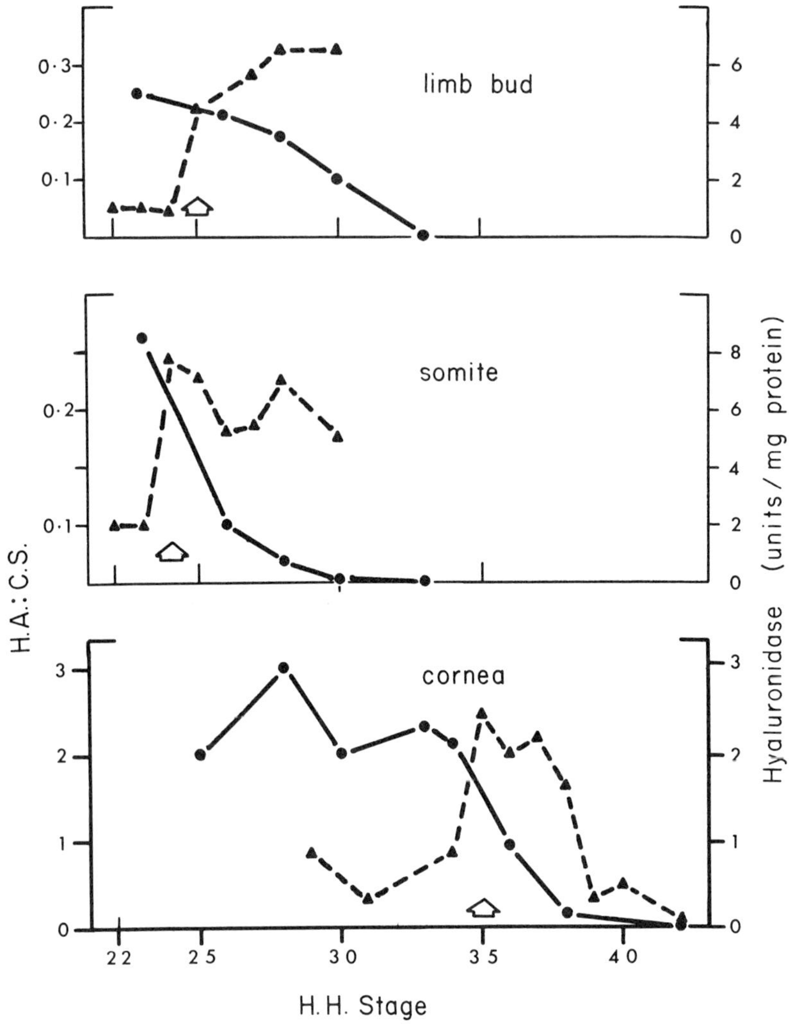

Fig. 20

sulfate that is synthesized has a shorter chain length than normal and is uncoupled from the hyaluronic acid backbone of the hyaluronic acid–proteochondroitin sulfate aggregate. When hyaluronic acid is added to cultures of chondrocytes obtained from either adult or embryonic mammals or birds, synthesis of chondroitin sulfate is depressed, suggesting that synthesis of chondroitin sulfate in part is controlled by the level of hyaluronic acid in the environment (Wiebkin and Muir, 1973, 1975; Solursh *et al.*, 1974). This effect is specific. Other glyco-

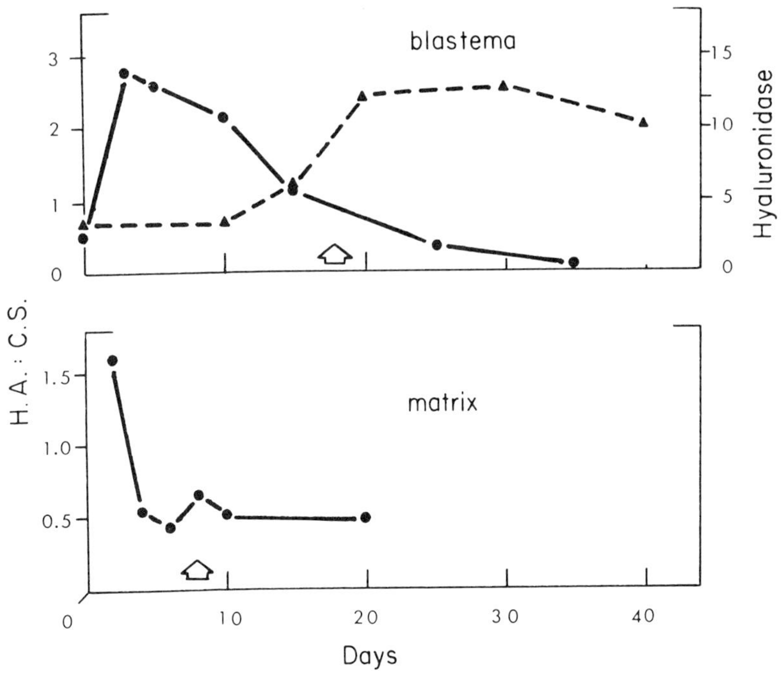

Fig. 20. The inverse relationship between the concentrations of hyaluronic acid (H.A.) and chondroitin sulfate (C.S., ●——●, plotted as the ratio of incorporation of radioisotope into H.A.:C.A.) and the concentration of hyaluronidase (▲---▲, units/mg protein, measured as the release of 1 μg *N*-acetylhexosamine under standard conditions) in five different skeletal systems. These are the developing avian limb buds, somites, and cornea; the blastema of the regenerating limb of adult *Diemictylus viridescens* (a urodele amphibian); and ectopic skeletogenesis in demineralized, morphogenetic bone matrix implanted intramuscularly in the rat (only data on hyaluronic acid are available for the latter system). Arrows indicate the onset of cytodifferentiation, which correlates with the removal of hyaluronidase and the increase in the amount of H.A. and C.S. in all systems. Based on data in Toole and Gross (1971), Toole and Trelstad (1971), Toole (1972), and Iwata and Urist (1973).

saminoglycans, such as chondroitin sulfate, keratan sulfate, or heparin sulfate, do not stimulate synthesis of chondroitin sulfate, whereas oligosaccharides derived from hyaluronic acid do.

Of course, the control of the synthesis of these extracellular matrix products is very complex. In chondrocyte cell cultures derived from avian embryos, synthesis of hyaluronic acid is unaffected by bUDR, whereas the synthesis of chondroitin sulfate is drastically reduced (Daniel *et al.*, 1973). The different chondroitin sulfates are also subject to independent controls. For example, there is increased synthesis of hyaluronic acid and chondroitin 6-sulfate, but not chon-

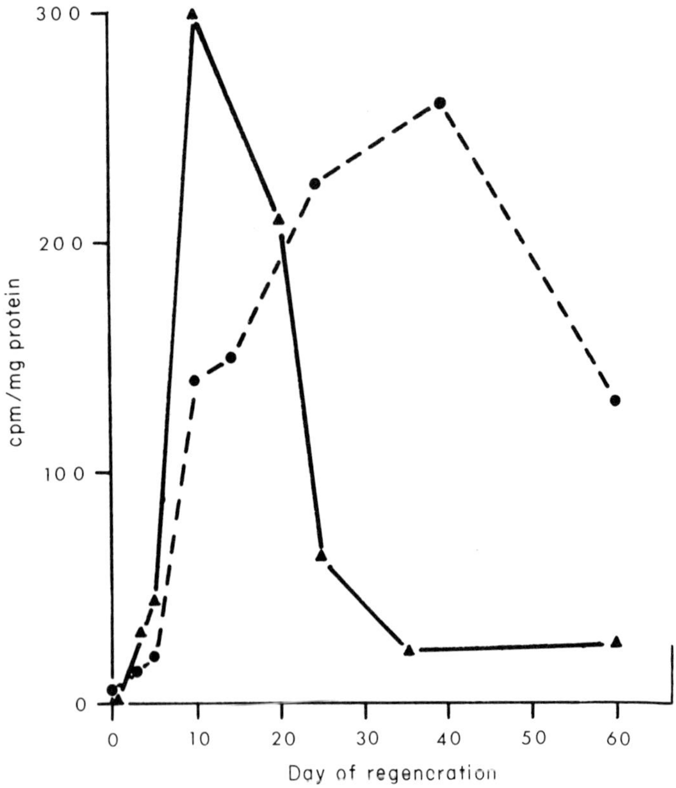

Fig. 21. Relationship between the concentration of hyaluronic acid (▲——▲) and chondroitin sulfate (●--●, both represented as cpm [^{3}H]acetate/mg protein) in the distal portion of the regeneration blastema and regenerating forelimb of the newt, *Diemictylus viridescens.* (cf. Fig. 19). Data taken from Toole and Gross (1971).

droitin 4-sulfate, when cells from rabbit aorta are stretched on elastic membranes (Leung *et al.*, 1976). In any event, it appears that the level of hyaluronic acid exerts some control on the synthesis of chondroitin sulfate, rather than the reverse. The high levels of hyaluronic acid that are found as cells condense act to depress the synthesis of chondroitin sulfate, while its removal by hyaluronidase aids in augmenting the synthesis of chondroitin sulfate, giving rise to the profiles seen in Figure 20.

Quite a different system, one which involves hyaluronic acid and its removal, has been explored by Iwata and Urist (1973). When allogeneic demineralized cortical bone matrix is implanted intramuscularly into rats, it first is invaded by circulating progenitor cells, which modulate to osteoclasts and begin to resorb the matrix. The implant then is invaded by surrounding progenitor cells, which accumulate within pores in the matrix. These host cells have synthesized and accumulated alkaline phosphatase by six days after implantation, have formed cartilage by seven days, and have formed woven bone by ten days. Marrow appears by 12 days. This is obviously a situation in which precursor cells migrate to a site, accumulate there, and then undergo cytodifferentiation. Peak levels of hyaluronic acid are found within the matrix two days after its implantation. As cytodifferentiation ensues, the level of hyaluronic acid within the matrix declines (Fig. 20). Whether the decline represents removal of hyaluronic acid by hyaluronidase or cessation of hyaluronic acid synthesis has not been investigated.

The cellular activity involving migration to this morphogenetic matrix parallels that exhibited by the cranial neural crest as its cells migrate, accumulate, and then differentiate (see Chapter 4, Section II,B,3). Two published studies implicate hyaluronic acid in the migration of neural crest cells. As assessed by autoradiography and by biochemical analysis, the cell-free space between the ectoderm and the mesoderm into which cranial neural crest cells migrate in the embryonic chick contains considerable hyaluronic acid (Pratt *et al.*, 1975). This suggests that hyaluronic acid may play a role in cell migration, as Toole (1972) has suggested for somite development. The addition of hyaluronic acid (30 μg/ml) or hyaluronidase (2 μg/ml) to a culture medium in which amphibian neural crest and pharyngeal endoderm are maintained prevents the chondrification of the neural crest that occurs in the absence of the hyaluronic acid (Corsin, 1975b). Controls for this system could provide a very useful model for the role of hyaluronic acid in cell migration and differentiation.

All of the above studies indicate a presumptive role for hyaluronic acid in the formation of condensations preceding chondrogenesis and for the removal of hyaluronic acid (or the presence of hyaluronidase) in the chondrification process. Several authors have sought more direct experimental verification of these presumptive roles.

Mayne *et al.* (1971) have provided support for the relationship between cell aggregation and synthesis of hyaluronic acid. Low-density cultures of amniotic cells from 10-day-old embryonic chicks produce hyaluronic acid almost exclusively. At higher cell densities, hyaluronic acid and chondroitin sulfate are produced in almost equal amounts. As the cell density increases, the synthesis of hyaluronic acid declines, whereas the synthesis of chondroitin sulfate rises. In fact, in fibroblasts from embryonic chicks maintained *in vitro*, there is a positive correlation between the concentration of hyaluronic acid and the rate of proliferation of the fibroblasts (Moscatelli and Rubin, 1975). Hyaluronic acid favors cell proliferation and aggregation and declines once critical cell densities have been attained.

Toole *et al.* (1972a) applied concentrations of hyaluronic acid as low as 1 ng/ml to limb bud mesenchymal cells and to somitic mesenchymal cells *in vitro*. Up to 90% of the colony formation and the subsequent differentiation of nodules into cartilage was blocked. The inhibition of cell aggregation into colonies prevented the initiation of chondrogenesis.

Several established nonskeletal cell lines (some fibroblastic, some epithelioid, and some lymphoblastic) produce a factor which is inactivated by hyaluronidase, induces cell aggregation, and whose action may be mimicked by the addition of hyaluronic acid (Pessac and Defendi, 1972).

Although we have far to go before we understand all of the factors involved in initiation of condensations preceding chondrogenesis or osteogenesis, these studies on the roles of hyaluronic acid and hyaluronidase clearly indicate that the microenvironment surrounding the cells holds the key.

6

Progenitor Cells and Their Differentiation

I. INTRODUCTION

If a summary of some of the concepts that were presented in Chapters 1 and 3 is used to introduce the discussion of the differentiation of precursor cells into chondroblasts or osteoblasts, one recalls that the skeleton-forming cells are set aside as separate races of progenitor cells some time before either the start of the cytodifferentiation or the time when they are found within the skeleton. For some cells, we can pinpoint the exact time when this determination occurs, but for others the timing is defined much less precisely. Although we do not understand the mechanism underlying the process of determination for any progenitor cell, we do have considerable information on the conditions that must be met if determination is to occur. Contact with an epithelium, migration, aggregation, and removal of hyaluronic acid from the aggregate have been presented as necessary conditions. We saw from the model proposed in Chapter 3 that this process of determination was progressive. The cells become more and more restricted in their potential with each successive developmental stage.

By the developmental stage when these progenitor cells can be recognized, by virtue of their position in condensations at sites of skeletogenesis, they have undergone considerable restriction. Whether particular progenitor cells have been restricted to only one fate at the time when cytodifferentiation is about to commence (e.g., differentiation into a chondroblast), or whether they still possess two or more possible fates (e.g., differentiation into chondroblasts or osteoblasts, or into osteoclasts) cannot be determined from their morphology or from their normal fate. All progenitor cells look alike and have more potential fates than they exhibit during normal development.

The existence of tissues intermediate between cartilage and bone implies a lability within particular scleroblasts for the production of differentiated products normally produced by different progenitor cells in separate tissues. This, however, does not prove that particular scleroblasts are capable of responding to different environmental conditions by differentiating into one or another cell type. There is, however, considerable evidence for the presence of bipotential progenitor cells even in the adult. The inducible osteogenic progenitor cells, which are capable of forming either fibroblasts or bone, would represent such a population (see Chapter 3). The literature on this ability of skeletal progenitor cells to differentiate into more than one type of differentiated cell is both vast and diverse. I have reviewed some reports previously (Hall, 1970, 1971); here, I shall present several examples to illustrate that at the stage when the progenitor cells are associated with ongoing skeletogenesis, they still possess the potential for forming more than one skeletal cell type and skeletal tissues. A description of these progenitor cells as they appear immediately preceding their cytodifferentiation into osteoblasts, chondroblasts, osteoclasts, or chondroclasts will be given first.

II. OSTEO- AND CHONDROPROGENITOR CELLS

In Chapter 3, I discussed the ingenious experiments of Friedenstein and his associates, whose studies indicated the presence of osteoprogenitor cells within a variety of tissues in the adult mammal. These experiments are remarkable, for there are no morphological criteria available for identifying these cells *in vivo*. Progenitor cells can be identified readily and visualized only when they are associated with the surfaces of preexisting bone or cartilage or when they are located in condensations in which osteogenesis or chondrogenesis is about to commence. In these cases, both the location and the knowledge of subsequent fate aid in identifying the precursor cells.

A variety of terms have been used to describe the proliferative progenitor cells at these stages. Those associated with osteogenic activity have been termed spindle cells or reticulum cells (Heller *et al.*, 1950); mesenchymal cells (Kember, 1960); preosteoblasts (Pritchard, 1956; Tonna, 1961; Owen, 1963; Owen and MacPherson, 1963; Scott, 1967); or osteoprogenitor cells (Young, 1962; Hall, 1971). I prefer osteoprogenitor cells and the companion term chondroprogenitor cells for the progenitors of osteoblasts and chondroblasts, respectively, although as will be seen below, because there are many sites where the precursor cells can become either chondrogenic or osteogenic, the term osteo-chondroprogenitor cells is really more appropriate, although clumsier. It should be reemphasized that these cells were committed as progenitor cells long before they first could be identified by virtue of their association with the skeleton.

Histochemically, these cells can be identified because of their high contents of DNA, RNA, alkaline phosphatase, and glycogen (Bevelander and Johnson, 1950; Pritchard, 1952; Fullmer, 1965; Hay, 1965; Milaire, 1965a, 1967; Hinchliffe and Ede, 1967; Searls, 1967; Hall, 1968a; Scott and Glimcher, 1971). Ultrastructurally, they are spindle shaped, having large nuclei, an extensive endoplasmic reticulum, small Golgi body, many free ribosomes, and numerous mitochondria. These features are characteristic, whether the cells are associated with endochondral or membrane bones or at fracture or ectopic sites (Cameron *et al.*, 1964, 1967; Scott, 1965, 1969; Aho, 1966; Silberberg *et al.*, 1966; Anderson, 1967; Hall and Shorey, 1968; Luk *et al.*, 1974; Thorogood and Craig Gray, 1975). Uptake of isotopically labeled precursors, response to hormones, and pattern of isozymes are additional means of identifying these progenitor cells (Searls, 1965a; Bingham *et al.*, 1969; Pawelek, 1969; Coffin and Hall, 1974; Rasmussen and Bordier, 1974; Thorogood and Hall, 1976, 1977).

The cell population kinetics of these progenitor cells have been examined in detail by Kember (1960, 1971, 1973), Walker and Kember (1972), Young (1962, 1963, 1964), Dixon (1970), Owen (1963, 1970, 1971), and Owen and MacPherson (1963). Kember (1971) has summarized and reviewed these data, and I have calculated the cell cycle data for an "average" progenitor cell of cartilage or bone as follows: S (DNA synthesis), 7 hr; G_2, 1.5 hr; M (mitosis), 1.25 hr, and T_c (total cell cycle time), 3.5 days.

Although these osteo- and chondroprogenitor cells have been described as a class of cells with uniform morphology, there is evidence that the potential for differentiation of all progenitor cells may not be the same, i.e., that races of cells exist within this broad class of osteo–chondroprogenitor cells. The determined and inducible osteogenic progenitor cells of Friedenstein and Owen (Chapter 3) represent two such races. The progenitor cells for osteoblasts and/or osteoclasts represent two additional races.

III. PROGENITOR CELLS FOR OSTEOBLASTS AND OSTEOCLASTS

Whether there is a single population of progenitor cells serving as the precursor to both osteoblasts and osteoclasts, or whether there are two separate populations of progenitor cells has been a controversial but important issue for many years. The adequate treatment of many metabolic bone diseases requires a speedy resolution of this question (Rasmussen and Bordier, 1974).

In some sites, such as the femur of the growing rabbit, bone deposition by osteoblasts and osteocytes occurs only on the periosteal surfaces, whereas bone resorption by osteoclasts occurs only on the endosteal surfaces. The progenitor cells on the periosteal surface are osteoblast precursors, those on the endosteal surface are osteoclast precursors. The study of these sites has led to the conclusion that there are separate precursors for osteoblasts and osteoclasts (Owen, 1963, 1970, 1971). At other sites, and more characteristic of osteogenesis, deposition and resorption of bone occur either side by side or sequentially on the same surface. While the osteoblasts can be distinguished readily from the multinucleated osteoclasts, their precursors cannot be distinguished readily from one another.

Alveolar bone is remodeled during the normal physiological drifting of teeth in their sockets. On one side, bone is deposited as the osteoprogenitor cells differentiate into osteo*blasts*; on the other, bone is resorbed as progenitor cells differentiate into osteo*clasts* (Baron, 1972, 1973; Markostamou and Baron, 1973). The appearance is of a spatial separation between the progenitor cells for osteoblasts and those for osteoclasts. If the direction of tooth drift is reversed by the application of orthodontic appliances, the pattern of progenitor cell differentiation also is reversed (Baron, 1973; Ten Cate, 1975). It would appear that the same progenitor cells can oscillate between osteoblast and osteoclast differentiation; however, there may be two intermingled subpopulations of progenitor cells surrounding the tooth socket. A similar possibility arises in estrogen-stimulated resorption of the innominate bones in the mouse (Chapter 7, Section III).

Several recent reviews detail the ultrastructural characteristics of osteoclasts (Cameron, 1972; Hancox, 1972a,b; Göthlin and Ericsson, 1976; Holtrop and King, 1977), while others have discussed the progenitor cells that produce them and whether osteoclasts arise by fusion of mononucleated cells (Owen, 1970, 1971; Rasmussen and Bordier, 1974; Hall, 1975c).

While Scott (1965, 1969), Göthlin (1973), and Luk *et al.* (1974) all have claimed that preosteoblasts could be distinguished from preosteoclasts on ultrastructural grounds, Young (1962, 1963, 1964) and Kember (1960) interpreted their pulse-labeling studies as indicating that the same progenitor cell could become either an osteoblast or an osteoclast. In my review (Hall, 1975c), I concluded that progenitor cells of osteoclasts arise as bloodborne and extraskeletal, rather than as sessile, intraskeletal cells, and that at least during osteogenesis within implanted morphogenetic matrix (Buring, 1975) and in the repair of fractured long bones (Göthlin and Ericsson, 1973), these migratory progenitors were a population *separate* from the local population of progenitors that differentiated into osteoblasts.

Despite the ingenious isotopic studies of workers such as Young and Kember, the question of the number of populations of progenitors within the skeleton during its *normal* development remains difficult to resolve. Ultrastructure is not sufficiently definitive to discriminate between progenitors of osteoblasts and of osteoclasts; in addition, as discussed by Hall (1975c), the isotopic studies are subject to various interpretations. To utilize the distinctive nuclear marker of the quail (see Chapter 4), implantation of tibiae from Japanese quail onto the chorioallantoic membrane of the embryonic chick (Kahn and Simmons, 1975) or of mouse calvaria onto the chorioallantoic membrane of the Japanese quail (Allyn and Minkin, 1976) indicates that progenitors from the host embryo migrate to the graft, invade it, and form osteoclasts. (Some osteoclasts also contain nuclei of both graft and host origin, confirming the growth of osteoclasts by cell fusion.) The osteoblasts of these grafts are all of graft, i.e., local, origin. These studies and the cell transfusion studies on grey-lethal and osteopetrotic mice (Barnicot, 1941; Barnes *et al.*, 1975; Walker, 1975a,b; Loutit and Sansom, 1976; Marks, 1976, 1978a,b), support the concept that during normal development of the endochondral and intramembranous skeleton, the osteoclasts arise from migratory progenitor cells, while the osteoblasts arise from a *separate* population of local progenitor cells. The possession of receptors for both calcitonin and parathyroid hormone on murine mononuclear phagocytes is consistent with their proposed role as migratory osteoclast precursors (Minkin, *et al.*, 1977). In fact, Kahn *et al.* (1978) now have provided direct evidence for the resorption of bone by transforming monocytes. These technical approaches and the separation of cell populations *in vitro* on the basis of their differential sensitivity to parathyroid hormone or to calcitonin (Wong and Cohn, 1974, 1975; Luben *et al.*, 1977; see Chapter 7, Section III), should be among the future methods of choice for identifying races of osteogenic precursor cells.

The evidence presented indicates that the progenitor cell for the osteoclast is not also a progenitor cell for the osteoblast. In fact, chondroclasts and osteoclasts may have a more common basis than osteoclasts and osteoblasts (Takuma, 1962; Schenk *et al.*, 1967; Savostin-Asling and Asling, 1973). Chondroclasts and osteoclasts represent two types of mineraloclasts (Knese, 1972) of the amebocyte series (Willmer, 1970). The fact that macrophages can be induced to synthesize and secrete collagenase (Wahl *et al.*, 1975) and that monocytes can resorb bone (Mundy *et al.*, 1977) is consistent with these amebocytes being a related series. While there is no direct evidence for the ability of osteoclast collagenase to degrade type II cartilage collagen [see Woessner (1973) for a review of mammalian collagenases], there is evidence that collagenase from a single source (cell type or tissue) can degrade more than one type of collagen. Rabbit corneal collagenase cleaves both rat and lamprey skin (type I) collagen, as well as calf articular cartilage collagen (Davison and Berman, 1973). Collagenase from cultured synovial cells from patients with rheumatoid arthritis degrades both skin and articular cartilage collagen (Leibovich and Weiss, 1973; Harris and McCroskery, 1974), and collagenase from human skin degrades lathritic type II cartilage collagen, although more slowly than type I collagen is degraded (Wooley *et al.*, 1973). The heterogeneous nature of synovium (see Janis *et al.*, 1967; Estabrooks and Schiff, 1972), as well as the fact that synovium only produces collagenase during arthritis, and the presence of both types I and II collagen in articular cartilage, are facts that make utilizing these data to support nonspecificity of collagenase action less than ideal. Robertson and Miller (1972) found that human gingival and rabbit leukocyte collagenase cleaved skin and bone type I collagen but did not cleave type II cartilage collagen, although *Clostridium* collagenase did act on both collagen types. They argued that the mechanisms of resorption of cartilage collagen differed from those of bone collagen. The specificity of the actions of cartilage and bone-derived collagenases remain, then, a problem for future investigation. The results of those investigations should provide useful information on the commonality of origin of osteoclasts, chondroclasts, osteoblasts, and chondroblasts and data on the resorptive (or protective) mechanisms present in cartilage and bone.

IV. PROGENITOR CELLS FOR OSTEOBLASTS AND CHONDROBLASTS

There is a very large body of literature supporting the concept that in many skeletal sites, the progenitor cell for the osteoblast (the osteoprogenitor) is also a progenitor for the chondroblast (the chondroprogenitor). Hall (1970) has reviewed much of the available evidence for the origin of precursor cells during endochondral ossification, during repair of fractures of long bones, and during the formation of secondary cartilage on membrane bones. Therefore, I shall concentrate on just a few examples and on the literature that has appeared since that 1970 review and shall discuss (1) the formation of secondary cartilage on the condylar process of the mammalian dentary; (2) the formation of secondary cartilage on the membrane bones of the head of the embryonic chick; and (3) the formation of callus cartilage during the repair of fractured bones.

A. Bipotential Progenitor Cells on the Mammalian Dentary

The condylar process of the dentary articulates with the temporal portion of the skull to form the temporomandibular joint. The body of the dentary and the basal portion of the condylar process develop by intramembranous ossification. However, the apical portion of the condylar process forms by the development of a secondary cartilage, the so-called condylar cartilage. (Similar, though more transitory cartilage is seen on the coronoid process.) Over the past 25 years, there have been numerous studies on the histodifferentiation of this condylar cartilage. Hall (1970), Durkin (1972), Durkin *et al.* (1973), and Beresford (1975b) have reviewed the present knowledge of the histogenesis of the condylar secondary cartilage. Koski (1975) provides an overview of 216 references on the development, growth, and function of this cartilage. The texts of Enlow (1975) and of Moore and Lavelle (1974) may be consulted for excellent reviews of the role of the cartilage in mandibular growth and morphogenesis (see Chapter 9).

Like all secondary cartilages, the condylar cartilage arises after intramembranous ossification has commenced [at the 48–50 mm CR length stage in man, according to Symons (1951)]. Several lines of evidence indicate that these chondroblasts and chondrocytes arise from the same population of cells as produces the bone. Histological studies of the developing condylar process give this impression (Baume, 1961, 1962a,b). Symons (1952) described progenitor cells at the apex of the process that differentiated into chondroblasts toward the cartilage but into osteoblasts toward the body of the process. In addition, the condylar cartilage contains both type I and type II collagen (Hirschmann and Shuttleworth, 1976). Under conditions of normal physiological activity, the progenitor cells of the condyle continue to proliferate and to differentiate into chondroblasts, which add to the bulk of the cartilage (Charlier *et al.,* 1969a,b; Meikle, 1973a,b; Petrovic *et al.,* 1973; Petrovic, 1974). The role of growth of this cartilage in the control of mandibular growth and in mandibular positioning is a controversial one and will be discussed in Chapter 9.

If the normal mechanical environment of the temporomandibular joint is disturbed by resecting the external pterygoid muscle, the progenitor cells differentiate into osteoblasts instead of differentiating into chondroblasts. Instead of depositing secondary cartilage, these progenitor cells deposit membrane bone (Stutzmann and Petrovic, 1974). Intracerebral transplantation of the joint for three weeks has the same effect (Meikle, 1973a,b). This population of progenitor cells is evidently a population having dual potential. Differentiation of the progenitor cells into chondroblasts requires a stimulus associated with normal joint function (shown in Section V,B,4 to be mechanical stimulation). In the absence of this environmental stimulus, the progenitors differentiate along an alternative pathway.

Do all the cells of the population possess the potential for both osteogenesis and chondrogenesis, or does the population consist of two cell pools, one of osteoprogenitor cells and one of chondroprogenitor cells? Stutzmann and Petrovic (1975a) maintain that there are two cell types within the progenitor pool. One line (their skeletoblast) is said to have a spontaneous tendency to form osteoblasts; however, *in vivo* this tendency is suppressed, and the cells form chondroblasts, i.e., these cells are bipotential for chondrogenesis and osteogenesis. The second cell line is said to form only chondroblasts, although the evidence for two cell lines with these differences is not overwhelming. It follows from their studies on the chalone-like effects of chondroblasts within the condyle (Stutzmann and Petrovic, 1975b).

It also has been suggested that the hypertrophic chondrocytes of the condylar cartilage survive calcification of their extracellular matrix and serve as osteoprogenitor cells (Silbermann and Frommer, 1972a,b,c, 1974). This concept is discussed in Section V,B,4.

B. Bipotential Progenitor Cells on Avian Membrane Bones

Secondary cartilages, similar in histological appearance to that found on the condylar process of the mammalian dentary, are found on various membrane bones of the mandible and skull of the embryonic chick. Murray (1963) has summarized the histogenesis of those bones that possess secondary cartilage. These are the quadratojugal, surangular, pterygoid, squamosal, and the palatine. Like the condylar cartilage, these secondary cartilages arise after the initiation of intramembranous ossification. For the quadratojugal, which is the first of the skull bones to ossify, intramembranous ossification commences at H.H. stage 30.5, with secondary cartilage arising some four days later at H.H. stage 37.

These cartilages depend upon mechanical stimulation for their formation from progenitor cells, which, in the absence of the mechanical stimulation, form osteoblasts instead of chondroblasts. If these membrane bones are removed from embryos younger than early H.H. stage 36 (ten days of incubation) and grafted to the chorioallantoic membrane, or maintained in organ culture, secondary cartilage does not form, and the progenitor cells differentiate into osteoblasts. If embryonic chicks are paralyzed with either curare or decamethonium, similar results are observed (Murray and Smiles, 1965; Hall, 1967a, 1968b, 1970, 1972b, 1975b). The progenitor cells are bipotential. Grafting or organ culturing these progenitor cells, or paralyzing embryos *in ovo,* does not suppress chondrogenesis if the cells are taken from, or the drugs injected into, embryos older than H.H. stage 36. This is a result of the progenitor cells either becoming determined at that age or only fully being able to express their chondrogenic bias by that age. The chondroprogenitor cells selectively divide and accumulate within mobile embryos. Paralysis removes some 60–75 cells from the progenitor cell pool during the tenth day, depriving the membrane bones of their source of chondroprogenitor cells (Hall, 1978d) [see Hall (1970, 1978b,c,d) and Ross and Hall (1978) for a discussion of the timing of determinative events in these cells].

C. Bipotential Progenitor Cells and Repair of Fractured Bones

The appearance of cartilage in the callus during repair of fractured bones is the result of the progenitor cells of the periosteum differentiating into chondroblasts instead of continuing to form osteoblasts, as they would have done in the absence of the fracture (Ham, 1930; Pritchard and Ruzicka, 1950; Murray, 1954; Tonna and Cronkite, 1962; Pritchard, 1965, 1972b, 1974; Hall, 1970; Ham and Harris, 1971; Bassett, 1972a; Becker, 1972; Tonna and Pentel, 1972; Kernek and Wray,

1973; White, 1975). These transformations are particularly striking when the fracture is in a membrane bone whose osteoprogenitor cells never before have been exposed to a stimulus evoking their chondrogenic potential (Pritchard, 1946; Girgis and Pritchard, 1958; Craft *et al.*, 1974; James *et al.*, 1974; Hall and Jacobson, 1975).

V. DEDIFFERENTIATION AS A SOURCE OF PROGENITOR CELLS

There is a further means by which the skeleton may obtain progenitor cells for subsequent differentiation, namely, through dedifferentiation of existing differentiated cells. The work of Silbermann and Frommer on the dedifferentiation of hypertrophic chondrocytes to osteoprogenitor cells during the calcification of condylar cartilage already has been mentioned (Section IV). Two systems in particular have been studied extensively, and accordingly, will be discussed here. These are the dedifferentiation of myoblasts and/or chondrocytes to progenitor cells during formation of the blastema following amputation of the amphibian limb and the possibility that hypertrophic chondrocytes survive calcification of the primary cartilaginous models of long bones and dedifferentiate to osteoprogenitor cells.

A. Dedifferentiation during Amphibian Limb Regeneration

In urodele amphibians, after amputation of the appendages, regenerative processes ensue that involve dedifferentiation of tissues exposed at the stump. These dedifferentiated cells accumulate as a blastema, from which the cytodifferentiation of cells as well as the histogenesis of the tissues, of the regenerate begin.

The concept that the tissues of the regenerate are derived from the dedifferentiation of local cells at the wound surface can be found in the work of Butler (1933), who postulated dedifferentiation as an alternative to the then prevailing theories of the origin of blastemal cells from either red blood cells, the epidermis, or a population of reserve cells mobilized to the side of regeneration.

The dedifferentiated cells in the blastema have the morphological features of progenitor cells. They are rounded, and have a high nucleo–cytoplasmic ratio, with prominent nucleoli, and many ribosomes; they lack the specialized morphological features normally associated with functional myoblasts, chondroblasts, or osteoblasts. They appear "embryonic" in form.

As well as undergoing morphological dedifferentiation, the blastemal cells also undergo functional dedifferentiation. They cease synthesizing the cell products that were associated with their prior differentiative activity. Thus, in *Ambystoma mexicanum* and in *Triturus viridescens,* muscle protein is not synthesized by blastemal cells (DeHann, 1956; Laufer, 1959). While ^{35}S-labeling of epiphyseal chondrocytes allows them to be followed into the blastema during dedifferentiation, as dedifferentiation proceeds, the level of incorporation declines until it is no different from that of neighboring fibroblasts at the stump (Mattson and Foret, 1973; McHenry *et al.*, 1974). Although sulfated glycosaminoglycans no longer are synthesized by these chondrocytes, the synthesis of hyaluronic acid *is* associated with the formation of the blastema (Toole and Gross, 1971; also see Chapter 5). No cartilage-specific (type II) collagen is synthesized by blastema cells during their dedifferentiation (Linsenmayer and Smith, 1976).

Chondrocytes divide at least five times as they undergo dedifferentiation (Steen, 1968), apparently in response to an interaction with the epidermis of the developing blastema (Mescher, 1976)—another example of epithelial control over mesodermal cell aggregation. Thus, both cytologically and functionally, the cells of the regeneration blastema become less specialized as they divide and accumulate.

Several problems have plagued the analysis of the subsequent fate of these blastemal cells. Since dedifferentiation involves the loss of specialized morphological and functional characteristics, it is difficult to follow the dedifferentiating cells for any length of time. And, because the dedifferentiated products of several mesodermal tissues of the stump (muscle, cartilage, bone, and connective tissue) are all morphologically equivalent, it is not possible to examine a blastema and to conclude which dedifferentiated cells were previously chondroblasts, which previously osteoblasts, which myoblasts, etc. Consequently, it is not possible to determine whether all of the blastemal cells that form the chondroblasts of the regenerate were chondroblasts before dedifferentiation, or whether some were myoblasts or fibroblasts.

But while the dedifferentiated blastema cells appear more "embryonic" (more like progenitor cells) than their differentiated ancestors, are they actually functioning as embryonic, totipotent cells within the blastema in the same way that blastomeres function within the blastula? Burgess (1974) attempted to answer this question by transplanting blastemal cell nuclei from the amputated hindlimbs of *Xenopus laevis* into enucleated *Xenopus* eggs. It is known that if totipotent nuclei are transplanted similarly, the enucleated ova are allowed to initiate normal embryonic development (Gurdon, 1970). Burgess found that blastemal cell nuclei (she was forced to inject many whole cells rather than a single nucleus; this is a criticism of the technique) would not support embryonic development of the enucleated eggs beyond the cleavage stage, while nuclei from gastrula-stage embryos would support complete development of a normal tadpole. Burgess concluded that blastemal cells are not totipotent and should not be equated with the blastomeres of the early embryo. The process of dedifferentiation is not equivalent to becoming embyronic again. However, some blastemal cells do possess more than one potential fate. For these cells, dedifferentiation is accompanied by the lability to modulate to progenitor cells having fates in the regenerate different from those they had in the original limb. This generalization is based on several lines of evidence, some of which are now classical studies in amphibian limb regeneration and in the study of differentiation, modulation, and metaplasia.

Weiss (1925), Thornton (1938), and Goss (1958), to name but a few workers, removed the skeleton (usually the humerus) from amphibian forelimbs and subsequently amputated the limb. In many cases, the regenerate formed a complete set of skeletal elements. The skeletal tissues of the regenerate could not have originated from preexisting skeletal tissues of the stump. They therefore must have arisen from blastemal cells whose source was muscle or connective tissue. (Blastemal cells are of local origin, except for the osteoclasts, which erode the skeletal stump and release cells that can then dedifferentiate. The osteoclast precursors migrate into the stump via the blood vascular system as shown by the [^{3}H]thymidine-labeling studies of Fischman and Hay (1962). (See also Section III.)

X-Irradiation of the limb prevents regeneration. Implantation of a muscle graft after X-irradiation allows regeneration of a limb complete with muscle, as well as with connective and skeletal tissues (Thornton, 1942). However, implantation of cartilage after X-irradiation of the limb produces a regenerate with only cartilage (Eggert, 1966). These implant studies suggested that, while dedifferentiated chondrocytes were stable in terms of determination, dedifferentiated muscle cells were not. This dichotomy between the fates of dedifferentiated myoblasts and chondroblasts now will be discussed.

Oberpriller (1967) approached this question by taking [^{3}H]thymidine-labeled blastemal cells from amputated intestines or tails of *Diemictylus viridescens* and grafting them into unlabeled limb blastema. Labeled chondrocytes were found in the cartilage of the regenerate limbs. After allowing for reutilization of label, Oberpriller concluded that the cells from the intestinal and tail blastemata had the potential to form chondrocytes and that they expressed that potential when transplanted to the blastema on the amputated limb.

Steen (1968, 1970) explored this question using the Mexican axolotl *Siredon mexicanum*. He implanted cartilage or muscle from triploid individuals that had been injected with [^{3}H]thymidine, into the limbs of diploid, unlabeled individuals, amputated the limbs, and examined the regenerates autoradiographically. At this point, he was working with a relatively small population of labeled cells in a limb that mostly consisted of unlabeled host cells. When cartilage was grafted into the limbs, only chondrocytes were found to be labeled in the regenerate. He therefore concluded that, in the regenerate, the chondrocytic cell type was stable through dedifferentiation, blastema formation, and redifferentiation. The source of the cartilage was not critical: appendicular, coracoid, or Meckel's cartilages all provided cells for the blastema and for the regenerate (the question of the specification of the cartilage cell type is discussed in Chapter 7, Section II,F). Using *Ambystoma maculatum,* Foret (1970) found that Meckel's cartilage was relatively inert to dissolution when grafted into the limb, and that it contributed few cells to the regenerate.

When Steen grafted isotopically labeled muscle into the host limb, the chondrocytes, myoblasts, and fibroblasts of the regenerate all were labeled, with both the triploid nucleolar marker and the [^{3}H]thymidine. He concluded that muscle cells were not stable and could alter their fate upon dedifferentiation. He was, however, aware that the muscle graft was not a homogeneous population of cells, as the cartilage grafts had been. (It would have included myoblasts, fibroblasts, satellite cells, neural cells, etc.) In following up this problem, he has provided perhaps the most convincing evidence for the ability of myoblasts to redifferentiate as either chondrocytes or fibroblasts during regeneration (Steen, 1973). Single myoblasts or fibroblasts were cultured clonally, and the cells were allowed to accumulate *in vitro* until a sufficiently large population of daughter cells was available for transplant to the amputated hindlimbs of larval *Xenopus laevis.* These cells were triploid and were labeled with [^{3}H]thymidine. The regenerates contained labeled fibroblasts within connective tissue, as well as within muscle; labeled muscle cells, and labeled chondrocytes. The conclusion that chondrocytes may arise from dedifferentiated myoblasts in hindlimb regeneration of *Xenopus laevis* seems inescapable. [In regeneration of hindlimbs of *Bufo viridis* tadpoles, considerable cartilage is found in the meager regenerate; some of this cartilage is inferred to have originated from cells of both muscle and connective tissue (Michael and Niazi, 1972).] Chondrocytes of anuran amphibians then can redifferentiate into other differentiated cell types.

In *Ambystoma mexicanum,* dedifferentiated chondrocytes produce perichondrial cells, connective tissues of the joints, fibroblasts of connective tissue and of muscle. The chondrocytes, however, do not produce myoblasts (Namenwirth, 1974).

The studies of Wallace's group (Wallace *et al.,* 1974; Maden and Wallace, 1975) on *Ambystoma mexicanum* contradict those of Steen on *Siredon* and of Namenwirth on *A. mexicanum.* To destroy host cells, *Ambystoma* were exposed to 2000 rads. Humeral cartilage was grafted into their limbs, which then were amputated and allowed to regenerate. Some reasonable instances of regeneration ensued, in which connective tissue and muscle were present. These investigators argue that the implanted cartilage consisted of a uniform population of chondrogenic cells, and that the fibroblasts of the connective tissue and the myoblasts of the muscle therefore must have arisen from dedifferentiated chondrocytes. More convincing is the finding that after approximately 16 weeks, the regenerate limbs were destroyed in an allograft reaction. This argues strongly that the regenerate consisted of graft and not of host cells.

There is, then, convincing evidence that, in both anuran and urodele amphibians, myoblasts can dedifferentiate to progenitor cells, which can redifferentiate into chondrocytes, and that anuran chondrocytes can redifferentiate into myoblasts. Whether chondrocytes of urodele amphibians show that same potential is not finally settled.

Little information is available on the factors that determine whether a particular dedifferentiated cell within the blastema will redifferentiate into a chondrocyte, a myoblast, or a fibroblast. The labeling studies indicate that the proportion of cells that redifferentiate into different cell types is small (probably less than ten percent). Type of redifferentiation is independent of influences from the stump, because isolated blastemata have been shown to undergo normal cytodifferentiation (Stocum, 1968, 1975a,b; Stocum and Dearlove, 1972). Summerbell *et al.* (1973) argued that the fate of individual blastemal cells depended upon their position relative to a labile "progress zone" at the tip of the blastema (see Chapter 8). Globus and Vethamany-Globus (1977) have provided evidence that the presence of ganglionic neurons is necessary before a sufficient mass of cells can accumulate to initiate chondrogenesis. The distribution of nerve fibers within the blastema thus may play a role in programming cells for particular differentiative events.

B. Dedifferentiation during Endochondral Bone Formation

The classical text book description of the normal replacement of primary cartilage by bone is as follows. The chondrocytes undergo hypertrophy and line up in columns whose transverse lamellae are dissolved but whose longitudinal lamellae remain intact. The removal of the extracellular matrix releases the hypertrophic chondrocytes, and they die. Blood vessels invade the lamellae; the bloodborne cells move in, line up on the now-calcified cartilaginous trabeculae, become osteoblasts, and deposit osteoid. This summary derives from the initial light microscopic study of Fell (1925), from the early electron microscopic studies of the epiphyseal zone by Scott and Pease (1956) and Trueta and Little (1960), and has been confirmed by later ultrastructural studies (Anderson and Parker, 1966; Schenk *et al.*, 1967, 1968; Engfeldt, 1969; Thyberg and Friberg, 1971). The reports of Ham (1974) and Serafini-Fracassini and Smith (1974) are good recent accounts of these events. It is not known whether the bloodborne cells were determined as osteoprogenitor cells while in the bloodstream, or whether they became osteogenic progenitors in the blood vessels at the invasion front or only after they settled down on the cartilage remnants. The latter, however, is probably the case (see Section III).

Some recent studies have cast doubt on whether the hypertrophic chondrocytes do indeed die after being released from their lacunae. I shall present evidence to indicate that some survive to become osteoprogenitor cells, presumably in a manner analogous to the dedifferentiation of chondrocytes in blastema formation during regeneration of the amphibian limb. This is not a new notion. Van der

Stricht (1890) and Brachet (1893) both proposed that the chondrocytes became osteoblasts either directly, i.e., by a metaplasia, or via immature connective tissue cells, i.e., by a dedifferentiation. Crelin and Koch (1967) provide an excellent account of the "turn of the century" views on whether chondrocytes survive the calcification of their matrices.

We should note that two types of hypertrophic chondrocytes must be considered: (a) those which are released from their lacunae when their matrices calcify, i.e., those at the ends of the columns; and (b) those in the columns in areas that are not invaded by capillaries, remaining as islands of cartilage within the bone matrix. These lie in uncalcified matrices and may have fates quite different from those that are released from calcified matrices (see Chapter I, Sections VI,B,C). Because of possible differences between different sites within the skeleton, each major site will be discussed separately.

1. The Ribs

The first modern attack on the notion that endochondral ossification meant automatic death for the hypertrophic chondrocytes came from the studies of Holtrop (1966, 1967, 1971, 1972a,b). She found (1966) that if rib fragments which included the zone of cell columns, but in which ossification had not commenced, were homotransplanted or cultured on solid media, the chondrocytes underwent hypertrophy and bone formed around them. The bone formed from progenitor cells of the perichondrium (periosteum). Ribs that had been deprived of their perichondria to exclude perichondrially located progenitor cells as the source of osteoprogenitor cells were labeled with [^{3}H]thymidine and were also homotransplanted. At the time of transplantation, many chondrocytes in the cell columns were isotopically labeled. Twelve days later, very few labeled chondrocytes were found; however, labeled periosteal cells, osteoblasts, and osteocytes were seen in the bone that had formed. Holtrop concluded that the chondrocytes had been the precursors of these bone cells, and that they had probably undergone dedifferentiation to progenitor cells equivalent to those of the normal perichondrium–periosteum (Holtrop, 1967).

Holtrop (1971, 1972a,b) has followed up these studies with an ultrastructural investigation of the ribs of three- to 14-day-old neonatal mice. As when the presence or absence of condensations was being interpreted (Chapter 5), fixation is very important in determining the viability of cells. She recommends a two-step procedure: two hours at 0°C in 2.5% glutaraldehyde:1% OsO_4 (1:2) in 0.1 *M* cacodylate buffer at pH 7.4, followed by postfixation in 0.25% uranyl acetate in 0.1 *M* sodium acetate (pH 6.3) for two hours at 0°C. As the hypertrophic chondrocytes mature (i.e., as one progresses down the cell column), the amount of endoplasmic reticulum increases, many free ribosomes appear, and electron-dense granules appear within the mitochondria. By electron microprobe analysis, the granules have been shown to be loaded with Ca and P (Sutfin *et al.*, 1971). Holtrop did not describe any process of degeneration or of chondrocyte lysis.

However, Brighton *et al.* (1973) have presented ultrastructural evidence that chondrocytes do degenerate in ribs from 21-day-old rats; however, the process is a rapid one and affects only the last four cells in the cell columns. These workers confirmed the increase in cytoplasmic constituents as the chondrocytes matured. Endoplasmic reticulum, mitochondria, lysosomes, and Golgi all increased, while lipid, vacuoles, and multivesicular bodies decreased. The last and penultimate cell in each column underwent fragmentation of its nuclear and cell membranes and lost its contents. Brighton *et al.* claim that the reason Holtrop did not describe these cells was because they were lost when the cell columns were dissected away from the bony metaphyses. However, it is possible that the rat and mouse ribs differ, or that the degeneration had not set in at the ages studied by Holtrop. The results of Brighton's group do allow for the possibility of survival of even the last cells under *in vitro* conditions in which oxygen tensions are high.

Shimomura *et al.* (1973) studied the costochondral junctions of 5-week-old rats autoradiographically and concluded that the preosteoblasts of the ossification front were derived from cells of the adjacent connective tissue, from the modulation of cells in Ranvier's groove (cells which can become either osteoblasts or chondroblasts), and *perhaps* from hypertrophic chondrocytes. They followed up this study with an experimental analysis providing strong presumptive evidence that hypertrophic chondrocytes induce local cells to become osteogenic, but themselves do not modulate to osteoblasts (Shimomura *et al.*, 1975). Chondroblasts or chondrocytes obtained from both the resting and the growth (hypertrophic) zones of rat ribs were placed in Millipore chambers and implanted intraabdominally into host rats. Provided that host cells invaded the chambers, bone was produced in those chambers containing hypertrophic chondrocytes, but not in those containing nonhypertrophic (resting) chondroblasts. The indication is that, in this system, while hypertrophic chondrocytes cannot modulate to osteoblasts, they can induce local progenitor cells (inducible osteogenic progenitor cells?) to become osteogenic.

2. Appendicular Long Bones

There is certainly evidence for species-specific differences in the fates of hypertrophic chondrocytes from embryonic long bones when they are maintained *in vitro*. Those from embryonic mouse tibiae survive, whereas those from chick tibiae cytolyze (Cooper, 1965).

In the rachitic metaphysis and epiphysis, the hypertrophic chondrocytes resume DNA synthesis (Mankin and Lippiello, 1969a; also see Chapter 7) to produce osteoblasts in lacunae previously occupied by chondrocytes (Dodds and Cameron, 1939).

Kuhlman and McNamee (1970) found that oxidative enzyme activity was very high in hypertrophic chondrocytes of rabbit femoral epiphyses. These workers concluded that the cells were physiologically active. Because the same phosphohydrolytic enzymes are present in both chondrocytes and osteoblasts, Arsenis and Huang (1977) have argued that chondrocytes could transform into osteocytes. However, similarity of enzymes does not provide presumptive evidence for cell transformations.

Bentley and Greer (1970) followed up Holtrop's observation on ossification in transplanted ribs and transplanted perichondrial-free rabbit metatarsal epiphyses intramuscularly after 3-hr labeling with [^{3}H]thymidine. Although bone developed at the tips of the cell columns, the osteoblasts were unlabeled, even though adjacent hypertrophic chondrocytes were labeled. They concluded that the latter cells were not the source of the osteoblasts; however, they made no comment as to the origin of the osteoblasts. Bentley and Greer reconcile their data with Holtrop's by invoking species specificity.

Kalayjian and Cooper (1972) found some interesting differences in the ossification of metaphyseal versus articular surfaces of epiphyses from guinea pig and mouse humeri, tibiae, and femora. On the metaphyseal side, where there was only slight calcification, many chondrocytes survived. Filaments accumulated within these surviving chondrocytes, and they came to resemble osteoblasts ultrastructurally, a metaplastic transformation similar to those described by Haines and Mohiuddin (1968) and by Hall (1972b). It has been shown by von der Mark and von der Mark (1977a) that hypertrophic chondrocytes contain Type I collagen at the time when endochondral osteoid is being deposited upon them. This ability of the mature hypertrophic chondrocyte to synthesize both collagen types is consistent with a possible role in osteogenesis.

Hanaoka (1976) examined the distal epiphyses of mouse femora ultrastructurally and found substantial cell death. At points where transverse lamellae were intact, the cells survived for longer periods than where lacunae were open, but they eventually died. [Previous studies of Knese and Knoop (1961a,b) had demonstrated intact hypertrophic chondrocytes in open lacunae in contact with primary spongiosa.] Hanaoka stressed the fact that cells transitional between chondrocytes and osteoblasts were not seen, and his point is well taken. He argued that, with the S phase in osteoprogenitor cells lasting eight hours, and with terminal lacunae in cell columns being opened up once every 4 hrs, there is considerable potential for transfer of label from degenerating chondrocytes to osteoprogenitor cells in the thymidine-labeling experiments.

3. The Interpubic Joint

Crelin and Koch (1965), culturing the mesenchymal primordium of the interpubic joint from 13-day-old mouse fetuses, found that a cartilaginous symphysis developed, and that endochondral ossification was initiated within the symphysis. It appeared that during ossification, the chondrocytes that were released from their matrices became either chondroclasts or osteoblasts. Crelin and Koch (1967) confirmed this histological impression with an autoradiographic analysis.

Pubes at this mesenchymal stage were labeled for two hours with [^{3}H]thymidine, and then were organ-cultured. Immediately prior to endochondral ossification (after approximately five or six days *in vitro*), only hypertrophic chondrocytes were labeled. With the appearance of bone, isotope label was seen in the osteoblasts and in the osteocytes. As the extracellular matrix surrounding the hypertrophic chondrocytes was removed, the cytoplasm of the chondrocytes became very basophilic, and mitoses of the chondrocytes were seen. Initial osteogenesis occurred within these cells—cells having large nuclei, very basophilic cytoplasm, and situated in clumps within lacunae. The cytological appearance was of hypertrophic chondrocytes modulating to preosteoblasts and to osteoblasts.

4. The Mandibular Condyle

Only part of the condyle of the mammalian dentary forms by endochondral ossification, and that by a secondary process after the condylar cartilage forms. In fact, as argued by Durkin (1972), if the chondrocytes survive, then the process is not really true endochrondral ossification. A similar situation is seen in the penile cartilages (Beresford, 1970, 1975a) and in the antlers (Chapter 4).

Silbermann and Frommer (1972a,b,c, 1974) studied the condyle of the mandible of the newborn mouse and provided ultrastructural, as well as isotopic (the cells take up [^{3}H]proline), evidence for chondrocyte survival. Many of the chondrocytes survived the calcification of their matrices. At the premineralization zone, many membrane-bound vesicles were seen within the chondrocytes. Meikle (1975), who observed similar bodies in the condyle of the 1-week-old rat, showed that they contained acid phosphatase and aryl sulfatase and concluded that they were lysosomes, as did Silbermann and Frommer (1973b). But Meikle also concluded that if these were lysosomes, then the cells would not survive, even though at the stage when they were found, the cells appeared very active: ultrastructurally, they appear intact, and still are synthesizing glycosaminoglycans (Silbermann and Frommer, 1973a). These bodies may not be related to cell death; rather, they may reflect the mineralization of the matrix and may indicate that the cells are participating actively in the mineralization process. Like the chondrocytes of the interpubic joint, these chondrocytes appear to be taking on the function of osteocytes.

5. Meckel's Cartilage

Melcher (1972) has shown that resorption of Meckel's cartilage in the fetal mouse is effected by the chondrocytes themselves, and that they survive the removal of the extracellular matrix. Savostin-Asling and Asling (1975) have presented histological evidence (see, e.g., their Figure 14) that released chondrocytes fuse with chondroclasts. Whether these or other surviving chondrocytes can modulate to progenitor cells of osteoblasts has not been investigated, but there is a substantial body of literature from the other skeletal sites, indicating that hypertrophic chondrocytes may survive removal of their extracellular matrices and become available as osteoprogenitor cells.

VI. THE DIFFERENTIATION OF PROGENITOR CELLS

Some time has been spent establishing the concept of the bipotentiality of the progenitor cells associated with the skeleton. Because an osteoprogenitor cell is situated on the surface of a membrane bone, it should not be assumed that its potential is limited only to differentiating into an osteoblast. Such a cell may differentiate into a chondroblast or into a fibroblast, depending on the conditions (Hall and Jacobson, 1975).

Stating that these progenitor cells can and do differentiate into either osteoblasts or chondroblasts describes their potentialities. We can state further, in the case of the formation of secondary cartilage on the condyle of the dentary or on avian membrane bones, or in cartilage formation during repair of fractured bones, that mechanical stimulation is the environmental influence to which these progenitor cells respond. While this describes the conditions under which a particular cytodifferentiation occurs, nothing is said about how the progenitor cells receive and translate the environmental message, nor as to which physiological and metabolic activities must be altered or set in motion for chondrogenic differentiation to occur. It is implicit that the potential to differentiate as an osteoblast or as a chondroblast existed in the progenitor cell *before* it was exposed to the environmental stimulus. Exposure to the environmental factor is a prerequisite to the expression of that determined state through differentiation.

Three major questions relating to the differentiation of progenitor cells have to be answered: (1) What are the environmental factors that can direct progenitor cells to differentiate along a particular pathway?, (2) How does the progenitor cell receive these messages? and (3) What biochemical, metabolic, and structural changes occur during the differentiation of progenitor cells into one differentiated cell type as opposed to another?

Cataloging the differences between a chondroblast and an osteoblast is not a very useful way of attempting to define the basis of their differentiation. I spent considerable time at the beginning of my own research career searching the literature for the biochemical features of chondroblasts and osteoblasts that would indicate the way in which their differentiation was controlled (Hall, 1968a,c) and have continued to gather information on how the two can be distinguished (Coffin and Hall, 1974; Thorogood and Hall, 1976, 1977; Ross and Hall, 1978). Many workers have used the differential synthesis and accumulation of sulfated glycosaminoglycans as the marker for chondroblast–chondrocyte differentiation (Bélanger, 1954; Amprino, 1955a–c, 1956; Johnston and Comar,

1957; Lash *et al.*, 1960; Holtzer, 1964a; Searls, 1965a,b; Holtzer and Abbott, 1968; Royal and Goetinck, 1977; also the review by Hall, 1977a), but osteoblasts also synthesize sulfated glycosaminoglycans, as do fibroblasts.

Until recently, no criteria were available to allow for the unequivocal separation of osteogenic form chondrogenic cell types at early stages in their differentiation. The discovery that collagen differs in its primary structure in a tissue-specific manner (see the reviews of Miller, 1972, 1973, 1976; also Trelstad, 1973), the characterization of their procollagens (von der Mark and Bornstein, 1973; Fessler *et al.*, 1973; Uitto *et al.*, 1977), and the translation of their mRNA's in cell-free systems (Benveniste *et al.*, 1973) provides very sensitive markers for osteogenic versus chondrogenic cells. But becoming a chondrocyte instead of an osteoblast must involve much more than the synthesis of type $[\alpha 1(II)]_3$ collagen instead of the synthesis of collagen of type $[\alpha 1(I)]2\alpha 2$. Specific collagens provide an excellent marker, but we just do not know what series of biochemical events is required in order for a progenitor cell to differentiate preferentially into a chondroblast rather than an osteoblast.

A more profitable approach to the study of cell differentiation might be to examine particular examples of specific environmental influences known to modulate the activity of progenitor cells. If we can define the conditions under which a particular differentiation occurs, we then might be better able to tackle the problem of analyzing the metabolic events involved in that differentiation. First, I shall examine the differentiation of chondroblasts and chondrocytes under the influence of mechanical stimulation as the example of the differentiation of progenitor cells under specific environmental conditions; then, I shall consider the differentiation of chondroblasts within the embryonic limb bud.

A. Mechanical Stimulation and the Differentiation of Chondroblasts and Chondrocytes

While Wolff (1892) is credited with originating the "Trajectory Theory," which relates the structure of bone to the mechanical forces imposed upon it, it was Ward (1838) who made the initial comparison of the architecture of the proximal portion of the femur with a bracket, and it was von Meyer (1867) who actually proposed the Trajectory Theory. Wolff was to von Meyer's Trajectory Theory what Huxley was to Darwin's theory of evolution—advocate, defender, and popularizer.

Murray (1936) provides what is still the best account of the early literature on the response of the skeleton to mechanical forces. Unfortunately, we still do not fully understand the mechanism of the skeleton's response. The translation of mechanical stimuli into bioelectrical activity and the response of skeletal cells to that bioelectrical activity are proposals that have received much recent support [see the reviews by Bassett, 1972b; Marino and Becker, 1977; Spadaro, 1977;

also the symposium volumes edited by Liboff and Rinaldi (1974) and by Attinger and Parakkal (1977)]. There is no doubt that electrical currents of small magnitude ($\mu\mu$amp range) can influence the proliferation, differentiation, and action of skeletal cells; however, the causal relationship between mechanical stimulation and these bioelectrical potentials has not been established. The direct application of electrical fields can accelerate the growth of epiphyseal cartilage (Watson *et al.*, 1975; Brighton *et al.*, 1976; Norton *et al.*, 1977) and of bone (Norton and Moore, 1972) and can alter local oxygen tensions around skeletal cells (Brighton and Friedenberg, 1974; Brighton *et al.*, 1975).

1. Intermittency of the Mechanical Stimulus

Respiration and locomotion produce cycles of bone deformation as tension and compression alternately are applied to the vertebrae and to the long bones (Lanyon, 1972; Lanyon *et al.*, 1975; Piekarski and Munro, 1977). J. Hert and his colleagues, in their studies on the response of rabbit tibiae to mechanical stress, have emphasized the importance of intermittent loading if the skeletal cells are to be activated (Hert *et al.*, 1969, 1971a,b, 1972a,b; Liskova and Hert, 1971; Chamay and Tschantz, 1972). Because mechanical loading of the skeleton is normally intermittent, Hert and Zalud (1971) argued that piezoelectric models based on the differential response of cells to either pressure or tension were inadequate (but see Bassett and Pawluk, 1972). They proposed local changes in potential and ion fluxes as the environmental changes actually sensed by the cell.

In order for progenitor cells to respond to mechanical stimuli by initiating chondrogenesis instead of osteogenesis, the mechanical stimulus must be an intermittent one. Rodbard (1970) has presented evidence that progenitor cells in the mesenchyme take up lipid and modulate to fat cells in the absence of mechanical stress, but in the presence of high rates of change in compression the synthesis of sulfated glycosaminoglycans is stimulated and differentiation into chondroblasts is initiated. Such a mechanism might explain the presence of bone and cartilage in the reptilian and mammalian heart (Torres, 1917; Kelsall and Visci, 1970; McConnell, 1970), and the formation of cartilage and bone from connective tissue cells of the endocardium of auxiliary heart grafts in the rat (Dittmer and Goss, 1972, 1974; Dittmer *et al.*, 1974).

If exposed to intermittent pressure and tension *in vitro,* the progenitor cells of the quadratojugal of the embryonic chick differentiate into chondroblasts. If exposed to constant pressure *in vitro,* or if allowed to differentiate within immobilized embryos, these cells differentiate into osteoblasts (Hall, 1968b, 1972b, 1978d; also see Section IV,B). Evidently, immobilization and constant pressure both are recognized as the same stimulus or as lack of stimulus.

Intermittent stresses favor chondrogenesis in the repair of fractures. This is the academic rationale for "walking casts" rather than immobilization in bed as therapy for those who are unfortunate enough to fracture a long bone. Defects

2–3 mm in diameter created in long bones and plugged with screws subjected to alternating motion either are filled with cartilage or are characterized by cycles of deposition and resorption of bone as cells modulate their differentiative activity (Schenk, 1973; Uhthoff and Germain, 1977). If fractured rabbit femora are completely immobilized, the periosteal progenitor cells differentiate into fibroblasts, and scar tissue develops. If allowed a normal range of movement, a mixture of fibrous tissue and fibrocartilage develops. In this case, healing takes about six months, whereas placing the fracture in continuous motion favors rapid differentiation of progenitor cells into chondroblasts. Intermittency, then, is important if the progenitor is to differentiate into a chondroblast. Proliferation of progenitor cells also is enhanced by mechanical stimuli, a topic that will be discussed in Chapter 9.

2. Mechanical Stimuli and Metabolic Activity

If progenitor cells differentiate into chondroblasts instead of osteoblasts or fibroblasts in the presence of intermittent mechanical stimulation, then we might expect particular cellular metabolic activities to be especially sensitive to changes in the mechanical environment surrounding the cell. Hyaluronate, a molecule already shown to play a role in cell aggregation (Chapter 5, Section IV), is very sensitive to mechanical stresses. Indeed, examining the correlations between cellular constituents and the biomechanical data from the same cells reveals the glycosaminoglycans often to be the only component whose levels correlate with the mechanical data (Åkeson *et al.*, 1974; Woo *et al.*, 1975).

The syntheses of collagen, hyaluronic acid, chondroitin 6-sulfate, and DNA are all enhanced when smooth muscle cells are stretched on membranes, particularly when that stretching is cyclical (Leung *et al.*, 1976, 1977). Subjecting cranial bones from embryonic chicks to high levels of compression *in vitro* slows the conversion of procollagen into collagen (Ehrlich and Bornstein, 1972). The filling of root canals stimulates the synthesis of collagen in the adjacent alveolar bone (Espie, 1975). Quite obviously, the levels of the major extracellular matrix components of cartilage and bone vary in accordance with the mechanical environment and reflect the response of the cells to alterations in that environment. How do these cells sense those changes?

3. Mechanical Stimulation Mediated by Cyclic AMP

Bioelectrical changes have been mentioned as one possible way whereby cells transduce mechanical stimuli into recognizable signals. Other possibilities include indirect sensing by changes in the blood vessels or nerves resulting from the liberation of metabolites, or by direct sensing by the membrane of the skeletal cell.

It is known that there are cell membrane receptors that act as binding sites for ions, hormones, and macromolecules (Roth, 1973; Edelman, 1976; Salomon and

Pratt, 1976). Are there also receptors that allow these cells to respond directly to changes in the mechanical characteristics of the environment? The potential of the fibroblast cell membrane varies from −8 to −17 mV, depending upon the amount of deformation to which the cells are exposed (Bard and Wright, 1974). If mechanoreceptors exist, and there is some evidence that they do, then there must be a mechanism for transducing the physical stimulus received by the receptor into a biochemical signal to which the cell can respond. The activation of adenylate cyclase and the alteration in the intracellular levels of cyclic AMP comprise an obviously attractive possibility.

A molecule with a ubiquitous distribution, adenosine 3′, 5′-monophosphate (cyclic AMP), mediates external signals received via a cell surface mechanism, thereby regulating intracellular metabolic activity in a great diversity of cell types and for a wide range of metabolic activities (Friedman, 1976). For example, the induction of both flagellar protein and sugar-digesting enzymes in bacteria is mediated by cAMP, a mechanism for which starvation is the external signal. Cyclic AMP also acts as the attractant (acrasin) in the aggregation of slime molds, with starvation again serving as the stimulus. It stimulates the aggregation of polymorphonuclear leukocytes. Levels of cAMP are high in the blastema during the first ten days of the regeneration of amputated amphibian limbs, possibly aiding in the accumulation of the blastemal cells (Jabaily *et al.*, 1975). The phosphorylation of glycogen to glucose 1-phosphate is catalyzed by cAMP. Synaptic transmission, activation of melanophores, the immune response, neural induction, cell division, the assembly of microtubules and microfilaments, and the release and activation of steroid hormones, all are mediated by cAMP.

The perception of a stimulus by the cell activates the synthesis of membrane-bound adenylate cyclase, which catalyzes the transformation of ATP into cAMP. The cAMP activates inactive enzymes and activates protein kinases, which in turn stimulate metabolic processes. In the presence of phosphodiesterase, cAMP is transformed into the inactive 5′-AMP (adenylic acid). Therefore, levels of cAMP may be increased by either activating adenylate cyclase or by inhibiting phosphodiesterase; cAMP levels are decreased by inhibiting adenylate cyclase or by activating phosphodiesterase.

Alterations in the levels of cAMP can affect the synthesis of the characteristic extracellular matrix products of skeletal cells. Dibutyryl 3′, 5′-cAMP has been shown to modulate the transformation of hamster ovarian cells into fibroblasts and to induce synthesis of collagen within these transformed cells (Hsie *et al.*, 1971); to increase the rates of synthesis and secretion of glycosaminoglycans from transformed fibroblasts (Goggins *et al.*, 1972), and to block hyaluronate-induced inhibition of chondrogenesis from somitic cells *in vitro* (Toole, 1973a).

Hormones (parathyroid hormone, thyrocalcitonin, thyroxine, and growth hormone) elevate the intracellular levels of cAMP in the whole skeleton and in isolated populations of osteoblasts (Toole, 1973a; Rodan and Rodan, 1974;

Shanfeld *et al.*, 1976; Davidovitch *et al.*, 1977b). The mechanisms of this elevation vary. Heersche and his co-workers (1971, 1974; Heersche, 1974) have shown that the elevated levels of cAMP in rat calvaria exposed to calcitonin are due to the activation of adenylate cyclase, but that cAMP is elevated following treatment of calvaria with dibutyryl 3′, 5′-cAMP because of phosphodiesterase inhibition.

Davidovitch and Shanfeld (1974) and Shanfeld *et al.* (1975) have assayed levels of cAMP in tibiae from cat and chick, in which levels range between 0.2–0.4 pmoles cAMP/mg wet wt and in the alveolar and mandibular basal bone of the cat, in which levels are somewhat lower (0.1–0.3 pmoles/mg wet wt). Orthodontic tipping of the cat canine (application of an initial force of 100 g by elastic and maintaining elastic in position for 7 or 15 days) increased the levels of cAMP by 50–130% (Davidovitch, 1973; Davidovitch *et al.*, 1974). With 80 g of initial force applied by coiled spring, an initial (24-hour) decrease in cAMP levels was observed, followed by a prolonged elevation (for 28 days) at about 50% above the levels recorded in untipped specimens (Davidovitch and Shanfeld, 1975). When cAMP was localized intracellularly by an immunohistochemical reaction (Davidovitch *et al.*, 1976a,b; Gustafson *et al.*, 1977), only a small number of cells were found to have elevated amounts of cAMP.

There is, then, evidence that cyclic AMP levels are stimulated at pressure sites when cat canines are tipped with an orthodontic force. What the precise cellular response is that causes this elevation, which cells produce the response, how the mechanical stimulus is perceived by these cells, and whether the alterations in levels of cAMP are related to the remodeling that follows this treatment, remain topics for future investigations.

As yet, little information is available on alterations in cAMP levels in response to electrical stimulation. Norton *et al.* (1976) exposed embryonic tibiae to electrical stimulation and found a 21% increase in cAMP levels, provided that the long axes of the tibiae were parallel to the applied electrical field. They speculated that the receptors on the cells or within the extracellular matrix were oriented along the long axis of the tibia. If so, this would provide an elegant mechanism for directing the response along the growing axis.

Rodan *et al.* (1975a) adapted a tuberculin syringe to deliver a known compressive force to embryonic tibiae *in vitro* and used this device to evaluate the effect of compression on glucose metabolism and on thymidine incorporation into DNA. Subsequently, Rodan *et al.* (1975b) used this device to show that whole tibiae, slices of tibial epiphysis, and isolated eiphyseal cells from 16-day-old chick embryos all could respond to a pressure load of 60 g/cm^2 by accumulating or by losing cAMP and cGMP. They were able to show a preferential decrease in cAMP in cells from the proliferative zone, a drop which matched the decrease in cAMP observed in the whole tibiae. It was evident that these prechondroblastic skeletal cells were able to relate pressure changes into altered levels of cAMP.

Bourret and Rodan (1976a,b) exposed the cells from the proliferative zone (i.e., chondroblasts and prechondroblasts) and hypertrophic chondrocytes from the tibiae of embryonic chicks to 60 g/cm^2 hydrostatic pressure for 15 min. Levels of cAMP activity were 21% lower in the pressure-treated proliferating cells than in non-pressure-treated cells. The level of cAMP in the hypertrophic chondrocytes was unaffected by the pressure treatment (see Table XII). The action of short exposure to pressure on the proliferative chondroblasts was mimicked by the calcium ionophore A23187 in the absence of pressure. The effects of ionophore and pressure were not additive, and the effect of pressure on cAMP levels was abolished in the presence of the chelating agent, ethylenebis-(oxyethylenenitrilo)tetraacetic acid (EGTA; see Table XII). They concluded that an increase in calcium ion concentration mediated the decrease in the level of cAMP within the proliferating cells. But pressure also increased calcium levels within the hypertrophic chondrocytes without altering levels of cAMP. This occurs because the adenylate cyclase of hypertrophic chondrocytes is not sensitive to changing calcium levels. Calcium inhibits adenylate cyclase in plasma membrane preparations of proliferating chondroblasts, but not in membrane preparations from hypertrophic chondrocytes. This loss of response evidently accompanies differentiation of chondroblasts into hypertrophic chondrocytes (Rodan *et al.*, 1977). Therefore, short duration (15 min) of physiological levels of pressure (60 g/cm^2) enhances calcium uptake into chondroblasts and chondrocytes. Proliferating chondroblasts respond to the increased calcium levels by a decrease in adenylate cyclase and by lowering the level of cAMP. That the effect of pressure will be localized on the immature cells is ensured by a loss of the

TABLE XII

A Summary of the Levels of cAMP (pmoles cAMP/10^6 Cells) in Proliferating Prechondroblasts from Tibiae of 16-Day-Old Chick Embryos after Treatment with Pressure (60 g/cm), the Calcium Ionophore A23187, and/or the Chelating Agent EGTA[a,b]

Treatment	cAMP[c]	Percent of control
Krebs–Ringer–glucose (control)	4.08 ± 0.37	
Pressure	3.21 ± 0.36	−21
A23187	3.47 ± 0.47	−15
Pressure + A23187	3.48 ± 0.42	−15
EGTA	6.93 ± 0.31	+70
Pressure + EGTA	6.79 ± 0.11	+66

[a] EGTA, ethylenebis(oxyethylenenitrilo)tetraacetic acid.
[b] Data adapted from Bourret and Rodan (1976a,b).
[c] $\bar{X}$ ± SEM, based on 6–12 replicates.

ability to respond to heightened calcium as the cells differentiate. The division of these proliferating cells also may be controlled by the changes in cAMP (Burger *et al.*, 1972; Rodan *et al.*, 1975a). Similar changes occur when these cells are exposed to oscillating electric fields (Rodan *et al.*, 1978). McMahon (1974), Whitfield *et al.* (1973), Rebhun (1977), and Rasmussen and Goodman (1977) provide overviews of the interplay between calcium and cAMP in regulating cell proliferation and differentiation.

The maintenance of a pool of proliferating chondroprogenitor cells and the ability of those cells to respond to changes in compression in their environment by altering the intracellular levels of cAMP make cAMP an attractive candidate for the role of transducer in those systems in which progenitor cells differentiate into chondroblasts. Additional facts support this model: namely, that direct application of dibutyryl cAMP to clonal myogenic cells depresses myogenesis and stimulates the synthesis of collagen and of glycosaminoglycans (Schubert and Lacorbiere, 1976), and that rabbit articular chondrocytes in suspension culture can be induced to switch from the synthesis of type II to the synthesis of type I collagen in the presence of dibutyryl cAMP or 1.8 m*M* $CaCl_2$ (Deshmukh and Sawyer, 1977). Ahrens *et al.* (1977a) and Solursh *et al.* (1978) have provided evidence for just such a model in their micromass cultures of aggregating and chondrifying limb mesenchymal cells.

I have not considered the possible role of cyclic GMP in this control. In hormone induction, the action of cGMP is antagonistic to that of cAMP. Only a few studies have examined cGMP in skeletal tissues; both found that levels of cGMP followed the same pattern as cAMP levels. Facial bones respond to parathyroid hormone by a rapid elevation of the levels of both cyclic nucleotides. Over a one-hour period, cGMP increases by 750% and cAMP by 150% (Shanfeld *et al.*, 1976; Davidovitch *et al.*, 1977a,b). In mechanically compressed tibiae, cGMP decreases in the hypertrophic chondrocyte zone, but increases in the proliferative and resting chondroblast zones, paralleling the changes in cAMP (Rodan *et al.*, 1975b). A possible involvement of cGMP in proliferation during amphibian limb regeneration has been documented by Liversage *et al.* (1977).

B. Chondrogenesis within the Limb Bud

The initial chondrogenesis of the limb buds of the embryonic chick has been studied intensively. This system is an excellent model for studying the segregation of myogenic and chondrogenic cells from one another and for the control of chondrogenic differentiation.

The process of cell condensation within the limb bud was discussed in Chapter 5, Section II. We saw that initially (until H. H. stage 22), both the central and the peripheral mesenchymal cells of the limb bud had a uniform histological appearance (Fig. 13). Between H.H. stages 22 and 26, the central cells formed a condensation. By H.H. stage 27, ultrastructural evidence of chondrogenesis within the central condensation (deposition of extracellular matrix, Fig. 16) and evidence of myogenesis within the peripheral mesenchyme were apparent. The appearance, from these light and electron microscopical studies, was of an initially uniform population of cells that progressively became separated into central chondrogenic and peripheral myogenic cells.

Also presented in Chapter 5 was evidence to indicate that the formation of a condensation might well be a mandatory prerequisite to chondrogenesis within limb mesenchyme (Ahrens *et al.*, 1977a; Solursh *et al.*, 1978). Umansky (1966) isolated mesenchymal cells from limb buds of 11-day-old mouse fetuses (the stage immediately preceding condensation), and upon culturing, obtained both clusters and monolayers. Chondrogenesis occurred within the clusters when the initial cell density exceeded 5000 cells/mm^2. Myogenesis took place in the monolayered cells. He concluded that the population of limb mesenchymal cells could produce more than one differentiated cell type depending upon the conditions with respect to cell-to-cell contacts. Dissociation and reaggregation of mesenchyme from young (H.H. stage 19) chick limb buds also favor chondrogenesis over myogenesis (Moscona, 1961; Medoff, 1967) and suggest bipotentiality of these cells. [Recall that fully differentiated myoblasts, and in some cases chondrocytes, can dedifferentiate to precursor cells with dual potentials (Chapter 6, Section V,A).]

1. Synthesis of Chondroitin Sulfate

The pattern of synthesis of chondroitin sulfate within the developing limb bud also supports the bipotentiality of these mesenchymal cells. Searls (1965a,b) showed that at H.H. stage 22, ^{35}S was distributed uniformly throughout the mesenchyme of the limb bud, and that these cells were synthesizing both chondroitin 4- and chondroitin 6-sulfates. Between H.H. stages 22 and 26, the uniform distribution of ^{35}S uptake changed, increasing within the central, condensing prechondrogenic mesenchyme and decreasing within the peripheral, premyogenic mesenchyme. He concluded that *all* of the mesenchymal cells of the early limb bud possessed not only the potential to synthesize chondroitin sulfate, but were synthesizing it. Furthermore, Searls (1965a) showed that ^{35}S was taken up uniformly by both flank and limb mesenchyme at H.H. stages 16 and 17, and Medoff (1967) showed that low levels of chondroitin sulfate were present as early as H.H. stage 15 at the site of the future limb bud. [Evidence also suggests that when the mesoderm that subsequently will form the wing buds is isolated from embryos as young as H.H. stage 7 (24 hours of incubation), it will chondrify *in vitro* (Pinot, 1969, 1970).]

Medoff (1967) demonstrated that, within the whole limb bud, the pattern of accumulation of UDP glucose dehydrogenase (1.1.1.22) and of UDP-*N*-acetylglucosamine-4-epimerase (5.1.37; two key and specific enzymes in the synthesis of chondroitin sulfate) rose, paralleling the accumulation of ^{35}S within the central mesenchyme. Chondrogenic and myogenic areas were not examined separately; however, mesenchyme from H.H. stage 19 limb buds was dissociated, reaggregated, and cultured for seven days, during which time chondrogenesis was initiated. A steady increase in the accumulation of the two enzymes was seen until chondrogenesis commenced (at the equivalent of H.H. stage 24), when there was a sharp increase in rate of accumulation. Medoff proposed that amplification of preexisting enzymatic activity occurred in the prechondrogenic mesenchyme but not in the myogenic mesenchyme (Medoff, 1967; Zwilling, 1968; Medoff and Zwilling, 1972).

If the synthesis of chondroitin sulfate is characteristic only of prechondrogenic and chondrogenic cells (as was thought during the 1960's), then all of the mesenchymal cells of the early limb bud, presumptive limb bud, and flank are prechondrogenic. In order for the peripheral mesenchymal cells within the limb bud to differentiate as myoblasts, a switch from their established chondrogenic orientation to a new myogenic basis would be necessary. However, the ability to synthesize chondroitin sulfate is neither an exclusive property of, nor diagnostic for, chondrogenic cells (see Chapter 3, Section II,B; also Daniel, 1976; Holtzer and Mayne, 1973; Kosher and Searls, 1973; Mattson and Foret, 1973; Ahrens *et al.*, 1977b); although the ability to accumulate chondroitin sulfate into an extracellular matrix is an exclusive property of chondrogenic cells (Abrahamsohn *et al.*, 1975). Differentiating chondrocytes also may synthesize a cartilage-specific chondroitin sulfate or proteochondroitin sulfate (Goetinck *et al.*, 1974; Pennypacker and Goetinck, 1976; Ho *et al.*, 1977; Wiebkin and Muir, 1977).

If, among other things, becoming chondrogenic involved amplification of the synthesis of chondroitin sulfate and deposition of that molecule into the extracellular matrix, then knowledge of the control of these processes might well illuminate the mechanism by which precursor cells differentiate into chondroblasts and into chondrocytes.

The control of the synthesis of chondroitin sulfate is very complex and as yet is understood incompletely. There may be several steps in the metabolic pathway at which synthesis and, hence, accumulation of chondroitin sulfate could be controlled. Suggested possible control steps are (1) the transformation of *N*-acetylglucosamine-1-phosphate to UDP-*N* acetylglucosamine by UTP and UDP-*N*-acetylglucosamine pyrophosphorylase (2.7.7.23) (Lash, 1968); (2) the transformation of UDP-*N*-acetylglucosamine to UDP-*N*-acetylgalactosamine by UDP-*N*-acetylglucosamine-4-epimerase (5.1.3.7) (Medoff, 1967; Weinstein *et al.*, 1971; Marzullo and Desiderio, 1972; Manasek, 1973; Ross, 1976; Ross and Hall, 1978); (3) transformation of glucosamine to glucosamine 6-phosphate by hexokinase (2.7.1.1) (Arrambide *et al.*, 1968; Winterburn and Phelps, 1970; Kim and Conrad, 1974); (4) the UDPxylose or xylosyltransferase (2.4.1.72) (Balduini *et al.*, 1973; Caplan and Stoolmiller, 1973; DeLuca *et al.*, 1975;

Schwartz and Dorfman, 1975); or (5) the core protein (Herington *et al.*, 1972; Schwartz and Dorfman, 1975; Ho *et al.*, 1977; Vasan and Lash, 1977). The possible control mechanisms as they relate to the initiation of chondrogenesis are reviewed by Thorpe and Dorfman (1967), Kleine (1972), Serafini-Fracassini and Smith (1974), Lamberg and Stoolmiller (1974), Levitt and Dorfman (1973, 1974), and by Levitt *et al.* (1974).

It is clear that the enzymatic machinery for chondroitin sulfate synthesis is established and functions at a low level in many, if not most, embryonic cells and tissues (Delbrück, 1970), even in the blastodisc (Abrahamsohn *et al.*, 1975) and certainly in the presumptive myogenic and chondrogenic mesenchyme of the limb buds. As they relate to initiation of chondrogenesis, the controls act on the rate of synthesis and on accumulation, not on initiation of synthesis. However, this does not appear to be the case for the synthesis of the so-called cartilage-type collagen, type $[\alpha 1(\mathrm{II})]_3$ collagen. Its appearance in the mesenchyme of the limb bud coincides with the differentiation of precursor cells into chondroblasts at H.H. stage 25 (Linsenmayer *et al.*, 1973a).

2. Cartilage-Type Collagen

At H.H. stages 23 and 24, limb mesenchyme produces only collagen of type $[\alpha 1(\mathrm{I})]_2\alpha 2$, and this same molecule (typical of adult bone, tendon, fibroblasts, ligament, dentine, and skin) is synthesized in the peripheral limb mesenchyme at H.H. stages 25 and 26. However, the mesenchymal cells in the central condensing prechondrogenic area begin synthesizing $[\alpha 1(\mathrm{II})]_3$ collagen at H.H. stage 25 (Linsenmayer *et al.*, 1973a) or at H.H. stage 24 (von der Mark and von der Mark, 1977b; H. von der Mark *et al.*, 1976). Later in development (H.H. stage 33 onward), this type of collagen, which is identical to that obtained from adult cartilage, is confined to the cartilaginous diaphyses of the long bones (Toole *et al.*, 1972b; Linsenmayer *et al.*, 1973a,b; Linsenmayer, 1974; H. von der Mark *et al.*, 1976; Oohira *et al.*, 1974; von der Mark and von der Mark, 1977b).

This appearance of typical adult-type cartilage collagen, i.e., $[\alpha 1(\mathrm{II})]_3$, in the limb bud coincides with the cytodifferentiation of the condensed mesenchymal cells as prechondroblasts and as chondroblasts. This collagen is not synthesized ubiquitously as is chondroitin sulfate. The onset of the synthesis of $[\alpha 1(\mathrm{II})]_3$ collagen is then a very sensitive marker for the initial differentiation of precursor cells into chondroblasts. Its appearance in central limb mesenchymal cells at H.H. stages 24 or 25 coincides with these cells' irreversible commitment toward chondrogenesis (Table VIII).

3. Regulation within the Limb Bud

When prechondrogenic mesenchyme from embryos younger than H.H. stage 24 is grafted to the premyogenic area of limb buds of similarly aged embryos, 94% of the grafts differentiate in conformity with the host and form muscle (Searls and

Janners, 1969). If the mesenchyme is taken from the central area of the limb buds of embryos of H.H. stage 25 or older and grafted into hosts aged between H.H. stages 22 and 27, 94% of the grafts produce cartilage, identified as nodules of ectopic cartilage within the muscle of the host limb (Searls and Janners, 1969). This elegant experiment tells us that the ability of the central cells to chondrify stabilizes between H.H. stages 24 and 25. It also implies, and perhaps even proves, that before H.H. stage 24 these central mesenchymal cells have the potential to differentiate into *either* cartilage *or* muscle. Several other pieces of evidence lead us to the same conclusion.

Zwilling (1966) isolated peripheral (premyogenic), central (prechondrogenic), and intermediate (partly within myogenic and partly within chondrogenic) mesenchyme from limb buds of embryos of H.H. stages 22 to 24. The mesenchyme either was cultured on embryo extract–plasma clots or was grafted to the chorioallantoic membrane. Mesenchyme from all three sites chondrified, and did so in similar proportions—67% of the "myogenic," 64% of the "chondrogenic," and 71% of the "intermediate" samples. Zwilling concluded that cells whose progeny would have differentiated into myoblasts if they had been left *in situ* differentiated into chondroblasts when transplanted.

Searls (1967) eliminated cell migration as a mechanism responsible for segregating separate populations of myogenic and chondrogenic cells within the limb bud. He grafted isotopically labeled prechondrogenic areas from limb buds of H.H. stages 22 to 24 into the prospective chondrogenic (humeral) region of wing buds of H.H. stage 20–24 embryos. There was no migration of labeled cells out of the grafts, nor was movement of unlabeled cells into the grafts observed.

The regulative capacity of the limb bud mesenchymal cells (the ability to compensate for loss of tissue by initiating normal and complete differentiation from the remaining cells) supports the concept of mesenchymal cells capable of either chondrifying or differentiating into muscle. Stark and Searls (1974) conducted two additional experimental procedures. In the first, they excised various prospective long-bone areas from the wing bud to determine whether adjacent premyogenic mesenchyme would regulate and replace them. In the second, they rotated the prospective elbow region and reimplanted it into the wing bud. Their general conclusion was that regulation occurred prior to H.H. stage 22; that between H.H. stages 22 and 24, the cells were more stable and less able to regulate; and that after H.H. stage 24, they were unable to regulate their activity, i.e., the limb bud was by then a mosaic (Table XIII). In the first experiment, three varying amounts of prospective chondrogenic tissue were excised: (a) the areas that normally form the humerus, radius, and ulna, representing about 60–70% of the limb mesenchyme; (b) the prospective radius and ulna mesenchyme, representing ca. 20–40% of the limb mesenchyme, or (c) the humeral, radial, or ulnal mesenchymes independently, representing some 10–30% of the total limb mesenchyme. Procedures (b) and (c) gave similar results, and have

TABLE XIII

Regulative Ability of Limb Mesenchyme As Assessed by the Ability to Replace Excised Prospective Humeral, Radial, or Ulna Mesenchyme and Produce a Morphologically Normal Limb[a,b]

Embryonic age at excision (H.H. stage)	Percent of normal limbs (*n*) 6 or 7 days postexcision: Prospective humerus, radius, and ulna all excised	Percent of normal limbs (*n*) 6 or 7 days postexcision: Prospective ulna and radius or humerus, radius, or ulna excised
19	67 (8/12)	
20	60 (3/5)	100 (9)
21	37 (3/8)	100 (18)
22	33 (4/12)	77 (27/35)
23	10 (1/10)	33 (11/33)
24	0 (0/2)	5 (2/40)

[a] Regulative ability decreases with embryonic age and is lost after H.H. stage 24.
[b] Based on data in Stark and Searls (1974).

accordingly been grouped together in Table XIII. From these results, we may conclude (a) that the greater the amount and the later that skeletal mesenchyme is removed, the greater the incidence of abnormal limbs and the less capable of regulating is the limb bud; (b) that regulative ability is lost at H.H. stage 24 (previously shown to be the point at which prechondrogenic mesenchyme becomes stabilized); and (c) that regulation occurs irrespective of the position of the piece excised along the proximodistal axis.

Stark and Searls' second experiment involved 180° rotation of the prospective elbow region and reimplantation into the wing bud. The adjacent ectoderm either was rotated with the mesenchyme or was left in place when the mesenchyme was rotated. Two criteria were used to assess regulation. Normal development of the joint after rotation would indicate regulation, whereas the presence of nodules of ectopic cartilage at the site of the joint would indicate failure to regulate. As shown in Table XIV; regulation occurred until H.H. stage 22, provided that the ectoderm retained its original orientation. Rotation of the dorsal ectoderm of the joint region alone was sufficient to prevent regulation [see Table XIV, also Barasa (1962) and Searls (1976)]. Limb mesenchyme grafted to the chorioallantoic membrane produced only jointed limb elements when ectoderm was included in the graft. The apical ectodermal ridge was not essential.

In other studies, Kieny (1977) and Kieny and Pautou (1976) grafted limb mesenchyme from Japanese quail embryos into various sections (stylopod, zeugopod, and autopod) of embryonic chick limb buds and found that regulation occurred until H.H. stage 23. Summerbell (1977b) removed slices along the

TABLE XIV

Regulative Ability of the Limb Mesenchyme As Assessed by the Ability to Compensate for Rotation of the Prospective Elbow Region (with or without Rotation of the Adjacent Ectoderm) and Produce a Normal Limb Bud[a,b]

Embryo age at rotation (H.H. stage)	Prospective elbow mesenchyme and ectoderm rotated		Prospective elbow mesenchyme rotated		Dorsal ectoderm rotated
	Percent of specimens with				
	normal joint	ectopic cartilage	normal joint	ectopic cartilage	normal joint
20	25	50			
21	17	67	100	0	60
22	5	80	76	23	36
23	0	100	0	100	19
24	0	100	0	100	

[a] Presence of nodules of ectopic cartilage at the normal site of the joint indicates inability to regulate. Regulative ability is lost after H.H. stage 22.

[b] Results are expressed as percent of specimens with normal joints or with nodules of ectopic cartilage. Based on data in Stark and Searls (1974) and in Searls (1976).

proximodistal axis of the chick wing bud and found that regulation ceased at H.H. stage 22.

The loss of regulative ability observed in these experiments is interpreted as the loss of the ability of the mesenchymal cells to modulate their differentiative activity in the same way that determination becomes more restricted in the whole embryo as it progresses from cleavage to gastrulation. But it may be that the cells become less capable of filling the larger wound in the older limb buds or less able to migrate across the bud to fill the gap [although Barasa (1962) found that regulation was more complete in large defects than in smaller ones].

4. Segregation from Precursors

Another approach to the origin of myogenic and chondrogenic diversity within the mesenchymal cell pool is to investigate the ability of chondrogenic cells to segregate from myogenic cells at various phases during their differentiation (Zwilling, 1968; Searls, 1971, 1972; Ede and Flint, 1972). In Searls' study, embryos of one developmental age were labeled *in ovo* with [^{3}H]thymidine, and mesenchymal cells from the limb buds were mixed in various proportions (4:1–1:4) with unlabeled cells from an embryo of a different developmental age. These mixtures then either were cultured on nutrient agar or were grafted to the chorioallantoic membrane of host embryos (in the latter case, the aggregates were enclosed in ectodermal jackets obtained from the limb buds of H.H. stages 20 to 22 embryos). Ede and Flint cultured their recombinations in liquid medium on a shaker.

One would expect terminally differentiated chondrocytes to segregate from terminally differentiated myoblasts. Indeed, chondrocytes from the limbs of H.H. stages 27 or 37 segregated from limb myoblasts of H.H. stage 38 and from ventricular myoblasts of H.H. stages 25 and 26 (Table XV) as well.

If the cartilage and muscle within the limb bud develop from a common cell population, there should be a stage during their differentiation when the two cell types would recognize one another as the same and not segregate. Knowledge of the time when the two cell populations first exhibit segregation from one another would aid in pinpointing the time of divergence of myogenic from chondrogenic populations. Mesenchyme from young (H.H. stages 20–22) limb buds (a mixture of prechondrogenic and premyogenic cells) segregates from differentiated myoblasts from H.H. stage 38 limbs, as well as from ventricular muscle from H.H. stage 25 embryos (Table XV). The mesenchymal cells do not recognize the myoblasts as the same cell type. These same cells do not segregate from chondrocytes until the latter have reached late stages of differentiation (Table XV).

Early limb mesoderm (H.H. stages 18 and 19) also segregates from dermamyotome from the same-aged embryos (Zwilling, 1968; also Table XV). Even though muscle will develop from these two sites, some cell specialization is present before overt differentiation of myoblasts. Flank mesoderm from H.H.

TABLE XV

Summary of the Results of Studies on the Segregation of Chondrogenic or Prechondrogenic Cells from Myogenic or Premyogenic Cells[a]

Source of chondrogenic cells	Source of myogenic cells	Segregation	Reference[b]
H.H. 31 limb chondrocytes	H.H. 38 limb myoblasts	+	1
H.H. 31 limb chondrocytes	H.H. 25 ventricular myoblasts	+	1
H.H. 27 limb chondrocytes	H.H. 24 limb mesenchyme	+	2
H.H. 20–22 limb mesenchyme	H.H. 38 limb myoblasts	+	1
H.H. 20–22 limb mesenchyme	H.H. 25 ventricular myoblasts	+	1
H.H. 18 or 19 limb mesenchyme	H.H. 18 or 19 dermamyotome	+	2
H.H. 14,15, or 18 limb mesenchyme	H.H. 14, 15, or 18 flank mesenchyme	+	2
H.H. 26 chondrogenic limb mesenchyme	H.H. 26 myogenic limb mesenchyme	−	3[c]

[a] Results are expressed as +, segregation or −, lack of segregation.

[b] 1, Searls (1971); 2, Zwilling (1968); 3, Ede and Flint (1972).

[c] These cultures were in liquid medium under conditions or agitation. All the others were on solid media or as grafts to the chorioallantoic membrane.

stages 14, 15, and 18 also segregates from presumptive limb mesoderm of the same ages (Table XV). Cell segregation between different populations of myogenic cells therefore develops very early in the ontogeny of these cells.

Whether chondrocytes that have attained terminal differentiation segregate from their precursors has been investigated by Searls (1971, 1972). No segregation was seen until after H.H. stage 27 (Table XVI). Chondrocytes isolated from H.H. stage 31 embryos segregated from H.H. stage 20–25 limb bud mesenchyme whether cultured or grafted to the chorioallantoic membrane. Segregation from H.H. stage 26 limb mesenchyme occurred only on the chorioallantoic membrane [cf. the observation of Ede and Flint (1972), Table XV].

Searls concluded that the ability of chondrocytes to segregate from their precursors occurred late in differentiation, later than the stabilization of prechondrogenic cells for chondrogenesis (H.H. stages 24–25) and after initiation of deposition of extracellular matrix (H.H. stage 27). The mesenchymal cells of the early limb bud recognize cartilage but not muscle as similar cell types. This is consistent with mesenchymal cells, whether in the central or the peripheral mesenchyme of the limb bud, all having chondrogenic potential. Vogel and Kelley (1977) have provided some information on the types of glycosaminoglycans found on the surfaces of human limb bud cells. However, since we do not fully understand the changes in the cell membrane that accompany differentiation, nor the basis of cell-to-cell recognition (Roth, 1973; Edelman, 1976; also see Chapter 7, Section II,F,3,c), it is difficult to interpret these results any further. But all of the evidence presented thus far in this section is consistent with the concept of the mesenchymal cells of limb buds as a homogeneous population of precursor

TABLE XVI

Summary of the Results of Studies on the Segregation of Chondrocytes from Their Precursor Cells within the Limb Bud[a]

Age and source of cells		Segregation	Reference[b]
H.H. 20 or 21 mesenchyme	with H.H. 24 mesenchyme	–	2
H.H. 20–22 mesenchyme	with H.H. 25 mesenchyme	–	2
H.H. 20 mesenchyme	with H.H. 26 mesenchyme	–	2
H.H. 20–22 mesenchyme	with H.H. 27 mesenchyme	–	2
H.H. 24 mesenchyme	with H.H. 27 mesenchyme	+	3
H.H. 20–22, 24, or 25 mesenchyme	with H.H. 31 chondrocytes	+	1,2
H.H. 26 mesenchyme	with H.H. 31 chondrocytes	+[c]	
		–[d]	2

[a] Results are expressed as +, segregation or –, lack of segregation.

[b] 1, Searls (1971); 2, Searls (1972); 3, Zwilling (1972).

[c] Cultured on an agar substrate.

[d] Grafted to the chorioallantoic membrane.

cells able to differentiate as myoblasts or as chondroblasts until H.H. stage 25. Two pieces of evidence that might argue against this concept will now be discussed.

5. Somitic Origin of Wing Myoblasts

In the discussion of limblessness in vertebrates (Chapter 4, Section III,A,1), it was seen that extensions from the ventral wall of the somites penetrate the developing limb bud and stimulate the proliferation of the limb bud som*a*tic mesoderm. Several studies indicate that somatic mesoderm stimulates the outgrowth of limb bud mesoderm in the embryonic chick, and that the inclusion of such somitic mesoderm along with limb bud mesenchyme *in vitro* enhances the chondrogenesis of the latter mesenchyme (Pinot, 1969, 1970; Kieny, 1971; Kieny *et al.,* 1972). Somitic cells penetrate the developing limb buds of the mouse (Houben, 1976; Milaire, 1976), and a stimulation of murine forelimb bud growth *in vitro* has been documented by Agnish and Kochhar (1977). Furthermore, when they cocultured [^{3}H]thymidine-labeled somites with unlabeled limb buds for nine days, these workers found labeled chondrocytes within the humeral cartilages of the limb skeleton, suggesting that the proximal limb skeleton in the mouse forelimb is of mixed somitic and somatic origin. Labeled chondrocytes were not found in more distal skeletal elements.

Until very recently, there has been no clear-cut evidence that the somitic cells *form* any of the tissues of the avian limb bud. However, Chevallier *et al.* (1977) has now shown that destroying the somitic mesoderm results in the absence of extrinsic musculature of the limb bud (because of the lack of a cellular contribution or a maintenance factor from the mesoderm), and that grafting Japanese quail somitic mesoderm into chick limb buds results in the development of limb myoblasts of donor (quail) origin. Fewer donor myoblasts were found when chick somitic mesoderm was grafted into chick limb buds, perhaps because of the slower invasion of chick cells relative to quail cells. The indication from his study is that both som*a*tic and som*i*tic mesoderm can contribute toward myogenesis of the limb bud. Christ *et al.* (1977) and Chevallier *et al.* (1977) also have obtained some compelling evidence to indicate that while the myogenic cells of the chick wing and leg bud are of somitic origin, the fibroblasts of the connective tissue of the muscles and the cartilage are of somatic origin. Their evidence also comes from the grafting of somites and/or somatic mesoderm between Japanese quail and chick embryos and from following the cells with the nuclear markers of these two species. The grafts were made at H.H. stages 12–14, considerably before any outgrowth of the limb mesoderm and before its segregation from flank mesoderm. When brachial level somitic mesoderm of the embryonic chick was replaced with brachial level somitic mesoderm of the quail, the somitic cells of the quail became localized within the peripheral premyogenic mesoderm of the chick wing bud. None appeared within the central prechondrogenic area. The somitic cells then differentiated into the myoblasts of the wing muscles. Christ and co-workers therefore argue that the myoblasts and chondroblasts of the wing bud have separate origins, with the former somitic and the latter somatic, and have demonstrated this to be the case. Christ and associates go on to argue that the mesenchymal cells of the wing bud are therefore not a "homogeneous population of pluripotential mesodermal cells," but, although they have shown that the population of cells is not homogeneous, in that it has two origins, their data do not provide information on the potentialities of the individual cells in isolation. In their grafts of somatic mesoderm, skeletal structures differentiated, while muscle did not. As a result, they claimed that the somatic (prechondrogenic) cells could not differentiate as myoblasts; however, their grafts were not innervated, and several lines of evidence indicate that innervation is necessary before myogenesis can be initiated. Paralysis of embryos prevents myogenic differentiation (Ahmed, 1966). Nerve fibers first enter the hindlimb bud at the same age (H.H. stage 23) that myosin can first be detected (Daneo and Filogamo, 1973; Fouvet, 1973), and limb buds from H.H. stage 22 to 26 embryos grafted to the chorioallantoic membrane form only myogenic clones when cografted with a nerve supply (Bonner, 1975). Nevertheless, the presence of two populations of cells within the wing and leg bud demands reconsideration of the concept of separation of one population into myogenic and chondrogenic pools by a process of restrictive determinations, i.e., the origins of the cells must be reconsidered. I do not believe that the potential for differentiation of those two cell pools into two or more differentiated cell types is negated by the findings of these experimenters, but these data do indicate that during normal limb development only one potential is probably expressed.

Newman (1977) has provided experimental evidence indicating that the cells at the distal tip of the limb bud at H.H. stage 25 can form cartilage but not muscle. This could be interpreted as the absence of somitic cells at the distal tip (*à la* Christ or Chevallier), the stabilization of cells for chondrogenesis after H.H. stage 24 (*à la* Searls), or as the continued lability of cells at the distal tip.

6. Clonal Cell Culture

A second line of evidence indicating that discrete and separate myogenic, chondrogenic, and fibroblastic cell lineages exist within the limb bud come from the cell culture analyses of Flower (1972), Mayne *et al.* (1972), Abbott *et al.* (1974), Dienstman *et al.* (1974), and Holtzer *et al.* (1975).

Flower (1972) utilized equilibrium density gradient centrifugation to isolate and clone mesenchymal cells from limb buds of H.H. stages 18 to 28. *Separate* subpopulations yielding colonies of chondrocytes, myoblasts, or fibroblasts were obtained, supporting the concept that the mesenchymal cells of the limb bud, at least as early as H.H. stage 18, are already a heterogeneous population. This is obviously in line with the concept (described in Chapter 3) of cell lineages set aside very early in development. Under conditions of *in vitro* cultivation, only a single developmental fate is expressed. Even so, Flower found that at H.H. stage 25, a population of cells appeared that formed colonies of chondrocytes much more readily than cells from younger limb buds or other cell populations from H.H. stage 25 limb buds. While lineages within the limb bud may be established early, their ability to differentiate into particular cell types is progressively modified and enhanced, as the studies of Searls (described earlier in this Chapter) indicated. A difficulty in the interpretation of cell cultures (and the possible limited repertoire of differentiated states expressed *in vitro*) is whether one can argue from the single expression of a differentiated product (e.g., chondrocytes rather than myoblast) to the conclusion that the cell population was limited, restricted in its determination only to chondrogenesis. The best that can be said is that under the culture conditions cloned cells express only one phenotype.

The second set of studies, also utilizing clonal analysis, established that myogenic cells from the leg muscle of H.H. stage 34 embryos yielded myogenic colonies, which on subculture produced fibroblasts (identified by the synthesis of $[\alpha 1(I)_2]\alpha_2$ collagen and hyaluronic acid) but not chondroblasts (Abbott *et al.*, 1974). It was concluded that cells are present within the muscle of the hindlimb of 8-day-old embryos that have the potential to form (are common progenitors for) myoblasts and fibroblasts but not to form chondroblasts. Of course, H.H. stage 34 is a much later stage than the developmental period during which cytodifferentiation is initiated within the limb; however Dienstman *et al.* (1974) found that this compartmentalization was already apparent at H.H. stage 21. These workers and Holtzer *et al.* (1975) have developed the concept of the limb bud containing distinct and multicompartmented lineages of cells rather than a homogeneous population of cells having an initial chondrogenic bias but able to shift to myogenesis.

It is clear that we no longer can regard the mesenchymal cells of the limb bud as a pool of chondrogenic cells, some of which later switch to myogenesis. The biochemical evidence on the ubiquitous distribution of glycosaminoglycans (especially chondroitin sulfate) confirms that the previously-used criteria of chondroitin sulfate synthesis cannot be invoked to conclude that all mesenchymal cells of the limb start out with a chondrogenic bias. However, these cells are capable of differentiating into chondroblasts when isolated at a very early stage (Pinot, 1970). Time spent within the limb bud is not a prerequisite for making these cells chondrogenic, only for allowing the ability to chondrify to be stabilized and expressed. Furthermore, additional studies indicate that clonally derived myoblast cultures can be induced to cease production of the cell products characteristic of myoblasts and to initiate synthesis of collagen and glycosaminoglycans, and also to undergo morphological transformations to cells indistinguishable from chondroblasts (Schubert and Lacorbiere, 1976; also see Chapter 6, Section V,A).

7. Homogeneous or Heterogeneous Population

We now have the grafting experiments of Christ *et al.* (1977) and of Chevallier *et al.* 1977) and the clonal analyses which indicate that myogenic and chondrogenic cell lines have separate origins within the wing bud (one somitic and the other somatic), and that myogenic colonies initiated from cells of stages later than H.H. 21, while not yielding chondrogenic colonies, do yield fibroblastic colonies. On the other hand, a substantial body of evidence indicates that at least until H.H. stage 25, limb bud mesenchyme can compensate for losses and transpositions of both myogenic and chondrogenic mesenchyme, and that cells from these mesenchymes can differentiate into both chondroblasts and myoblasts. Are these two bodies of evidence compatible? The simplest superficial union is that the clonal culture conditions in the above studies did not allow full expression of the potentials of these cells, and that despite their disparate origins from two mesodermal sources, both populations are able to chondrify or to initiate myogenesis under appropriate conditions. They arise as separate populations, but until H.H. stage 25 or later, retain a bipotentiality for myogenesis and for chondrogenesis. One difficulty is that much of the earlier work, which asked the question, "when does cell determination occur and how is it environmentally regulated?" produced data that really answered the question, "what allows cytodifferentiation to occur?" The determination of these cells has its basis in the origin of the cell lineages during the very earliest stages of embryonic development. Most studies have been directed toward asking what is required to stabilize, enhance, or allow the expression of, that potential.

Accordingly, if one is convinced that the evidence obtained from the various *in vivo* analyses indicates that mesenchymal cells from the peripheral zones can become chondrogenic until a certain stage of limb bud development, then it is appropriate to explore the conditions that might allow such an event to occur. However, it must be borne in mind that these data probably will not yield any information on the process of determination. The studies of Arnold Caplan of Case Western Reserve University provide considerable data on the conditions within the limb bud that allow myogenesis or chondrogenesis to be initiated; these data hold true whether they are interpreted as allowing a homogeneous population of cells to differentiate selectively as chondroblasts, or as allowing a committed cell lineage to differentiate as chondroblasts.

8. Metabolic Regulation of Chondrogenic and Myogenic Differentiation

The introduction of 3-acetylpyridine at levels of 400–600 μg/egg into eggs incubated for at least 96 hours produces severe muscular hypoplasia later in development (Landauer, 1957). The direct action of 3-acetylpyridine on the myoblasts is shown by the inhibition of myoblast proliferation and the slowing of myoblast fusion into myotubes, a process which follows direct treatment of myogenic cell cultures with 3-acetylpyridine. Cotreatment with nicotinamide, both *in vivo* and *in vitro*, counteracts 3-acetylpyridine (Caplan, 1971, 1972a). McLachlan *et al.* (1976) maintain that 3-acetylpyridine does not influence myoblast differentiation, but rather, that it acts to destroy peripheral nerves, thereby slowing muscle growth. However, this objection, which is based on the injection of large doses of 3-acetylpyridine (2000 μg) into embryos beyond the stage of mesenchymal differentiation into myoblasts (6 days of incubation), does not counteract the observations of the direct action of 3-acetylpyridine on myoblasts *in vitro*. Its neurological action may well be an additional effect.

Taking advantage of the fact that the mesenchymal cells are not irreversibly committed to chondrogenesis before H.H. stage 25, Caplan and his associates asked: "What was the fate of the cells that 3-acetylpyridine prevented from becoming myogenic?" Did they die, become fibroblastic, or become chondrogenic? As the experimental model to test these questions, Caplan utilized high-density cell cultures (5–30 $\times$ 10^6 cells/plate). These were obtained by trypsinization of limb buds from embryos of H.H. stages 20 to 32 (Caplan *et al.*, 1968; Caplan, 1970). At 30 $\times$ 10^6 cells/plate and with limb buds from embryos younger than H.H. stage 26, 100% of the area of the plates chondrified within seven days. Chondrification was assessed as the production of metachromatic extracellular matrix. Nodules of chondrocytes invested with a perichondrium-like membrane were interspersed with flatter zones of chondrocytes. Under these conditions, mesenchymal cells from limb buds of H.H. stage 26 or older did not chondrify (Fig. 22).

Addition of 3-acetylpyridine (1330–1500 μg/ml) increased the number of H.H. stage 25 limb bud cells producing metachromatic matrix such that at densities as low as 12 $\times$ 10^6 cells/plate, 100% of the cells exhibited chondrogenesis. Furthermore, 3-acetylpyridine allowed mesenchymal cells from limb buds of embryos older than H.H. stage 26 to chondrify (Fig. 22). Chondrogenesis was not observed without the addition of 3-acetylpyridine.

Thus, in the cells of this culture system already able to undergo chondrogenesis (those from embryos younger than H.H. stage 26), chondrogenic expression was stimulated by the addition of 3-acetylpyridine. In addition, the cells that did not express chondrogenesis at all (older than H.H. stage 26) were able to form cartilage with the addition of 3-acetylpyridine. Caplan speculated that 3-acetylpyridine acted as an exogenous derepressor or inducer.

This action of 3-acetylpyridine was prevented by the addition of equivalent amounts of nicotinamide to the culture medium (Caplan, 1970). He therefore argued that 3-acetylpyridine was acting by depleting endogenous levels of either nicotinamide or of NAD, and he was able to show that 3-acetylpyridine does increase the incorporation of nicotinamide by mesenchymal cells of H.H. stage 24 limb buds when they are cultured at moderately high densities (12.5 $\times$ 10^6

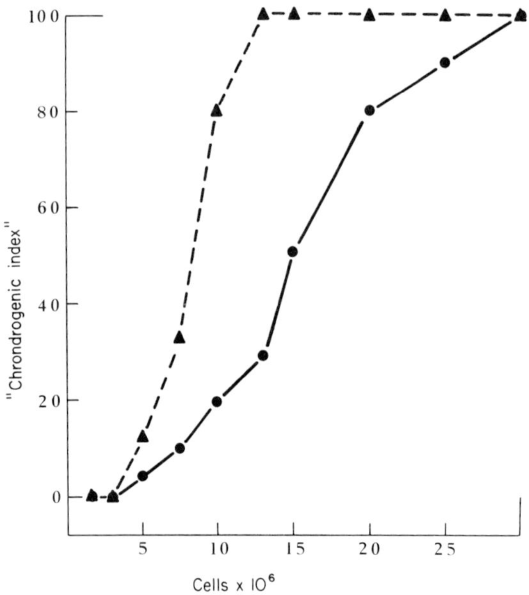

Fig. 22. The "chondrogenic index" (percent of the area of cultured cells that have differentiated into chondrocytes) of cultured H.H. stage 24 limb bud cells plotted against the initial cell count (cells $\times$ 10^6). All of the untreated cells (●——●) chondrify only if the initial cell density is high. Addition of 4000–5000 μg of 3-acetylpyridine/culture dish (▲--▲) allows chondrogenesis to be expressed at lower initial cell densities. Redrawn from Caplan (1970).

cells/plate), Caplan (1972b). The sizes of the total pools of nicotinamide and NAD were reduced, as were the syntheses of RNA, protein, and glycosaminoglycans. The levels of three of the key enzymes involved in the synthesis of chondroitin sulfate (UDPglucose pyrophosphorylase, xylosyltransferase, and *N*-acetylgalactosaminyltransferase) were all sharply increased (Caplan, 1972b; Caplan and Stoolmiller, 1973), reflecting perhaps the increasing number of cells involved in chondrogenesis.

There is a general increase in total NAD with *in vivo* limb bud development and correlation between high levels of NAD and myogenesis, and low levels and chondrogenesis *in vitro* (Rosenberg and Caplan, 1974) and in 6-AN-induced micromelia (Chapter 7, Section II,D,7). These authors proposed that high levels of nicotinamide inhibited chondrogenesis, and that low levels favored chondrogenesis while also inhibiting myogenesis. Further, they proposed that the levels of nicotinamide were influenced by exogenous 3-acetylpyridine, and that these in turn influenced the internal pool size of NAD. High intracellular levels of NAD would favor myogenesis, and low levels would favor chondrogenesis.

Although cells are not permeable to NAD, they are permeable to nicotinamide. Elevated levels of nicotinamide external to the cell would lead to elevated intracellular levels of NAD. Preliminary indications are that NAD may act via conversion to poly(ADPribose), which is bound to both histone and to nonhistone chromosomal proteins (Caplan and Rosenberg, 1975; Caplan, 1977). Synthesis of poly(ADPribose) parallels the differentiation of mesenchymal cells into chondrocytes *in vitro* and is suppressed by both 3-acetylpyridine and by nicotinamide. They suggest that altering the internal pool size of NAD may influence transcription by altering the levels of the NAD-dependent ADP-ribosylation of chromatin, thereby enhancing or initiating chondrogenic expression. Holliday and Pugh (1975) and Strom and Dorfman (1976a,b) have provided additional evidence that moderately repetitive, cartilage-specific sequences of DNA are amplified and modified during the differentiation of chondrocytes *in vitro*. As myoblasts differentiate, there is a *decrease* in transcriptional diversity (Ordahl and Caplan, 1976).

If NAD also exerts this control over chondrogenesis *in vivo,* then one would expect to find gradients of NAD across the limb bud. NAD would be low in the central condensed mesenchyme, where chondrogenesis occurs, and would be higher in the peripheral myogenic mesenchyme. While these gradients have not been detected, Caplan and Koutroupas (1973) propose that differential vascularity across the limb bud provides the morphological transport system necessary for the establishment of such gradients. It now seems apparent that such gradients would not be responsible for the *determination* of *somatic* mesenchymal cells for myogenesis within the limb bud. Whether gradients play any role in the determination of mesenchymal cells for chondrogenesis or myogenesis *in vivo* is not known. The *expression* of differentiation *in vivo* clearly is influenced by these metabolic changes. Under *in vitro* conditions, in which presumptive myogenic cells become chondrogenic, these metabolic factors influence both the expression of differentiation and the ability of presumptive myogenic cells to chondrify (i.e., expression of an alternate pathway of differentiation). Even if myogenic and chondrogenic populations are distinct and separate *in vivo,* it is clear that chondrogenic expression can be elicited from presumptive myogenic cells *in vitro*.

7

Maintenance of the Differentiated State

I. INTRODUCTION

There is a vast literature on the factors that enhance or retard the expression of the differentiated state of chondroblasts or myoblasts *in vitro*. The majority of these studies have involved the analysis of populations of cells that were already terminally differentiated at the outset of the culture period. As a result, these studies provide very valuable information on the *expression* of the differentiated state and on the ability of differentiated cells to modulate their phenotypes. They do not, however, aid our understanding of the *acquisition* of the differentiated state or the process of differentiation from precursor cells. Those studies on the expression and maintenance of the differentiated state *in vitro* are relevant to the maintenance of a differentiated population(s) of cells within the skeleton *in vivo*.

We know considerably more about the factors involved in the maintenance of differentiated chondrocytes than we do about the maintenance of differentiated osteocytes, perhaps because in most skeletal sites the chondrocytes are not permanent, but degenerate, to be replaced by osteoblasts and by bone. The permanent mineralized matrix surrounding the osteocytes makes the study of their continued activity so difficult. I then, shall spend some time discussing the maintenance of the chondrocyte and then briefly discuss maintenance of the osteocyte.

II. MAINTENANCE OF THE DIFFERENTIATED CHONDROCYTE

We can distinguish chondroblasts and chondrocytes from the other cells of the connective and skeletal tissues on the basis of morphology and specialized functions. The distinction between chondroblast and chondrocyte is often an arbitrary one—the chondrocyte is more mature than the chondroblast. The characteristic surface projections of the rounded chondrocyte resting within its lacuna (Fig. 23a), the synthesis of chondroitin sulfate and its deposition into an extracellular matrix, the synthesis of type $[\alpha 1(\mathrm{II})]_3$ collagen, and the ability to yield chondrogenic clones, all serve to identify these cells.

Understanding how precursor cells differentiate into chondroblasts requires knowledge of the synthesis and deposition of the products of the extracellular matrix. Hyaluronic acid must be removed before differentiation can be initiated (Chapter 5, Section IV), and the synthesis of chondroitin sulfate must be enhanced, and that of type II collagen initiated (Chapter 6, Section VI,B). The maintenance of that differentiated state requires the continued synthesis and deposition of this specialized extracellular matrix. Thus we might expect differentiated chondrocytes to be sensitive to this extracellular environment and to use it to monitor and thereby regulate their synthetic activity.

Hyaluronic acid as a component of the extracellular matrix already has been discussed (Chapter 5). Here, I shall concentrate on chondroitin sulfate and collagen, especially with respect to their synthesis by the same chondrocyte, the positive and negative feedback to the chondrocyte to regulate rates of synthesis, and their deposition and interaction within the extracellular matrix.

A. Synthesis of Collagen and Chondroitin Sulfate by the Same Chondrocyte

Prockop *et al.* (1964) set out specifically to determine whether collagen and chondroitin sulfate were synthesized by the same or by different chondrocytes. Uptake of $[^3H]$- and $[^{14}C]$-labeled proline into hydroxyproline confirmed that isolated vertebral chondrocytes from 10-day-old (H.H. stage 36) embryos were synthesizing collagen *in vitro*. Double labeling with $^{35}SO_4$- and ^3H-labeled proline produced incorporation of the labels into glycosaminoglycans and collagen, respectively. Synthesis of both products stopped under conditions that allowed chondrocytes to dedifferentiate (growth on fibrin clots). It was argued that since the inability to synthesize both products was lost simultaneously in these dedifferentiated chondrocytes, their syntheses might be closely interrelated. However, Bhatnagar and Prockop (1966) went on to show that the two could be inhibited independently (see Section C).

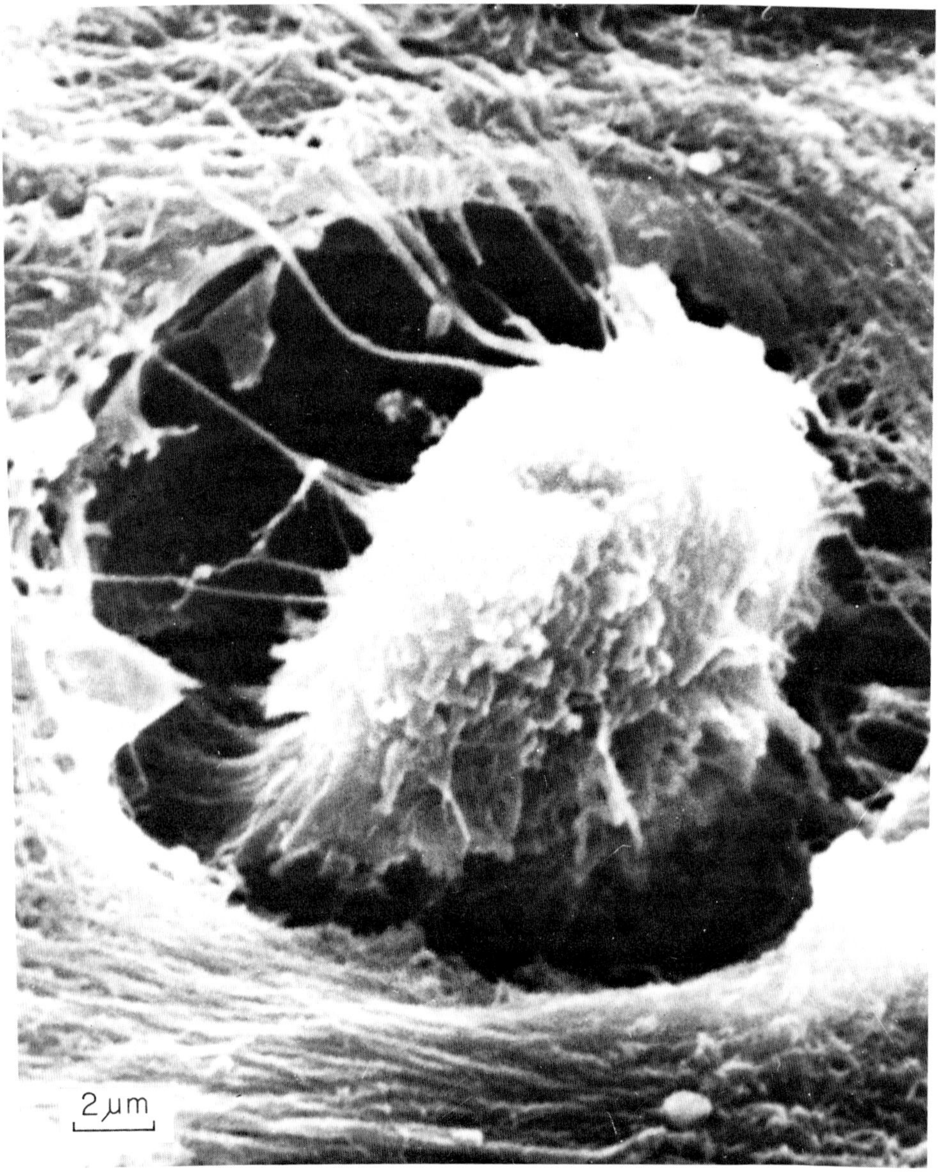

Fig. 23a. A chondrocyte from the articular cartilage of the patella of a normal patient. Note the prominent lacuna and the dense meshwork within the extracellular matrix. Reproduced with the permission of the author and publisher from Redler (1974).

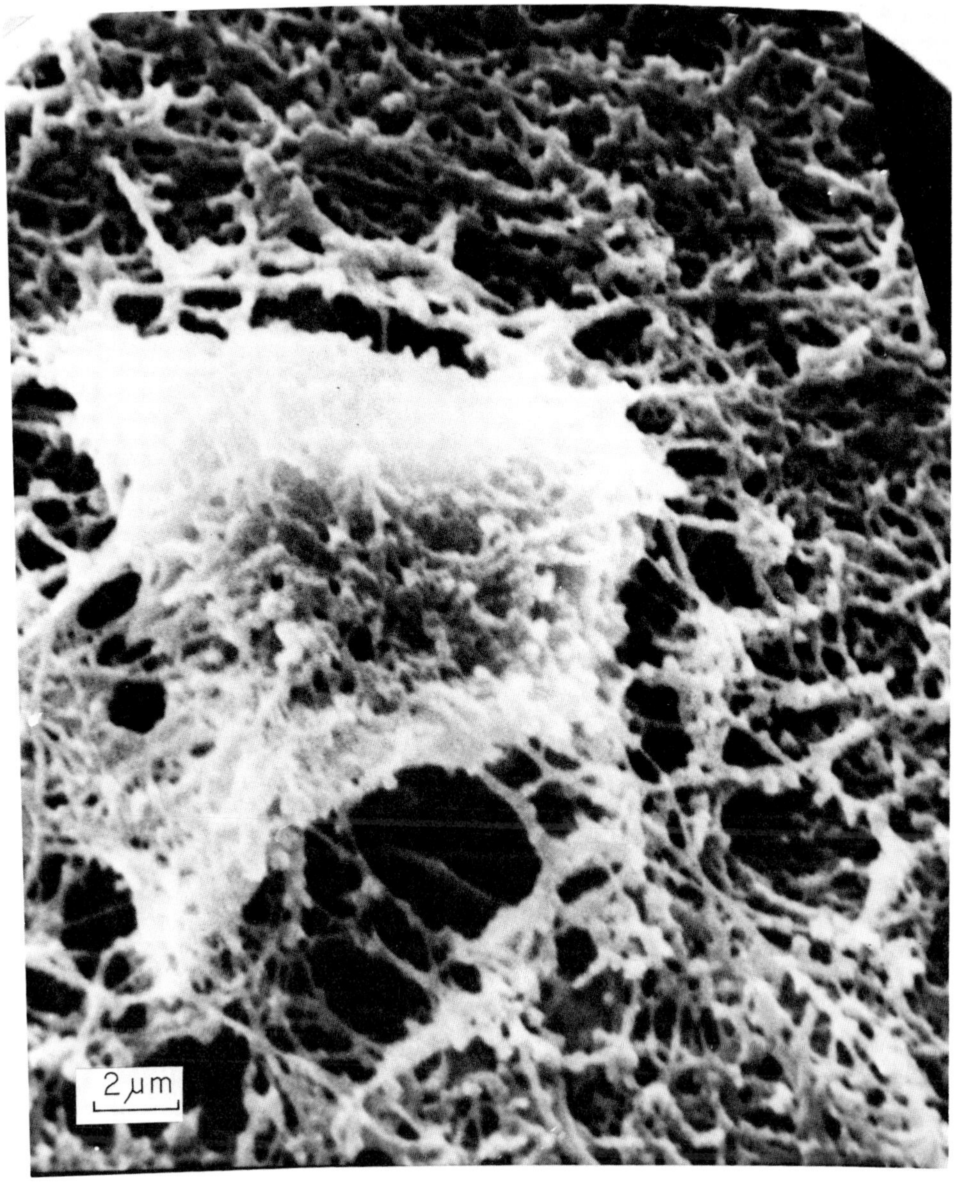

Fig. 23b. A chondrocyte from the articular cartilage of the medial tibial condyle from an osteoarthritic patient. Note the collapse of the lacuna, the distortion of the chondrocyte, and the open and haphazard meshwork within the extracellular matrix (cf. Fig. 23a). Photomicrograph kindly supplied by Dr. Irving Redler.

Smith (1972), using an autoradiographic analysis of double-labeled sternal chondrocytes from 13-day-old embryos, showed that all of the chondrocytes in his preparation incorporated both labels over the same time period. Short pulses of label were used to overcome the problems of reutilization or exchange of label. The visualization of the chondrocytes was such that he could state that the same cells were synthesizing collagen and glycosaminoglycans at the same time. Nimni (1973) and Bornstein (1974) have provided excellent reviews on the routes by which these products are synthesized.

B. Feedback Control of the Synthesis of Glycosaminoglycans

Over the past few years, there has been an accumulation of reports indicating that the amount and nature of the glycosaminoglycans in the extracellular matrix of chondrocytes play a role in regulating the rate of synthesis of glycosaminoglycans by those same chondrocytes. Whether that feedback is positive or negative depends upon whether the three-dimensional architecture of the cartilage has been retained (as in whole skeletal rudiments or in organ culture), or whether the chondrocytes have been liberated from their matrices and established as cell cultures.

1. Organ Culture

Depletion of glycosaminoglycans from organ-cultured cartilaginous tibiae by agents such as testicular hyaluronidase results in removal of glycosaminoglycans from the extracellular matrix and produces a compensatory increase in the synthesis of glycosaminoglycans. Collagen synthesis is unaffected (Fitton Jackson, 1970, 1976). The recovery of synthesis depends upon the dosage of hyaluronidase used to deplete the matrix. This is indicative of regulation via the level of glycosaminoglycans within the matrix. The glycosaminoglycans that are synthesized are smaller than normal; as well, less chondroitin sulfate is synthesized, either because the synthesis of other glycosaminoglycans is preferentially increased, or because of decreased glycosylation of the glycosaminoglycans (Hardingham *et al.*, 1972). This stimulation of glycosaminoglycan synthesis by negative feedback is in contrast to the stimulation by positive feedback in cultured *isolated* chondrocytes (Nevo and Dorfman, 1972).

The production of glycosaminoglycans by H.H. stage 17 somites is increased two- to threefold by the addition of glycosaminoglycans derived from embryonic cartilage (Kosher *et al.*, 1973). The magnitude of this stimulation indicates that either chondroblast precursors or chondroblasts in the early stages of differentiation when little matrix is present are especially sensitive to regulation of glycosaminoglycan synthesis by end-product stimulation. This also has been shown by Takeichi (1973). Stimulation may also come from other intermediates in the synthetic pathway of chondroitin sulfate. Kim and Conrad (1974) organ-cultured vertebral chondrocytes from 8-day-old embryos in the presence of exogenous D-glucosamine and found that glycosaminoglycan synthesis increased as the concentration of added D-glucosamine was increased up to 2 m*M*, but that at 4 and 25 m*M* synthesis was suppressed. Ehrlich *et al.* (1974) found that glucosamine, UDP-*N*-acetylglucosamine, and UDP (at concentrations of 1 m*M*/ml) stimulated glycosaminoglycan synthesis by rabbit articular cartilage *in vitro*. UDP was the most effective, with the chondrocytes exhibiting differential responses depending on the concentration of added UDP (Fig. 24). It appears that chondrocytes can respond differentially, depending on the local concentration of extracellular matrix components. Exogenous glucose does not exert a regulatory role (Delcher *et al.*, 1973).

The mechanism whereby organ-cultured cartilage both senses and regulates the synthesis of glycosaminoglycans is not understood. As indicated in Chapter 6, there may be several points of regulation within the metabolic pathway. Even an increase in the Ca^{2+} concentration of the medium from 0.5 to 1 m*M* can stimulate glycosaminoglycan synthesis (Shulman and Opler, 1974).

Other matrix products also may be regulated by end-product stimulation. For example, the concentration of lysozyme within the extracellular matrix regulates the rate at which lysozyme is synthesized (Kuettner *et al.*, 1972).

2. Chondrocyte Cell Cultures

Nevo and Dorfman (1972) maintained epiphyseal chondrocytes in suspension culture, added chondromucoprotein and other polyanionic glycosaminoglycans,

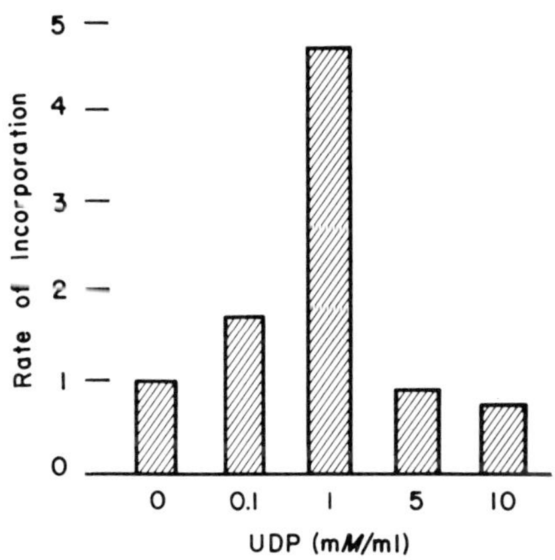

Fig. 24. The effect of the concentration of added UDP (m*M*/ml culture medium) on the rate of incorporation of $^{35}SO_4$ into slices of rabbit articular cartilage *in vitro*. Control medium without addition of UDP is given a rate of incorporation of 1. From data in Ehrlich *et al.* (1974).

and obtained enhancement of glycosaminoglycan synthesis over a 2.5-hour period (Table XVII). Enhancement was dose dependent. The synthesis of collagen, and that of total protein, were not altered, indicating specificity of the response to glycosaminoglycan synthesis. Enzyme induction was ruled out, as xyloslytransferase and *N*-acetylgalactosaminyltransferase activities were not enhanced. Nor does initiation of core protein synthesis appear to be the mechanism. Schwartz and Dorfman (1975), adding chondroitin sulfate or heparin to monolayered chondrocytes, were able to stimulate the synthesis of chondroitin sulfate. When β-D-xyloside and chondroitin sulfate were added together, a stimulation of synthesis was observed, an increase exceeding that obtained with either agent alone. They therefore concluded that the stimulation did not involve *stimulation* of core protein synthesis: although this synthesis is required, puromycin stops the synthesis of chondroitin sulfate. Nevo and Dorfman (1972) speculated that cell membrane receptors were involved, and that these receptors sensed the extracellular environment. There is evidence for cell-to-cell interaction under the influence of carbohydrates (Turner and Burger, 1973), and Wiebkin and Muir (1975) have indicated that hyaluronic acid interacts with glycosaminoglycans at the chondrocyte surface to regulate the synthesis of glycosaminoglycans.

Using sternal chondrocytes that synthesize glycosaminoglycans at rates proportional to the rate of accumulation of matrix, Solursh and Meier (1973, 1974) and Solursh *et al.* (1973) found that a nondialyzable, trypsin-, and heat-sensitive molecule of molecular weight 30,000 to 150,000 was liberated into the medium by these chondrocytes, and that it stimulated the syntheses of both chondroitin sulfate and collagen. Meier and Hay (1974a,b) and Solursh and Karp (1975) provide excellent discussions of the possible mechanisms of these stimulations, for both chondrocytes and corneal epithelium.

TABLE XVII

Results of the Stimulation of the Synthesis of Glycosaminoglycans by Cultured Chondrocytes as a Function of the Amount of Exogenous Chondromucoprotein Added to the Medium[a]

Chondromucoprotein (μg/ml)	Glycosaminoglycan synthesis (cpm ^{35}S)
0	5790
20	8207
200	22,896
2000	26,521

[a] Based on data in Nevo and Dorfman (1972).

We saw that organ-cultured cartilage responded to enzymatic depletion of extracellular matrix by initiating the synthesis of glycosaminoglycans. Cultured chondrocytes respond to hyaluronidase, chondroitinase, or trypsin by decreased incorporation of ^{35}S (Huang, 1974, 1977; Solursh and Karp, 1975). The reason for the differential response of rudiments versus isolated cells is not known. Nevertheless, it is clear that chondrocytes are sensitive to the amount and quality of their extracellular matrices, and that they respond by regulating the rates of their glycosaminoglycan syntheses, thereby aiding in the maintenance of their differentiated phenotype.

C. Interaction between Glycosaminoglycans and Collagen within the Extracellular Matrix

The interaction that exists between the glycosaminoglycans and the collagen fibrils and fibers of the extracellular matrix is very complex. The integrity of both the matrix and the cartilaginous rudiment or tissue is dependent upon this interaction (Bazin *et al.*, 1962). Complex formation occurs between collagen and chondroitin sulfate and between collagen and hyaluronate. These complexes are highly ordered and provide the biochemical basis for the integrity not only of cartilage, but of the entire early embryo (Partridge, 1948; Mathews, 1965; Wiebkin and Muir, 1973; Campo, 1974). According to Serafini-Fracassini and Smith (1974): "In cartilage, the main function of collagen is probably that it determines the ordering within the matrix of other physiologically important components, such as proteoglycan macromolecules."

The syntheses of collagen and chondroitin sulfate are regulated independently. Bhatnager and Prockop (1966) inhibited collagen synthesis with α,α^1-dipyridyl and found that the rate of chondroitin sulfate synthesis was unaffected. When they inhibited this latter synthesis with 6-diazo-5-oxonorleucine (a glutamine analogue), collagen synthesis was unaffected (see also Bhatnagar and Rapaka, 1971). Thus, while chondroitin sulfate can be synthesized in the absence of collagen, the latter is required if the polysaccharide is to be deposited into an insoluble matrix (Lavietes, 1971). Continuous RNA synthesis is required for maintenance of synthesis of chondroitin sulfate, but not for collagen synthesis (Solursh and Meier, 1972), indicating that control for chondroitin sulfate is transcriptional, whereas control of collagen synthesis is translational or posttranslational (see Nimni, 1973, for a review). BuDR inhibits chondroitin sulfate synthesis but not the synthesis of collagen (Levitt and Dorfman, 1972). Dondi and Muir (1976) used the glycosaminoglycan initiator, 4-methylumbelliferyl-β-D-xyloside, to inhibit synthesis in pig laryngeal cartilage and found that synthesis of collagen was unaffected. Insulin stimulates the synthesis of glycosaminoglycans but has little effect on collagen synthesis (Hajek and Solursh, 1975).

This sampling of experimental approaches indicates that the syntheses of collagen and chondroitin sulfate are under independent control. Nor are interactions between the two molecules required for transport through, or secretion from, the cell (Nimni, 1973; Ross, 1975). However, the concentrations of collagen and glycosaminoglycans within the extracellular matrices of cartilages show very close correlations, even after a variety of quite different treatments (Hall, 1973; also Fig. 25).

Certain interactions between collagen and chondroitin sulfate within the extracellular matrix are necessary before either can be deposited into an insoluble complex. The chemistry of this interaction has been reviewed by Steven (1972), Walton (1974), and by Scott (1975). Glycosaminoglycans, such as chondroitin sulfate and dermatan sulfate, and glycoproteins interact with tropocollagen to aid in collagen fibrillogenesis (Toole and Lowther, 1968a,b; Toole, 1969; Öbrink and Wasterson, 1971; Lowther and Natarajan, 1972; Neméth-Csóka, 1974). The thermal stability of collagen and, hence, its triple helical structure, are stabilized by the interaction with chondroitin 6-sulfate (Gelman and Blackwell, 1973). These latter experimenters added chondroitin 6-sulfate to collagen in concen-

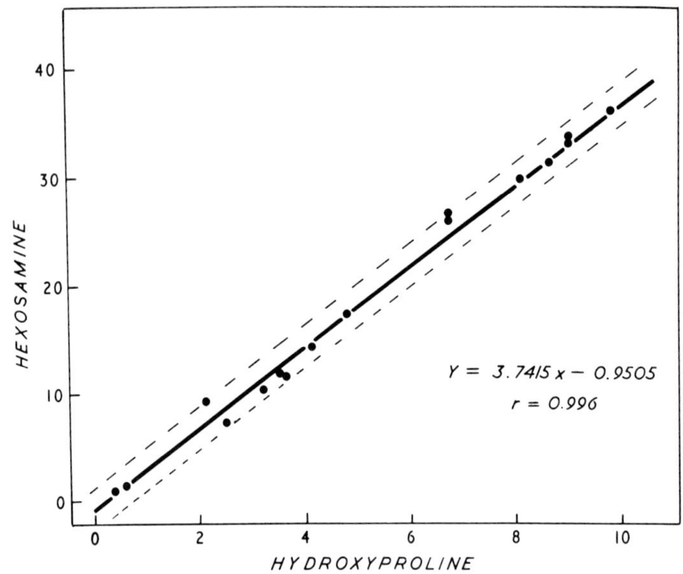

Fig. 25. The concentration of hexosamine (μM/pair of tibiae) plotted against the concentration of hydroxyproline (μM/pair of tibiae) for tibiae of 12-day-old control embryos and for tibiae from embryos previously treated with thallium and vitamin C. [See Hall (1973) for details of the treatments.] The concentrations of hexosamine and hydroxyproline are highly correlated. Reproduced with permission of the National Research Council of Canada from Hall (1973).

trations of 5.5 disaccharide residues/ 100 amino acids and found that the "melting" temperature of calf skin collagen could be raised from 38 to 46°C, thereby stabilizing the collagen fiber.

Cartilage-type collagen renatures slowly (Igarashi *et al.*, 1973) and is particularly capable of binding to glycosaminoglycans, hence, maintaining the integrity of the extracellular matrix. Nimni and Deshmukh (1973) showed that type II collagen contained four or five times more hydroxylysine and glycosidically linked glycosaminoglycans than type I collagen, although other areas of the collagen chains are homologous and highly conservative (Butler *et al.*, 1974). Hydroxylysine provides binding sites for hexose sugars. Any reduction in glycosylation or hydroxylation could potentially disrupt the integrity of the cartilage matrix, and a number of agents known to disrupt cartilage (thyroxine, corticosteroids, growth hormone, β-aminopropionitrile) in fact do reduce the glycosylation of cartilage collagen (Blumenkrantz and Prockop, 1970). This biochemical characteristic of cartilage collagen also is thought to be responsible for the characteristic morphological appearance of the cartilage collagen fibers (see Stark *et al.*, 1972; Bruns *et al.*, 1973). The more glycosaminoglycans bound within the extracellular matrix, the more able cartilage is to resist stresses (Lee-Own and Anderson, 1975). Nimni and Deshmukh (1973) noted that osteoarthritic human articular cartilage contained considerable type I collagen and concluded that it therefore was less able to bind glycosaminoglycans and more liable to destruction from abrasion.

Ultrastructural evidence demonstrates the spacing of glycosaminoglycans along the collagen fibers in bovine and rabbit articular cartilage (Smith and Serafini-Fracassini, 1967; Shepard and Mitchell, 1977), on epiphyseal cartilage (Eisenstein *et al.*, 1971a,b; Campo and Phillips, 1973; Thyberg, 1977), and on condylar secondary cartilage (Silbermann and Frommer, 1974). Figures 26 and 27 illustrate the periodic spacing along the collagen fibers seen in rabbit articular cartilage. However, Goel and Jacob (1976) have cast doubt on the nature of these matrix granules. Observing that treatment of newborn mouse epiphyseal cartilage with collagenase before fixation removed the granules, they concluded that the granules were artifacts. But there is such an abundance of evidence for their reaction with staining procedures for glycosaminoglycans that their existence and nature are well established. Eisenstein *et al.* (1971b) used ruthenium red, protamine, and lysozyme to distinguish three proteoglycans in the epiphyseal growth cartilage of the dog and mouse, each having different associations with collagen fibers.

D. Achondroplasia (Chondrodystrophy)

The importance of the interaction between extracellular collagen and glycosaminoglycans for the maintenance of the differentiated state and the integ-

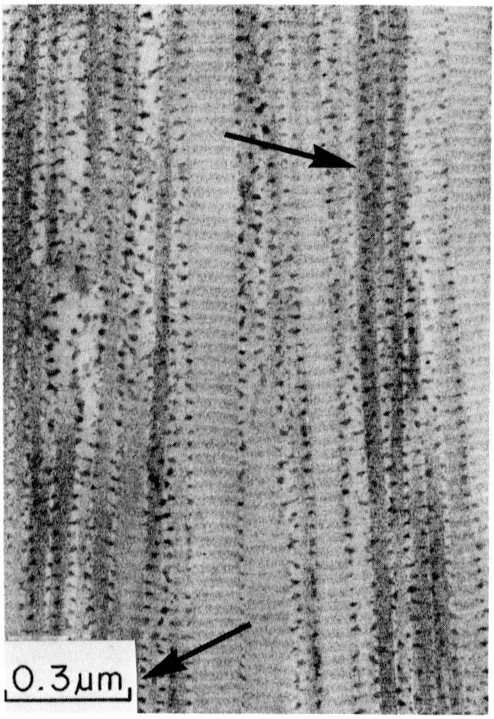

Fig. 26. Extracellular matrix from articular cartilage of the rabbit stained with 0.2% ruthenium red and counterstained with uranyl acetate and lead citrate. Ruthenium red-positive proteoglycan in a repeated pattern (arrows) links adjacent collagen fibers. Reproduced with permission of the author and publisher from Shepard and Mitchell (1977).

rity of the chondrocytes can be illustrated very well by an analysis of a series of achondroplastic mutations. In each of these mutations, the abnormality involves some aspect of collagen and/or glycosaminoglycan synthesis, transport, deposition, or interaction. While the bases of these changes differ from mutant to mutant, the gross morphological manifestation is similar.

Achondroplasia is an abnormality in the conversion of cartilage into bone that results in asymmetrical dwarfism. Histological changes usually accompany achondroplasia, but only in some types are they diagnostic, and some may show no histological abnormality at all (Rimoin *et al.*, 1970, 1976). Chondrocyte necrosis is a feature of many of the genetic and teratological achondroplasias that have been described in the achondroplastic human, mouse, and rabbit (Shepard *et al.*, 1969; Nardi *et al.*, 1974; Silberberg *et al.*, 1976), in the crooked-neck dwarf chick (Allenspach and Babiarz, 1975), and in insulin, 6-aminonicotinamide-, and thallium-induced micromelia (Seegmiller *et al.*, 1972a; Daimon, 1973a,b; Hall, 1972a,c, 1977b; Hinchliffe, 1974). This necrosis may be re-

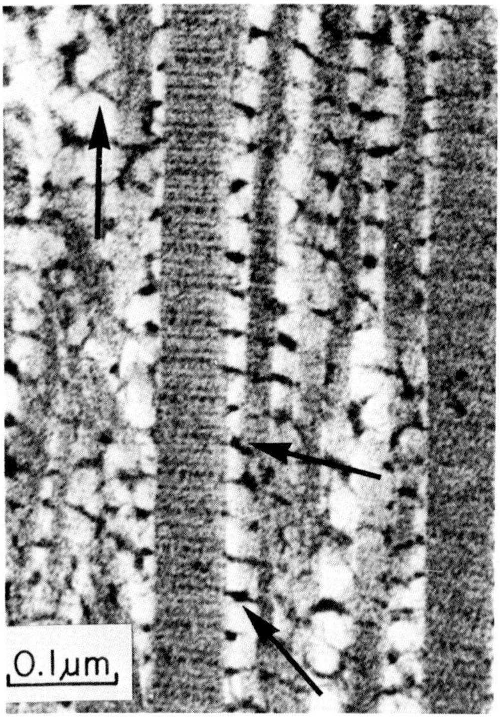

Fig. 27. Extracellular matrix from the articular cartilage of the rabbit stained with 0.1% Safranin O and counterstained with uranyl acetate and lead citrate. Safraninin O-positive proteoglycan (arrows) forms regularly repeated attachments to the collagen fibers. Reproduced with permission of the author and publisher from Shepard and Mitchell (1977).

duced or reversed with either ascorbic acid or sodium ascorbate (Hall, 1972c; Overman *et al.*, 1976), suggesting an impairment in collagen metabolism. Achondroplasia may be inherited as an autosomal recessive, as in the "bulldog" calves of the Dexter breed, in the Dachshund, and in mice, or as an autosomal dominant, as in birds and man.

The literature on achondroplasia, as seems to be the case with all of the topics thus far, is enormous (for review, see Grüneberg, 1963; McKusick, 1972; Rimoin, 1975). Landauer (1969b) indexed 1755 references. Here, I shall discuss only several types, and shall concentrate on those for which some knowledge of the mechanism of action of the gene is known.

1. Cartilage Anomaly (*can*) in the Mouse

Johnson and his colleagues (Johnson and Wise, 1971; Johnson, 1974; Johnson and Hunt, 1974) have been investigating the recessive mutant *can,* which arose spontaneously in their stock in 1965. All of the symptoms of classical achondroplasia are exhibited by these mice. Their growth is retarded, and their skulls, axial, caudal, and appendicular skeletons are disproportionately shortened. The homozygote (*can/can*) is lethal at about ten days after birth. The primary defect lies in the cartilage.

The first sign of the biochemical alterations in these skeletons is reduced protein synthesis at 17 days of gestation (this is before the mutant can be identified morphologically). However, protein synthesis rises to above-normal levels after birth, when the defect is manifested. The biochemical changes are quite complex (Johnson and Hunt, 1974), although both the ultrastructure and the synthesis of collagen are normal. There is less extracellular matrix within the cartilage as assessed by light and electron microscopy and by incorporation of [^{14}C]glucose. Although the levels of UDPG dehydrogenase and of UDP-*N*-acetylglucosamine-4-epimerase are reduced, glucose-6-phosphate dehydrogenase is not—consistent with a specific reduction in the synthesis of glycosaminoglycans. Oxidative phosphorylation is normal (cf. the *ac* rabbit below). *Can* therefore involves decreased protein synthesis and a mechanism that results in reduced deposition of extracellular matrix by chondrocytes.

2. Achondroplasia (*ac*) in the Rabbit

Shepard *et al.* (1969), Shepard (1971), and Bargman *et al.* (1972) have studied this autosomal recessive achondroplastic mutant. Synthesis of cartilaginous extracellular matrix is reduced, and there is considerable cell death and necrosis, especially in the center of the cartilaginous primordia, a fact which prompts the suggestion that isolation from vascularity is involved. Incorporation of both glucose and galactose is increased, in comparison with normal littermates, indicating defective glucose metabolism. Oxidative phosphorylation in liver mitochondria is also defective. Other than these data, we have no information about how the defective extracellular matrix is produced in this mutant.

3. Achondroplasia (*cn*) in the Mouse

Lane and Dickie (1968) described the genetics and growth characteristics of this achondroplastic recessive mutant in the mouse. As in human achondroplastics, the main effect on growth occurs in early postnatal life. In the mouse, the growth rate is below normal only during the first three weeks of life (Konyukhov and Paschin, 1970). The growth zone of the long bones is normal, but the hypertrophic zone is reduced (Silberberg and Lesker, 1975). Bonucci *et al.* (1976, 1977) and Silberberg *et al.* (1976), using histochemical and ultrastructural criteria, claim that the difference in cytodifferentiation between *cn* and normal cartilage is premature aging of the chondrocytes. There is shortening of the cartilage columns, decreased deposition of glycosaminoglycans into the extracellular matrix, accumulation of glycogen within chondrocytes, and consequent delays in calcification. Many of the hypertrophic chondrocytes survive the delayed matrix calcification. Subperiosteal osteogenesis is unaffected. This is probably because the gene acts directly on the chondrocytes—as shown by the reciprocal transplantation of cartilage between normal and *cn* mice (Konyukhov and Paschin, 1967).

Interstitial growth of the sphenooccipital, midsphenoidal, and nasal cartilages is inhibited; however, the mandibular condylar cartilage, whose growth is appositional, grows normally in these achondroplastic mice, as it does in human achondroplasia (Shepard *et al.*, 1969; Jolly and Moore, 1975; Brewer *et al.*, 1977; also see Chapter 9).

Little biochemical information is available for this mutant. The levels of both hydroxyproline and sialic acid are reported to increase (Silberberg and Lesker, 1975) and to lie within normal limits (Kleinman *et al.*, 1977). The mutant acts primarily to inhibit the maturation and growth of the chondrocytes. Such mutants, which inhibit intrinsic regulation of skeletal growth, are notoriously difficult to understand (see Chapter 9, Section II,C,2).

4. Chondrodysplasia (*cho*) in the Mouse

Chondrodysplasia is a recessive lethal mutant in the C57 mouse that produces shortened but wide long bones, a shortened face, cleft palate, and collapsed tracheae (Seegmiller and Fraser, 1977). A decreased amount of extracellular matrix is present; there is formation of large (600–2000 Å diameter), banded collagen fibers, and an increase in the size of the Golgi body, especially in the number of Golgi vacuoles (Seegmiller *et al.*, 1971, 1972b). [Similar thick collagen fibers are seen in the cloverleaf skull syndrome, a chondrodystrophy of the skull base (Bonucci and Nardi, 1972).]

In *cho*, the chondroblasts fail to align in columns within the long bones and instead spread laterally, contributing to the expanded appositional growth of the epiphyses. Seegmiller *et al.* (1971, 1972b) proposed that glycosaminoglycans were not released from the chondrocytes, but were retained within Golgi vacuoles. In two of the chondrodystrophies affecting man (metaphyseal dysostosis and pseudoachondroplastic dwarfism), there is evidence of retention of material within distended cisternae of the endoplasmic reticulum: granular (glycosaminoglycans?) in the former, and lamellar (protein–glycoprotein?) in the latter (Cooper and Ponsetti, 1973; Cooper *et al.*, 1973).

The reduced glycosaminoglycan content of the extracellular matrix of *cho* would enhance polymerization of tropocollagen into collagen fibrils. A matrix having abnormally large collagen fibers but with reduced glycosaminoglycan levels thereby would be produced and, as a result, would be weakened. This would affect the alignment of the chondroblasts and would cause collapse of the tracheal cartilage, resulting in asphyxiation. Kochhar *et al.* (1976) applied the glutamine analogue, 6-diazo-5-oxonorleucine, to mouse limb buds *in vitro* and observed the formation of a dense meshwork of unusually wide collagen fibers.

Collagen synthesis was unaffected. Thus, reduced deposition of glycosaminoglycan into the extracellular matrix of cartilage can affect collagen fibrillogenesis and the integrity of the cartilage. However, in the nanomelic chick (see below), synthesis and deposition of glycosaminoglycans are severely retarded, yet collagen fibrillogenesis is normal (Pennypacker and Goetinck, 1976).

Stephens and Seegmiller (1976) have provided further biochemical data on rib and limb cartilage from *cho* mutants that indicate normal levels of incorporation of $Na_2{}^{35}SO_4$, glucosamine and glucose, total uronic acid, and protein, and that glycosaminoglycans of normal molecular weight are synthesized. Less glycosaminoglycan was deposited into the extracellular matrix (more ^{35}S is washed out of *cho* cartilage than from normal cartilage), but the collagen content also was recorded as below normal. A similar failure of deposition of glycosaminoglycans into the extracellular matrix is seen in one of the human achondroplastic mutants (Pedrini-Mille and Pedrini, 1971). So, whether the defective collagen fibrillogenesis in *cho* results from a primary defect in the deposition of glycosaminoglycans into the extracellular matrix, or whether from a primary defect in the collagen itself, remains to be determined.

5. Brachymorphic (*bm*) in the Mouse

Another mutant producing disproportionate dwarfism in the mouse is the brachymorphic. Its basic genetics have been described by Lane and Dickie (1968); however, there has been little investigation into the mechanism of action of this mutant. The collagen produced is normal cartilage-type II collagen, but the chondroitin sulfate is undersulfated and fails to aggregate into proteochondroitin sulfate granules within the extracellular matrix (Orkin *et al.*, 1976, 1977). A similar failure to aggregate is seen in the nanomelic chick, although the mechanism is quite different (see below).

Another recently discovered chondrodystrophic mutant in the mouse is Stumpy (*stm*). In Stumpy, although the rate of chondroblast proliferation is normal, the chondrocytes fail to separate from one another (Johnson, 1977, 1978), perhaps indicating a cell membrane dysfunction, as is seen in the brachypod (bp^H) mouse (Section F,3,c).

6. Nanomelia (*nm*) in the Chick

Nanomelia is a recessive lethal chondrodystrophic mutant in the chick that causes extreme (homozygote) or moderate (heterozygote) micromelia of the appendicular skeleton (Landauer, 1965). It was shown by Mathews (1967b) that synthesis of chondroitin sulfate in the nanomelic limb and sternal cartilage was depressed 90% below the normal level, while synthesis in the skin was normal, i.e., *nm* does not involve a systemic alteration in the synthesis of chondroitin sulfate. Fraser and Goetinck (1971) and Palmoski and Goetinck (1972) confirmed reduced synthesis of chondroitin sulfate in sterna both *in vivo* and *in vitro*. The level of synthesis was similar to that found in the H.H. stage 18 normal limb bud mesenchyme, a finding which points toward a block in the enhancement of synthesis in the mutant.

Cultured normal sternal vertebral and epiphyseal chondrocytes synthesize two major peaks of proteochondroitin sulfate. The major of these is completely missing from the medium in which *nm* sterna are cultured and is severely reduced in the chondrocytes and may represent a cartilage-specific proteoglycan (Palmoski and Goetinck, 1972; Pennypacker and Goetinck, 1976; also see Okayama *et al.*, 1976; Wiebkin and Muir, 1977). The number of glycosaminoglycan granules in the extracellular matrix, and the amount of matrix itself, are severely reduced below normal. Xyloside was found to stimulate the synthesis of chondroitin sulfate in *nm* chondrocytes. This indicated that the glycosyltransferases for chondroitin sulfate were present and functional (Pennypacker and Goetinck, 1976). The glycosaminoglycan that is produced fails to aggregate with hyaluronic acid (Goetinck and Pennypacker, 1977). The rate of collagen synthesis and the ultrastructure of collagen fibers are normal in *nm* chondrocytes. The action of the *nm* gene is then to suppress (or not allow) the augmentation of the synthesis of the proteochondroitin sulfate that is characteristic of chondrocytes. Failure to aggregate with hyaluronic acid leads to absence of matrix granules. This reduced content of glycosaminoglycans in the extracellular matrix has a minimal effect on collagen fibrillogenesis (P. I. Goetinck, personal communication).

7. 6-Aminonicotinamide-Induced Micromelia

In chick embryos, micromelia can be produced by a variety of teratological, treatments, including administration of 6-aminonicotinamide (6-AN) (Landauer, 1957). 6-AN is converted to 6-aminonicotinamide adenine dinucleotide (6-AN-NAD), which displaces endogenous NAD, reducing oxidative phosphorylation and impairing the dehydrogenase-linked reactions involved in the utilization of glucose (Caplan, 1972c; Seegmiller and Runner, 1974).

Cotreatment with NAD, or with sodium or calcium ascorbate or ascorbic acid, *in vivo* and *in vitro*, prevents the teratological action of 6-AN (Caplan, 1972d; Overman *et al.*, 1972, 1976). The possible role of NAD in the regulation of cartilage differentiation was discussed in Chapter 6, Section VI,B,8).

Clonally cultured myogenic cells synthesize both collagen and glycosaminoglycans in the presence of 6-AN (Schubert and Lacorbiere, 1976). Within chondrocytes, the sulfation of chondroitin sulfate is reduced, although the incorporation of glucosamine into glycosaminoglycans and that of proline into collagen are normal (Overman *et al.*, 1972; Seegmiller and Runner, 1974). The number of Golgi vesicles in the chondrocytes is greatly reduced, and there is some reduction in the rough endoplasmic reticulum, indicative of reduced sites of synthesis and slowed transport of collagen and glycosaminoglycans (Seegmiller *et al.*, 1972a).

The deposition of extracellular matrix is inhibited, especially in the chondrogenic core of the limb bud (Caplan 1972c,d; Seegmiller, 1977).

This induced micromelia and the existence of the achondroplastic mutants indicate that any impairment in the interaction between glycosaminoglycans and collagen in the extracellular matrix will lead to an instability of the differentiated phenotype of the chondrocyte and to abnormal chondrogenesis.

In osteoarthritis and other degenerative joint diseases, similar disruptions are seen. Disruption of collagen–glycosaminoglycan interactions leads to deposition of swollen and disorganized collagen fibers in osteoarthritic patients (Redler, 1974; also see Fig. 23b). The glycosaminoglycan that is present is bound less firmly to the extracellular matrix (Brandt, 1974) and correlates with later alterations in the structure of the weight-bearing portions of the skeleton (Pugh *et al.*, 1974).

E. Modulation of the Chondrocytic Phenotype

A very profitable way of studying the factors responsible for the maintenance of the differentiated state of chondrocytes is to perturb them in cell culture, and a very considerable literature is available on the *in vitro* cultivation of avian chondrocytes. Excellent reviews and discussions of the literature can be found in Lasher (1971), Green (1971), Searls (1973a,b), Sokoloff *et al.* (1973), Levitt and Dorfman (1974), Levitt *et al.* (1974), and Wigley (1975).

Coon (1966) established that differentiated chondrocytes could produce clones in culture that were stable over many generations, and that while phenotypic expression was not lost with morphological dedifferentiation, it could be reexpressed in a permissive medium. Depletion of the extracellular matrix initiates a compensating synthetic response from the chondrocytes (Section A). Alteration of the extracellular matrix or of the external environment of the chondrocyte also can lead to modulation of its phenotype so that it expresses differentiative characteristics that normally are associated with other cell types.

During the past ten years, attention has been directed toward investigating the properties and responses of chondrocytes that have been isolated from mammalian articular cartilage and maintained *in vitro*. The impetus for such studies comes from the need to understand the metabolism and maintenance of articular cartilage, so that the degenerative bone diseases and osteoarthroses that involve articular cartilage may be treated. Of particular significance in this regard are the differences between the behavior of mammalian articular chondrocytes and avian chondrocytes *in vitro* that have emerged. These might make transference of the information gained from avian studies to mammals, and especially to man, hazardous. Sokoloff (1974) offers a useful review in the context of osteoarthritis. Thus, for these and related reasons—because of these differences, because these studies are not as well known to developmental biologists as are the avian studies, and because I wish to discuss reinitiation of cell division in mammalian articular cartilage in the context of cartilage maintenance and the reparative response to degenerative joint diseases (Section 2)—I shall discuss briefly the data available on the behavior of mammalian articular chondrocytes *in vitro*.

1. Mammalian Articular Chondrocytes *in Vitro*

Manning and Bonner (1967) were the first to culture adult mammalian articular chondrocytes. They isolated chondrocytes from individuals up to 84 years of age by collagenase digestion and mechanical disruption. Although they appeared "fibroblastic" in monolayer culture, the chondrocytes continued to produce extracellular matrix if pelleted by centrifugation prior to culturing [obviously a parallel of the situation described by Abbott and Holtzer (1966) for avian vertebral chondrocytes]. Collagenase digestion of femoral articular cartilage from mature female rabbits produced chondrocytes that formed colonies containing metachromatic matrix when cultured for 14 days in serum-supplemented Ham's F12 medium (Ham and Sattler, 1968; Ham *et al.*, 1970). The proliferative ability of articular chondrocytes when released from their matrices thus was established (see also Section 2).

The synthesis of glycosaminoglycans continues *in vitro* even under monolayer conditions unfavorable to this synthesis by avian chondrocytes (Sokoloff *et al.*, 1970; Srivastava *et al.*, 1974a). Glucose-metabolizing enzymes have been examined by Marcus and Srivastava (1973), and their activities have been found to be typical of fully functional chondrocytes. They also examined lactate/malate dehydrogenase ratios, finding these to be greater than those seen in fibroblasts. It was also learned that the levels of both enzymes increased with increasing oxygen tension [see Thorogood and Hall (1976, 1977) for a discussion of the usefulness of LDH/MDH ratios for the identification of chondrogenic cell types]. These mammalian articular chondrocytes produce A and B arylsulfatases and modify their rates of synthesis in the presence of hormones and ascorbic acid (Jones and Addison, 1975; Schwartz and Adamy, 1976). Obviously, the physiological activity and the responses of these chondrocytes are maintained *in vitro*.

The oxygen consumption of articular cartilage *in vivo* is low (6 μl O_2/g dry wt/hr) in comparison to that of epiphyseal cartilage, whose oxygen consumption ranges from 500 to 900 μl O_2/g dry wt/hr (Serafini-Fracassini and Smith, 1974). Nevertheless, like avian chondrocytes, human articular chondrocytes, when maintained in organ culture, will respond to lowered oxygen tensions (20% O_2) by increasing their synthesis of glycosaminoglycans and will react to higher oxygen tensions (50% O_2) by decreasing their synthesis (Jacoby and Jayson, 1975; Lemperg *et al.*, 1975). They can continue to produce glycosaminoglycans at a high rate for prolonged periods of time (Lane and Brighton, 1974). However, Marcus (1973) found that hypoxic conditions did not promote chondrogenesis from monolayered articular chondrocytes and questioned the general role of hypoxia in chondrogenesis. Numerous literature reports have indicated that hypoxia favors both initiation and maintenance of chondrogenesis (Bassett and Herrmann, 1961; Bassett, 1962; Hall, 1969, 1970; Pawelek, 1969; Brighton and Heppenstall, 1971; Hadházy *et al.*, 1971; Brighton and Kregs, 1972a,b; Nevo *et al.*, 1972; Brighton *et al.*, 1974; Heppenstall *et al.*, 1975). But Marcus points out the alarming fact that plastic culture dishes (on which most cells are cultured) contain considerable amounts of diffusible oxygen!

Cultured articular chondrocytes respond to environmental perturbations by modifying their synthetic activity. Under clonal conditions, differentiated hyaline chondrocytes can be maintained through many cell generations (Green and Dickens, 1972). If, after some time in culture, they are placed on gelfoam or on decalcified bone matrix, chondromyxoid or chondroid is produced (Green, 1971; Green and Ferguson, 1975). The type of extracellular matrix, and of cartilaginous tissue produced thus can be modified by varying the substrate on which the cells are cultured.

Under other conditions, synthesis of cartilage-type collagen ceases, and synthesis of type I collagen is initiated. [Lane and Weiss (1975) have reviewed the organization of collagen within articular cartilage.] Treatment with liver lysosomes, vitamin A, cAMP, or $CaCl_2$, PTH or CT, cell culture under monolayer conditions, or prolonged culture, will initiate this shift, which is reversible (Layman *et al.*, 1972; Sokoloff *et al.*, 1973; Deshmukh, 1974; Schindler *et al.*, 1975; Gay *et al.*, 1976b; Deshmukh and Kline, 1976; Benya *et al.*, 1977; Desmukh and Sawyer, 1977; Desmukh *et al.*, 1977; Norby *et al.*, 1977). Explorations of these shifts may point up a relationship to the osteoarthritic state, for articular cartilage from osteoarthritic patients synthesizes type I collagen (Nimni and Deshmukh, 1973; Gay *et al.*, 1976b; Müller and Kühn, 1977).

Not all mammalian articular chondrocytes behave similarly under *in vitro* cultivation. Srivastava *et al.* (1974b) found human articular chondrocytes to be quite different from those of the rabbit, in that their growth was poorer, and they produced a wide range of glycosaminoglycans. However, these samples were obtained from a variety of human sarcomas, so one might expect there to be abnormalities and poor growth. Nonarticular chondrocytes (e.g., from rabbit sternum) tend to produce a predominantly type II collagen, while the articular chondrocytes produce a mix of types I and II (Schindler *et al.*, 1975). Webber *et al.* (1977) examined the behavior in monolayer culture of articular chondrocytes from eight subprimate species and found the cell proliferation, the synthesis of matrix products, and the response to exogenous vitamin C all to be species related, with rabbit chondrocytes maintaining the best functional activity.

In summary, the differentiated chondrocyte is sensitive to its extracellular environment and maintains its terminally differentiated state by initiating appropriate responses to changes in that environment. Derangement in the extracellular matrix can lead to abnormal expression of the differentiated state when chondrocytes are maintained *in vitro,* or in such *in vivo* situations as achondroplasia. The abnormal expression is stable and maintained in both cases. The chondrocyte may respond to its altered environment by altering the nature of its differentiated products. Synthesis of abnormal glycosaminoglycans may be initiated, or the type of collagen synthesized may be switched from type II to type I. *In vitro,* such changes are often accompanied by phenotypic "dedifferentiation" of the chondrocyte, so that it appears more like a fibroblast. Indeed, such shifts may represent a true modulation to the fibroblastic state. When these changes occur *in vivo*, foci of abnormally differentiating cells may appear among the normal cells. These foci are typical of the histopathological changes in articular cartilage during the early stages of osteoarthritis and degenerative joint diseases.

2. Mechanisms of Articular Cartilage Repair

The release of chondrocytes from their extracellular matrices does not necessarily imply the death of those cells. The liberated hypertrophic chondrocytes from the cartilaginous models of long bones and of the mandible can dedifferentiate to osteoprogenitor cells (Chapter 6, Section V). As long as the mineralized extracellular matrix is intact, the hypertrophic chondrocytes are maintained. Once the calcified matrix is removed, the cells resume DNA synthesis and divide. Parallel events occur *in vitro*. Removal of the extracellular matrix and the culturing of chondrocytes under conditions of low cell density allow them to resume DNA synthesis and to proliferate (Holtzer *et al.,* 1960; Abbott and Holtzer, 1966, 1968). When hypertrophic chondrocytes from Meckel's cartilage of the mouse are exposed to hydrocortisone *in vitro,* or when similar cells from the embryonic chick tibia are exposed to complement-sufficient antiserum, reinitiation of DNA synthesis within these terminally differentiated cells is evoked (Fell *et al.,* 1968; Melcher, 1971a,b).

Reinitiation of both DNA synthesis and mitosis within chondrocytes is a compensatory response initiated to maintain the population of differentiated chondrocytes in the face of environmental influences seeking to destroy it. Mammalian articular chondrocytes provide a useful example, since, in the early stages of osteoarthritis, a proportion of them resume DNA synthesis and mitosis, thereby maintaining, at least for a short time, the population of differentiated chondrocytes. I should like, then, to consider the articular chondrocytes during either osteoarthritis or degenerative joint disease as an example of a population of differentiated cells maintained by a retention within at least some of them of the ability to divide [for reviews of the biology of osteoarthritis, see Bollet (1969) and Jaffe (1972)].

Many cells lose the ability to divide upon undergoing terminal differentiation, although a mutual exclusion of division and differentiation is not universal. It was thought for many years that mammalian articular chondrocytes completely lost the ability to divide upon reaching adulthood (Elliott, 1936; Crelin, 1957a,b; Barnett *et al.*, 1961). (An interesting exception were the articular chondrocytes of the pubic symphyseal cartilage of adult mice, in which mitosis frequently was observed (Crelin, 1957a,b; Crelin and Koch, 1965, 1967; also Section 3.) Consequently, no attention was paid to the possible regenerative or reparative abilities of articular cartilage as a means of overcoming degenerative conditions. Theories of "wear and tear" predominated (see Ali *et al.*, 1974). However, considerable evidence now indicates that articular chondrocytes, taken from osteoarthritic patients or from animals in which degenerative conditions of the joint have been produced experimentally, can resume DNA synthesis and *perhaps* even divide. I emphasize *perhaps* because many authors have assumed that evidence of DNA synthesis (incorporation of [^{3}H]thymidine into DNA) was evidence that the cell was about to enter mitosis (Cahn and Lasher, 1967; Cameron and Jeter, 1971). However, several lines of evidence indicate that in periosteal osteogenesis, in fibroblasts of the periodontal ligament, and in the blastema of amputated amphibian appendages, DNA synthesis need not necessarily be followed by initiation of mitosis.

Owen and MacPherson (1963), on the basis of grain counts of labeled cells, indicated that some osteogenic precursor cells labeled with [^{3}H]thymidine still had not divided 72 hours after labeling. These cells could have a very long S phase or could be the precursors of cells whose activity was not required during that period. The same grain-counting procedure indicates that only 37% of the fibroblasts of the orthodontically stimulated periodontal ligament of the rat have divided 27 hours later (Roberts and Jee, 1974). *In vitro,* the periodontal ligament behaves similarly (Melcher and Turnbull, 1974, 1976), so that it is probably not the orthodontic treatment that blocks cell division. This uncoupling of DNA synthesis from cell division is also seen in the regeneration of the newt limb after denervation. In the normal regenerate, the labeling index and the mitotic index both increase in the blastema. In the denervated animal, the labeling index rises to 54% of the control value, but mitosis is not seen. The cells remain blocked in G2 of the cell cycle (Tassava *et al.*, 1974; Mescher and Tassava, 1975). One must be aware, then, of the possibility that incorporation of [^{3}H]thymidine and mitotic activity need not show a one-to-one correspondence.

The fact that chondrocytes from adult mammalian articular cartilage can proliferate *in vitro* obviously indicates that the ability to initiate mitosis is not irreversibly lost with terminal differentiation (Section 1).

Mankin has carried out a series of studies on the proliferative abilities of articular cartilage *in vivo*. He applied [^{3}H]thymidine autoradiography to the study of the distal femoral articular cartilage of immature rabbits and showed that a small percentage of chondrocytes near the articular surface incorporated labeled thymidine (Mankin, 1962a). While trauma to these cartilages increased the number of labeled cells, these cells, however, made little contribution to repair of the defect (Mankin, 1962b)—because mitosis was blocked or because of a long cell cycle in these cells (Mankin, 1964)?

The metabolic activity of articular cartilage from adult rabbits is considerable, and turnover of glycosaminoglycans is high (Mankin and Lippiello, 1969b; Ali *et al.*, 1974; Hadházy *et al.*, 1975). In osteoarthritic human articular cartilage and in rheumatoid arthritis, the synthesis of glycosaminoglycans, especially chondroitin 4-sulfate (normally characteristic of immature cartilage), increases parallel to the incorporation of [^{3}H]thymidine until arthritic degeneration is severe, when both decline in parallel (Mankin, 1973). It is tempting to speculate that the chondrocytes produced by proliferative activity (although no mitotic index data are available) synthesize chondroitin 4-sulfate in order to repair the defective matrix (Ali and Evans, 1973a; Muirden *et al.*, 1974), and that this continues until their proliferative ability is exhausted. Synthesis of type I collagen also is initiated in osteoarthritic chondrocytes (Nimni and Deskmukh, 1973; Gay *et al.*, 1976b; Müller and Kühn, 1977). The fact that cultured articular chondrocytes also produce type I collagen under monolayer conditions prompted Gay *et al.* to propose that it was the presence of a normally functioning extracellular matrix that allowed continued synthesis of type II collagen and prevented initiation of synthesis of type I collagen. The initiation of DNA synthesis by these osteoarthritic chondrocytes is evidence of the ability of chondrocytes to revert to a prior differentiated state. "The chondrocytes invoke their ontogenetic prerogative to become chondroblasts and reinstitute DNA synthesis" (Mankin, 1973).

Hulth *et al.* (1972) have also shown incorporation of [^{3}H]thymidine into chondrocytes from osteoarthritic patients and have noted that some of these cells were present in nests—presumptive evidence of localization of daughter cells. Recently, Telhag (1976) provided quantitative data showing that, while total DNA and RNA decrease below the levels seen in articular cartilage from normal patients (the decrease is on the order of 50% and represents cell loss), the synthesis of DNA/μg DNA *increases* in the osteoarthritic cartilage (Table XVIII). Telhag concluded that the remaining cells were metabolically very active [cf. Mankin (1973) above and Mayor and Moskowitz (1974)]. Normally, synthesis and total DNA and RNA *decrease* with age in healthy articular cartilage (Telhag and Havdrup, 1975).

These authors also have devised an experimental model for osteoarthritis using rabbits (Telhag, 1972, 1973; Havdrup *et al.*, 1975). If prolonged degeneration of rabbit knee joints is induced by surgical excision or by division of ligaments of the joint, [^{3}H]thymidine-labeled chondrocytes can be seen as early as five days postoperative. Labeled chondrocytes also are seen within the tibial and femoral condyles, indicating that uninjured but nearby tissues also respond to the trauma. Total immobilization of the joint is not a sufficient stimulus to initiate mitosis. The cells die and the cartilage degenerates. Slight movements may induce some

TABLE XVIII

Total DNA and Rates of Synthesis of DNA in Articular Cartilage from Healthy (Control) and Osteoarthritic Individuals[a]

		Rates of DNA synthesis	
	Total DNA (μg DNA/mg wet wt)	(cpm DNA/mg wet wt)	(cpm DNA/μg DNA)
Control	2.2	49.7	24.1
Osteoarthritic	1.0	33.8	44.0
Osteoarthritic/control	−54%	−32%	+82%

[a] Based on data in Telhag (1976).

chondrocytes to divide, but the number doing so is negligible, and must be accumulated with colchicine arrest (Crelin and Southwick, 1960, 1964).

Articular chondrocytes respond to other stimuli by initiating mitotic activity. Local freezing of rabbit femoral articular cartilage produces degenerative changes accompanied by the production of nests of new chondrocytes, especially at the margins of the defect (Simon *et al.,* 1976). Similar nests are seen at the chondrosynovial junction, where fibrous tissue is transformed into chondroid tissue after intraarticular injection of anti-chondromucoprotein immunoglobulin (Eguro and Goldner, 1974). Labeled chondrocytes also were observed, but not after injection of either anti-muscle globulin or saline. These authors propose that, early in degeneration, chondromucoprotein is released by the articular chondrocytes, but is recognized as foreign, initiating antibody formation and further cycles of degeneration. Implantation of active electrodes produces an electrochemical environment that also favors regeneration of fibrocartilage, especially from marginal chondrocytes (Baker *et al.,* 1974a,b).

Mammalian articular chondrocytes, then, normally are maintained in a postmitotic state, but can respond to degenerative changes by proliferation and by reinitiating DNA synthesis. While this does represent a loss of the differentiated state of individual chondrocytes, it obviously represents a means of maintaining the differentiated state of the cell population as well as the cartilaginous tissue, and of repairing defects within that cartilage. If a greater proportion of these chondrocytes could be brought into a proliferative and metabolically active pool, would larger defects be repaired or would repair be more permanent? Obviously, these are questions that can be pursued experimentally.

3. Formation of the Interpubic Ligament

The sexual dimorphism of the pelvic girdle is evidenced either in the size and shape of the ilium, ischium, and pubic (innominate) bones, or in the nature of the tissues present within the pubic symphysis, i.e., the juxtaposition of left and right pubic bones that forms the interpubic joint.

Within the vertebrates, there are three basic types of pubic symphysis (Ruth, 1932): (a) those in which the innominate bones are fused at the ischiopubic or pubic symphysis, forming a synostosis, as in some rodents; (b) those in which the innominate bones are connected by an interpubic cartilage, forming a synchondrosis, as in some rodents, carnivores, primates, and the Artiodactyla; and (c) those in which the innominate bones are connected by a ligament, forming a syndesmosis, as in some rodents and in bats. Here, I shall concentrate primarily on those vertebrates in which the interpubic joint is a synchondrosis, i.e., group (b).

Within this group, several types of interpubic joint may be distinguished on the basis of the histological changes observed to occur within the interpubic cartilage during pregnancy. In the primates, there is some softening of the interpubic cartilage such that by the time of parturition, the joint is flexible (Gingerich, 1972; Rawlins, 1975). Exceptions are species such as the African lorises, in which the fetus is very small even at birth (Leutenegger, 1973). In the rat and the guinea pig, the interpubic cartilage is converted into a flaccid interpubic ligament, and there is some resorption of the medial edges of the innominate bones during pregnancy (Ruth, 1935, 1936a,b; Crelin and Brightman, 1957; Bernstein and Crelin, 1967). In the guinea pig, the modulation of the interpubic cartilage to an interpubic ligament is controlled by relaxin (Talmage, 1947), and collagenase is produced by symphyseal tissues only during, and immediately before and after, parturition (Wahl, 1971). In the mouse, the formation of an interpubic ligament is accompanied by extensive resorption of the medial edges of the innominate bones. The bulk of the experimental work on the control of the transformation of cartilage to ligament in the female, and on the inability of the male to initiate this transformation, i.e., the presence of sexual dimorphism for this ability, has been performed on the mouse by E. S. Crelin and his colleagues at Yale University (see Crelin, 1960a). Their studies provide considerable insight into the hormonal control of the development of this sexual dimorphism and of its expression in the female.

The predetermined, agonal bony pelvic type in the mouse is the female type (Crelin, 1960a). If animals of either sex are gonadectomized at birth, in all cases the female type of pelvis is present at sexual maturity. Administration of testosterone to these gonadectomized mice results in the production of a male-type bony pelvis. The sexual dimorphism becomes evident between 17 and 20 days after birth (Crelin and Haines, 1955). Muscle function does not influence the type of pelvis that develops (Crelin, 1960b), but testosterone transforms the agonal (female) type into a male pelvis.

The chronology of the histogenesis of the female interpubic joint is as follows. One week after birth, cartilage is present at the medial edges of the innominate bones. After three weeks, an immature pubic symphysis is present, and by four weeks fibrocartilage begins to appear within the interpubic cartilage. By six weeks postnatally, the ligament formation and the sexual dimorphism of the pubic symphysis are complete (Crelin and Levin, 1955). The conversion of the symphyseal cartilage into an interpubic ligament involves the release of enzymes from the chondrocytes and the liberation of the chondrocytes from their extracellular matrices. These chondrocytes, after modulating to a fibroblastic state, deposit collagen and align it across the interpubic joint (collagen is randomly arranged in the cartilage; Hall, 1947; Crelin, 1969). Other possible fates of these chondrocytes were discussed in Chapter 6, Section V. These changes to ligament formation are reversible, for the ligament may be transformed back into a cartilage. Perhaps this occurs because the ligament retains some features of cartilage—it is flaccid, hydrated, and contains considerable amounts of glycosaminoglycans.

Gardner (1936) was the first to confirm experimentally that interpubic ligament formation and resorption of the medial edges of the innominate bones were controlled by estrogen, and that these events could occur in males if the animals were gonadectomized and given estrogen. He thought that these differential responses were attained only at sexual maturity, but in fact they occur earlier. When administered alone to virgin female mice, estrogen produces a specific hydration and swelling of the interpubic cartilage. If estrogen is administered in association with relaxin, the cartilage swells and is converted into an interpubic ligament (Crelin and Levin, 1955). The same changes are observed after administration of hormone to gonadectomized males or females, even when the pelves are transplanted to an ectopic site (Crelin, 1954a,b, 1958a, 1959). Such hormone-induced changes are rapid, because after one day, dissolution of cartilage is apparent with fibrocartilaginous cells present after two days, and a complete ligament formed by the fifth day (Crelin, 1956).

If females possessing an interpubic ligament are ovariectomized, the ligament reverts to cartilage, indicating that maintenance of the ligament, as well as its initiation, depends upon a continued supply of estrogen and/or relaxin.

Implantation of immature pelvic joints to ectopic sites outside the pelvic area (usually into the pectoral musculature), or cultivation of embryonic pelvic primordia for a time before their transplantation back to the pelvis of 5-day-old females, shows that the unique hormonal responses of the interpubic cartilage are acquired by precursor cells of the cartilage during development (Crelin, 1959, 1960b, 1963; Crelin and Koch, 1965).

Mitosis is reinitiated within the chondrocytes as they transform to fibroblasts (Crelin, 1957a,b; Linck *et al.*, 1975). However, the role played by hormones in this process has not yet been determined.

The response of interpubic cartilage to estrogen and relaxin cannot be initiated simply by placing the tissues in direct contact with the hormones, either by implantation of pellets (Crelin and Haines, 1955), intrasymphyseal injection (Crelin, 1958b), or by cultivation *in vitro* in the presence of hormone (Harkey and Crelin, 1963a,b). The longer the ovary is present before puberty, the greater the enhancement of ligament formation; accordingly, the longer the testis is present, the greater the inhibition of ligament formation (Crelin and Grillo, 1957). Conditioning by both growth hormone and thyroxine is also necessary (Horn, 1958; Steinetz and Beach, 1963; Steinetz *et al.*, 1965). Calcitonin inhibits this resorption (Steinetz *et al.*, 1973), although the precise roles of these other hormones on the interpubic cartilage have not been determined.

In addition, the interpubic cartilage may in part serve a mechanical function. If the symphysis in nulliparous females is severed, the pubic bones separate, and the cartilage disappears. The ligament produced when the interpubic joint is grafted back to a pelvic site is longer than that produced when the joint is transplanted to an ectopic site, indicating a mechanical involvement in the determination of the length of the ligament (Crelin, 1960b, 1963). However, since paralysis of the pelvic musculature *in situ* does not affect the length of the interpubic ligament produced (Crelin, 1958d), muscle action may not play a vital role in its normal development.

Several other instances of transformation of cartilage to ligamentous tissue during development are also known. In the mammal, a portion of Meckel's cartilage transforms to the sphenomandibular (tympanomandibular) ligament; however, tension rather than hormonal stimulation is the initiating factor (Richany *et al.*, 1956; Burch, 1966; Savostin-Asling and Asling, 1973; Asling, 1976). The myelohyoid bridge, an inherited deficiency in Meckel's cartilage, involves the conversion of cartilage to ligament (Ossenberg, 1974). Removal of chondroitin sulfate unmasks the fibrous component of articular cartilage, and an avascular connective tissue on the human ulna is produced during postnatal growth (Haines, 1976).

The several examples cited in this section should serve to illustrate some of the conditions under which the phenotype of the chondrocyte can be altered. Also stressed is the fact that such alterations may lead to modulation of both the phenotype and function, or to dedifferentiation and subsequent redifferentiation.

F. Maintenance of Cartilage Diversity

In the discussions of cartilage, chondrocytes, and chondrogenesis, there up to now have been peripheral indications that all cartilages may not be equivalent to one another. In Chapter 1, I briefly discussed the different histological types of cartilage. These reflect the synthesis and accumulation of differential products into the extracellular matrix. Cartilage may be either mesodermal or ectomesenchymal in origin. Cartilage can precede osteogenesis and have a temporary existence, as does the primary cartilage of endochondral bones, or, as exemplified by the secondary cartilage on membrane bones, it can succeed osteogenesis and have a more permanent existence. Although most primary cartilages undergo chondrocyte hypertrophy, in many species Meckel's cartilage does not. Cartilage is usually avascular, but in antlers it is highly vascular. In land vertebrates, calcification of the primary cartilages of the long bones enables these cartilages to be replaced by bone, whereas in sharks, calcification of cartilage inhibits osteogenesis. Quite obviously, cartilages vary—in their origins, their structure, and in their physiological and developmental roles within the body.

We have very little systematic information on the basis of this heterogeneity of cartilage as a tissue, or of chondrocytes as differentiated cells. Does this heterogeneity have its origin in the initial determination or setting-aside of precursor cells for chondrogenesis, or is it associated with later stages of chondrocyte differentiation? We do not know whether cell lines become determined initially to form cartilage or to form particular types of cartilage, although the information available points towards some specification of heterogeneity (nonequivalence) early in development.

Weiss and Moscona (1958) cultured mesenchymal cells from limb buds of 3½- to 4-day-old embryos and also cultured ectomesenchyme from periocular (scleral) regions of 6- to 7-day-old embryos, i.e., before any differentiation into chondroblasts. They found that the patterns of reaggregation of the differentiated chondrocytes *in vitro* were distinctive and type specific (see also Weiss and Amprino, 1940). The fibulae of brachypod mice fail to undergo the chondrocytic hypertrophy typical of fibulae from normal littermates. This defect can be traced back at least to the time of condensation of limb mesenchyme, i.e., at twelve days of gestation (Selleck and Elmer, 1973; Elmer and Selleck, 1975). The pattern of morphogenesis and the ability to undergo hypertrophy seem to be specified in the precursors of the cells that actually exhibit these behaviors. The condylar cartilage from 6-month-old calves has some antigenic determinants in common with nasal and epiphyseal cartilages, but also has at least one determinant that is distinctive (Brigham *et al.*, 1977). A developmental study using immunological markers would be a very profitable way of determining the timing of the specification of such cartilage types.

In Chapter 6 (Section VI,B,4), I presented evidence indicating that, as chondrocytes differentiate, they acquire the ability to segregate from their own precursors. Similar studies indicate that chondrocytes from one portion of the skeleton (e.g., the limb) recognize and reaggregate with chondrocytes from another portion of the skeleton (e.g., the vertebra), but that the precursors of the two sets of chondrocytes segregate from one another (Moscona, 1957; Levak-Svajger and Moscona, 1964; Zwilling, 1968; also see Table XIX). The heterogeneity is present early in the life of the cell lines and is lost with differentiation—perhaps concomitant with the deposition of the extracellular matrix, isolating the cells from one another. Provided that the chondrocytes have the same embryological origin, they do not segregate from chondrocytes from other sites or from other species (Table XIX).

TABLE XIX

Summary of the Results of Studies on the Segregation of Avian Limb Bud Chondrocytes or Their Precursors from Avian Vertebral Chondrocytes or Their Precursors, or from Murine Limb Bud Mesenchyme

Age of limb bud cells	Age of avian vertebral or murine limb bud cells	Segregation[b]	Reference[a]
H.H. 34 chondrocytes	H.H. 34 vertebral chondrocytes	−	1
H.H. 22 mesenchyme	14-day mouse mesenchyme	−	2
H.H. 22 mesenchyme	12-day mouse mesenchyme	−	3
H.H. 18 mesenchyme	H.H. 18 scleratomal mesenchyme	+	1
H.H. 18 mesenchyme	H.H. 13 somitic mesenchyme	+	1

[a] Results are expressed as +, segregation or −, lack of segregation.
[b] 1, Zwilling (1968); 2, Levak-Svajger and Moscona (1964); 3, Moscona (1957).

1. Cartilages of Different Embryological Origins

Cartilages of different embryological origins show distinctive metabolic and differentiative properties. Chondrocytes from Meckel's cartilage having an ectomesenchymal origin have been compared to tibial or femoral chondrocytes of mesodermal origin. Levenson (1969, 1970) has shown the differentiative behaviors of these two types of chondrocytes when they are maintained *in vitro* to be distinctly different.

Chiakulas (1957) utilized fusion versus nonfusion of cartilaginous rods of ectomesenchymal or mesodermal origin to investigate the distinctiveness of cartilage of neural crest origin. Cartilage was excised from larval newts (*Ambystoma maculatum*) and ectopically grafted into the tail fin in contact with a second cartilaginous rod. When both cartilages were of a common embryological origin, e.g., femur against femur, humerus against humerus, Meckel's against Meckel's, the two fused. When the embryological origins of the partners differed, e.g., femur against Meckel's cartilage, fusion was not observed. This was taken as evidence of heterogeneity based on embryological origin. Foret (1970), implanting humeral or Meckelian cartilage into the upper forearms of *A. maculatum,* amputated the limb and graft, and followed the regeneration of the host limb. Some fusion of Meckel's with the host humerus was observed. Both of these studies were complicated by the presence of other host cells, by differential vascularity, by graft survival, etc. Cultivation *in vitro* would overcome these objections.

The Meckelian symphysis in the lower jaw of the hamster consists of a fibrocartilage that forms by the fusion of Meckel's cartilage with two secondary cartilages that arise within the joint (Trevisan and Scapino, 1976a). Both of these cartilages are of neural crest origin. The symphysis would provide a good *in vivo* site for testing the cartilage fusion or lack of fusion after transplantation of non-neural crest-derived cartilage into the joint.

Silberzahn (1968) cultured Meckel's and femoral cartilages from 8-day-old chick embryos and from 14th-day-of-gestation mice in various combinations on a semisolid medium for 14 days. Incidences of fusion between adjacent rods of cartilage were as follows: femur against femur, 77% fused; Meckel's against Meckel's, 56% fused; Meckel's against femur, 37% fused. The fusion involved elaboration of extracellular matrix and the proliferation and chondrification of perichondrial cells at the contact points. In many instances, a common perichondrium was found. It was noted that, while fusion of homologous cartilages of either embryological origin had begun by the third day, fusion of heterologous cartilages did not commence until the 6th day *in vitro*. When Meckel's and femur were dissociated with trypsin and then reaggregated, a common cartilage was produced, with intermingling of chondrocytes from both sources. Segregation may well be a property of the extracellular matrix.

Fyfe (1977) in our laboratory has cultured Meckel's cartilage in contact with tibial cartilage on both semisolid and liquid media, and has obtained segregation of the two rods from one another. Autoradiographic analysis of cultures in which one partner was prelabeled with [^{3}H]thymidine established the demarcation between adjacent cartilaginous rods with some precision.

There is obviously a conflict between these studies and those of Chiakulas and Foret. The species differed, the ages differed, and the experimental approach (culture vs. graft) differed. These traditional ways of explaining differences between different studies merely indicate that we do not understand the basis of the phenomenon of segregation of cells of the "same" phenotypes but with differing developmental histories.

2. Resistance to Vascular Invasion

During its early development and at those sites where it is retained as a permanent tissue, cartilage is avascular and unmineralized. Cartilage will develop from mesenchyme in the absence of a vascular supply, as for example, in organ culture, and it can be maintained as a differentiated tissue without vascularization, obtaining metabolites by diffusion through the extracellular matrix. During embryonic development, cartilage may be invaded by vascular channels as a prelude to endochondral ossification, but in these cases the cartilaginous matrix undergoes mineralization and the chondrocytes either degenerate or become osteoprogenitor cells. Cartilage is also a tissue that is rarely subject to tumors. When it is, the tumors do not invade the cartilage, but rather encapsulate nodules of cartilages.

Does this lack of invasion of cartilage by blood vessels and by tumors represent a distinctive capability of unmineralized cartilage, or is it merely fortuitous? It recently has become evident that grafted cartilage actively resists invasion by blood vessels, and that an active component that confers this ability (and the ability to resist invasion by tumors?) can be isolated from cartilage. Such a factor might well be useful for containing the growth of a variety of tumors by inhibiting their vascularization. It might also provide a way of understanding why some cartilages remain as permanent unmineralized tissues.

One body of data, derived from the grafting of cartilage from one species to another, indicates that *intact* cartilage is immunologically privileged. Homografts of cartilage survive longer than any other grafted tissue except cornea. Chondrocytes that have been separated from their extracellular matrices and then grafted do sensitize the host and elicit an immunological response (Hadházy *et al.*, 1972; Elves, 1974, 1976; Langer and Gross, 1974). The glycosaminoglycans of the extracellular matrix prevent exposure of the chondrocytes to tissues of the host or to antisera *in vitro* (Heyner, 1969; Dingle *et al.*, 1975; Fell, 1975). Antibodies against cartilage glycosaminoglycans have been used to suppress chondrocyte hypertrophy and to reduce replacement of cartilage by bone (Jain and Sabet, 1974), a finding consistent with the concept that it is the intact cartilage which resists replacement by bone. The type II collagen of articular cartilage has antigenic specificity associated with the glycine–proline–alanine sequences of one fraction (Faulk *et al.*, 1975; Hahn *et al.*, 1975; Furthmayr and Timpl, 1976), but this also may be masked by the glycosaminoglycans of the extracellular matrix.

Evidence for a more active resistance to vascular invasion comes from the study of Eisenstein *et al.* (1973). These workers tested various tissues for their resistance to invasion when grafted on Millipore filters to the chorioallantoic membrane of host embryonic chicks. Tissues normally are vascularized rapidly on this membrane by blood vessels from the chorionic mesenchyme. Eisenstein *et al.* found that hyaline cartilage, whether live or devitalized, resisted vascular invasion, provided that it was not mineralized and provided that the extracellular matrix was intact (Fig. 28). Calcified cartilage and bone were rapidly invaded by blood vessels (Fig. 29). If the uncalcified cartilage was extracted with 1 *M* GuHCl (a procedure whereby glycosaminoglycans are extracted from the extracellular matrix) and then grafted to the chorioallantoic membrane, resistance to vascular invasion was reduced (Sorgente *et al.*, 1975; also see Fig. 29).

Brem and Folkman (1975) and Langer *et al.* (1976) utilized extraction with guanidine hydrochloride to isolate a multiprotein fraction from calf scapular cartilage that inhibited vascular invasion of tumors induced in the cornea of the rabbit. Such tumors are normally vascularized at a maximum rate of capillary growth of 0.6 mm/day. In the presence of the extract, capillary growth is reduced to 0.2 mm/day, after which regression of these vessels sets in. Folkman (1976) has reviewed these data on tumor vascularization.

Evidence for the proteinaceous nature of the inhibitory factor lies in its inhibition of protease activity. Eisenstein *et al.* (1975, 1976) and Kuettner *et al.*

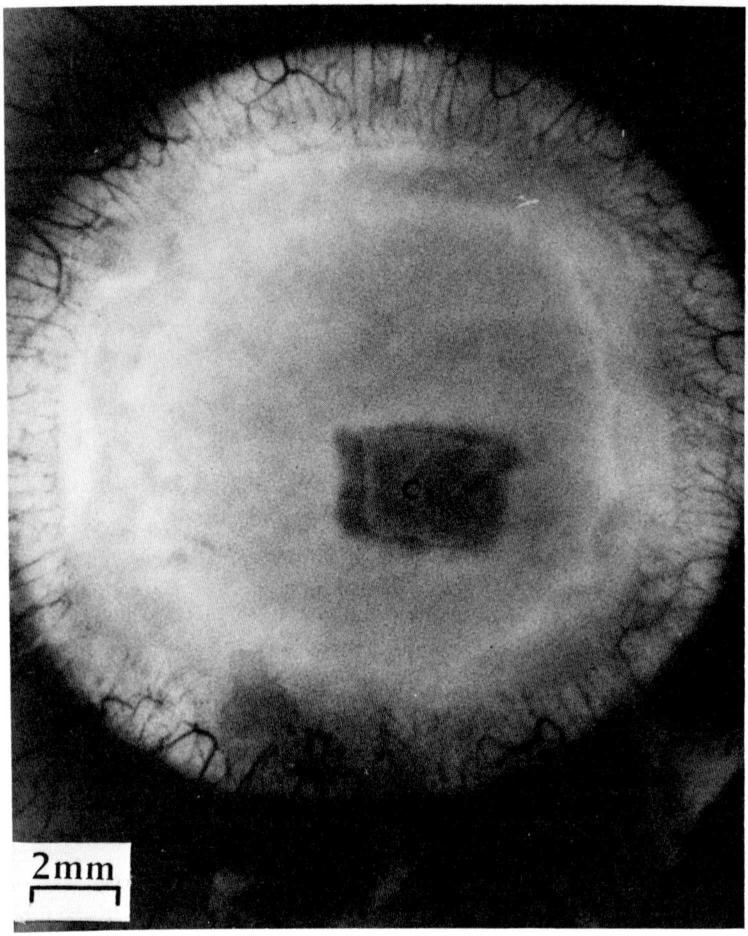

Fig. 28. Uncalcified cartilage (c) on a Millipore filter when grafted to the chorioallantoic membrane of the embryonic chick prevents ingrowth of chorionic blood vessels and remains unvascularized. Reproduced with permission of the author and publisher from Kuettner *et al.* (1976b).

(1976a, 1977a), using bovine nasal septum, also have isolated an inhibitor fraction which does not contain collagen, has less than 10% glycosaminoglycans, and which also may be a protease inhibitor. This fraction inhibits the growth of endothelial cells *in vitro* (so that the possibility of cartilage actively retarding blood vessel growth is very real) and is mimicked by known protease inhibitors. The inhibition of endothelial cell growth is dose dependent (Table XX). This

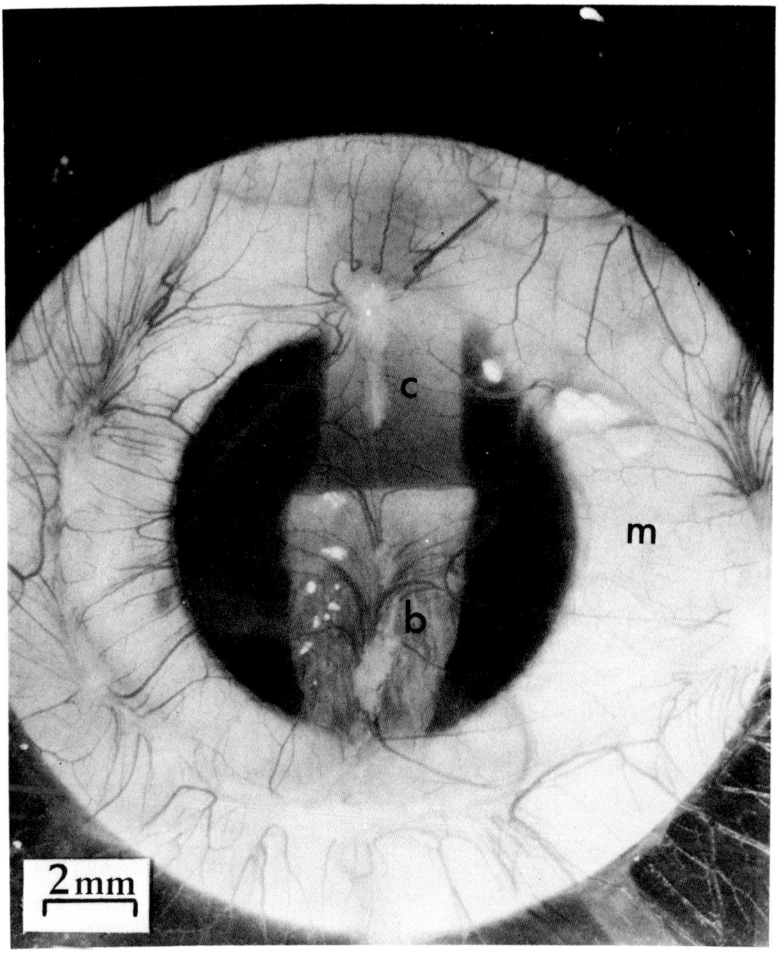

Fig. 29. Uncalcified cartilage (c) and mineralized bone (b), extracted with 1.0 *M* GuHCl and grafted to the chorioallantoic membrane on a Millipore filter (m) are both invaded by chorionic blood vessels (cf. Fig. 28). Reproduced with permission of the author and publisher from Kuettner *et al.* (1976b).

inhibitor also suppresses the action of collagenase from human skin and from osteosarcomas (Kuettner *et al.*, 1976b, 1977b). Cells derived from human osteosarcomas secrete collagenase *in vitro* if challenged with heparin, but not when low molecular weight (less than 50,000 daltons) cartilage extract is added. One can readily see how inhibition of collagenase activity would enable the cartilage to resist vascular invasion, to maintain its differentiated, unmineralized state, and

TABLE XX

Dose-Dependent Inhibition of the Growth of Endothelial Cells *in Vitro* in the Presence of Extracts Isolated from Bovine Nasal Septal Cartilage[a]

	Concentration of cartilage extract added (μg/ml)	Number of endothelial cells/dish[b]
Crude extract	0	700,000
	20	510,000
	100	285,000
	500	52,000
<50,000 MW fraction	5	300,000
	20	260,000
	100	65,000
	500	22,000
>50,000 MW fraction	500	710,000[c]

[a] Based on data from Eisenstein *et al.* (1975), who should be consulted for the method of preparation and purification of the extract.
[b] Cell number counted four days after seeding 40,000 cells/dish.
[c] This fraction was inactive.

to resist replacement by bone. It would be necessary to inactivate or remove the inhibitor before the cartilage could either produce its own collagenase (none has yet been found) or be invaded (Ehrlich *et al.*, 1977; Wolley *et al.*, 1977). When human osteosarcoma or metastatic mammary carcinoma cells were cocultured with human growth plate, the tumor cells invaded the bone, but not the unmineralized cartilage of the growth plate (Kuettner *et al.*, 1977b). With the onset of aging, or in pathologies such as Addison's disease, in which cartilages that normally do not ossify (such as the trachea, epiglottis, and elastic cartilages) do undergo ossification (Beneke *et al.*, 1966; Andrew, 1971; Siebenmann, 1977), this resistance to vascular invasion may be lost. Acquisition of that ability would be a welcome replacement for the distress of prolonged corticosteroid treatment that these patients must suffer.

3. The Hypertrophic Chondrocyte

Nonhypertrophic chondrocytes within an unmineralized matrix resist vascular invasion (Section 2). The attainment of chondrocyte hypertrophy is usually accompanied by mineralization of the extracellular matrix (for a review of cartilage calcification, see Alcock, 1972). Coincident with this mineralization, the ability to resist vascular invasion is lost, and with vascular invasion, breakdown of the cartilaginous extracellular matrix sets in. Without chondrocyte hypertrophy, calcification is rare, but can occur if vascular ingrowth is artificially initiated (Riede *et al.*, 1971).

The ability of vascular buds to penetrate the perichondrium may be related to the death of the chondrocytes, an event which accompanies mineralization of the extracellular matrix. Kaminski *et al.* (1977) found that viable—but not dead—chondrocytes could inhibit lymphocyte-induced formation of blood vessels *in vivo*.

a. Matrix vesicles. Mineralization of extracellular matrix is initiated within the membrane-bound vesicles of the extracellular matrix. Such vesicles have been observed in calcifying cartilage and bone (both intraskeletal and ectopic), and in dentine, antlers, and calcifying aorta. These vesicles contain calcium, phosphate, lipids, glycosaminoglycans, alkaline phosphatase, pyrophosphorylase, ATPase and 5′-AMPase (Anderson, 1969; Ali *et al.*, 1970; Anderson *et al.*, 1970; Anderson and Reynolds, 1973; Katchburian, 1973; Thyberg *et al.*, 1973; Landis *et al.*, 1977a,b).

Matrix vesicles first were isolated by Ali *et al.* (1970). When incubated with ATP under alkaline conditions, they accumulate calcium ions (Ali and Evans, 1973b).

From the top to the bottom of the cell columns in the growth plates of long bones, one sees a gradual shift from intramitochondrial calcium at the top of the columns to intravesicular calcium at the bottom (Brighton and Hunt, 1974; Anderson *et al.*, 1975). However, matrix vesicles are not mitochondrial in origin, for they do not contain any respiratory enzymes (Person *et al.*, 1977).

b. Hypertrophic chondrocytes and subperiosteal ossification. In such cartilages as Meckel's cartilage and those of the ear and trachea, in which chondrocytes do not normally undergo hypertrophy and where extracellular matrix remains unmineralized, the cartilage may remain intact throughout the life of the individual. Subperiosteal ossification is not seen within the perichondria of such cartilages (cf. exceptions such as Addison's disease, above). When subperiosteal ossification is seen, it is found adjacent to intact hypertrophic chondrocytes. Initially, it is recognized as being (a) a collar of bone surrounding the diaphyses of embryonic long bones (Fell, 1925, 1928); (b) adjacent to areas of hypertrophic chondrocytes in cartilages such as Meckel's cartilage (Friant, 1964, 1968; Rajtova, 1971; however, Trevisan and Scapino (1976b) claim that Friant misidentified the secondary cartilages of the angular and condylar processes of the hamster mandible as Meckel's cartilage); or (c) adjacent to hypertrophic chondrocytes in ectopic sites (Copher, 1935; Lacroix, 1951; Shimomura *et al.*, 1975). Although its life span as a cartilaginous type is short, hypertrophic cartilage plays an important role in eliciting osteogenesis from the adjacent perichondrium, converting the latter into a periosteum.

The possibility that liberated hypertrophic chondrocytes might play a role in providing osteoprogenitor cells during endochondral ossification has already been discussed (Chapter 6, SectionV,B). Of course, superiosteal ossification is intramembranous, and membrane bone makes up a much greater proportion of the bulk of so-called endochondral bones than is usually thought. Virtually all of the bone deposited during embryonic life in avian long bones is in fact subperiosteal and not endochondral (Rinaldi *et al.*, 1974). According to the early view, bone growth was entirely appositional, and occurred from the periosteum: "as an exogenous stem grows from the inner layer of the bark" (Duhamel de Monceau, 1742). It was not until 1912 that MacEwen finally showed that bone formed from origins other than the periosteum, i.e., endochondrally.

Growth of endochondral bone is dependent on maintenance of the growth of the primary cartilaginous model. Since this model requires functional stimuli such as mechanical stress, continued deposition of endochondral bone depends secondarily upon such biomechanical stimuli. Deposition of membrane bones of the skull occurs independent of such stimuli. Smith and Abramson (1974), who graft rabbit ilium or frontoparietal bones either subcutaneously or subperiosteally, found one year later that the endochondral bone was being resorbed and had been reduced to 25% of its original volume; the membrane bone, however, was found to have increased in volume. Patterns of bone deposition in the calvarium of the rat are similar *in vitro* to those observed *in vivo* (Lewis and Irving, 1970), and during the repair of fractures of long bones, intramembranous ossification occurs in fixed (immobilized) fractures, while endochondral ossification occurs in free fractures (Jarry and Uhthoff, 1971). Intramembranous ossification is also more sensitive to disruption of collagen biosynthesis (Diegelmann and Peterkofsky, 1972; Klein and Zika, 1976), perhaps because intramembranous is more rapid than endochondral ossification (Zika and Klein, 1975). What of subperiosteal intramembranous ossification?

Copher (1935) found that when bladder mucosa was transplanted in contact with hyaline costal cartilage in the dog, hypertrophy of the costal chondrocytes was induced and then ossification initiated. The experiments of Shimomura *et al.* have already been mentioned (Chapter 6). Cartilage within diffusion chambers implanted into mammals will induce osteogenic activity from host cells, provided that the cartilage cells are hypertrophic. Cooper (1965), who found that chondrocytes undergoing hypertrophy would act inductively on somitic mesoderm to evoke chondrogenesis, presents a good discussion of the literature on the association between chondrocyte hypertrophy and osteogenesis. Mareel (1967) sought experimental verification of this association by culturing intact hypertrophic chondrocyte zones of 9-day-old embryonic tibiae in contact with the outer surface of same-aged perichondrial tibial surfaces. He obtained a rapid build-up of "contact cells," which he thought were young osteoblasts; however, no osteoid or bone was deposited. The tibiae were probably too old (the perichondrium becomes a periosteum at H.H. stage 29) and the culture period too short to obtain osteogenesis. Peter Scott-Savage of this laboratory has repeated and extended Mareel's experiments without obtaining initiation of osteoblasts. His current experiments involve isolated perichondria from 6-day-old embryos cocultured with hypertrophic chondrocytes to determine whether this association is involved in the initiation of subperiosteal osteogenesis within these embryonic long bones.

Retardation of chondrocyte hypertrophy *in vitro* will suppress subperiosteal osteogenesis and is discussed in Chapter 9 in the context of the Creeper (*cp*) gene in the chick. Failure of chondrocyte hypertrophy is also seen in the brachypod mutant in the mouse. Since some detailed information is available on the origins of this defect, it merits discussion in some length.

c. Brachypodism (bp^{H}) *in the mouse.* The autosomal recessive mutant brachypod has a mode of action that makes it very valuable for the study of the relationship between chondrocyte hypertrophy and subsequent osteogenesis. Tibiae and femora of brachypod mice are essentially normal, while the fibulae fail to undergo chondrocyte hypertrophy and do not initiate osteogenesis until the third week after birth (Konyukhov and Ginter, 1966; Grüneberg and Lee, 1973).

Evidence to date indicates that brachypod acts on limb mesenchyme condensation at a very early stage (12 days of gestation or earlier). Grüneberg and Lee showed that the size of the condensation for the elements of the hindlimb was reduced at 12 days. Selleck and Elmer (1973) and Elmer and Selleck (1975), culturing mesoblast cells from 12-day normal and brachypod mice, found that the formation of cartilage nodules was both delayed and quantitatively reduced in the mutant. Cell division and DNA synthesis were prolonged in the cells from the mutant but declined with chondrogenesis in normal cells. Cells with a fibroblastic morphology were much more evident, and confluency was not attained in the mutant cells. This evidence was all taken to indicate a cell membrane dysfunction that retarded condensation of cells both *in vitro* and *in vivo* (Elmer, 1977). Confirmation of the absence of the changes in mesenchymal cell membranes that normally occur between 11 and 12 days of gestation in the *bp*^{H} mutant was obtained by Hewitt and Elmer (1976, 1978). These workers found that agglutination to concanavalin A or to wheat germ agglutinin did not take place in mutant cells, but did occur in normal cells, and Duke and Elmer (1977) also found that the rate of decline of single cells in rotation culture was reduced when *bp*^{H} cells were cultured. In the latter study, only half of the *bp*^{H} cells chondrified. Clearly, then, an early defect in cell membrane properties affects cell-to-cell contact and the ability of *bp*^{H} cells to accumulate and to chondrify.

A growth inhibitor (a 76,000 MW protein) that inhibits the growth of tissues from normal limb buds is present in 13-day-old brachypod embryos (Pleskova *et al.*, 1974). The production of this inhibitor and the failure of cells to pack into condensations have not been correlated experimentally. However, the failure of the cells to chondrify normally *in vitro* and the delay *in vivo* are evidently responses to an earlier event.

The fibulae of these mutants contain chondrocytes that do not undergo hypertrophy (Fig. 30). The possibility that this failure stems from alterations in cell behavior at the precondensation phase of limb mesenchyme development has important implications for studies on the origin of specificity of cartilage types. Specification of the ability to undergo hypertrophy may occur very early in development and not during cytodifferentiation of the chondroblast or chondrocyte.

The synthesis of matrix products in the fibulae of newborn mutants *in vivo* has been studied by Rhodes and Elmer (1972, 1975). The collagen produced by cultured bp^{H} cells is of the cartilage type (Elmer and Selleck, 1975). In brachypod fibulae, the levels of collagen and glycosaminoglycans are increased over those in normal fibulae. This is due to the dissociation of the synthesis and degradation processes, which are normally highly correlated (see Section II,A; also Hall, 1973). A general slowdown in protein synthesis is responsible for reducing collagen synthesis to half the normal rate. Collagen degradation is also slowed, the net result being an increase in collagen content. Hydroxylation of proline is retarded in the mutant fibulae and collagen synthesis (incorporation of [^{3}H]proline) is independent of ascorbic acid stimulation. A proline:hydroxyproline ratio of 28:1 versus the normal 1.4:1 is the result.

The rate of glycosaminoglycan synthesis is normal, but, as with collagen, degradation is slowed and glycosaminoglycans accumulate. Whether these alterations in the products of the extracellular matrix of the differentiated chondroblasts are related to the initial defect observed at 12 days of gestation, or to an independent later action of the mutant, is not known.

Because brachypod fibulae fail to exhibit chondrocyte hypertrophy and to initiate subperiosteal osteogenesis, and because alkaline phosphatase has been implicated in cell hypertrophy preceding osteogenesis (Fell and Robison, 1929, 1930), Krotoski and Elmer (1973) examined levels of alkaline phosphatase between 14 and 17 days of gestation in tibiae, femora, and fibulae from both brachypod and normal mice. In the normal fetus, there is a slow rise in the accumulation of alkaline phosphatase in the tibiae and femora between 14 and 16 days, and an accelerated accumulation between 16 and 17 days (Table XXI). Both phases of this accumulation were depressed in brachypod tibiae and femora. Normally, the level of alkaline phosphatase is low in normal fibulae at 17 days,

---

Fig. 30. The fibula from a 6-hr-old brachypod (bp^{H}) mouse lacks chondrocyte hypertrophy and consequently lacks subperiosteal bone. Hematoxylin and eosin. Reproduced from Rhodes and Elmer (1975) with permission of the author and publisher.

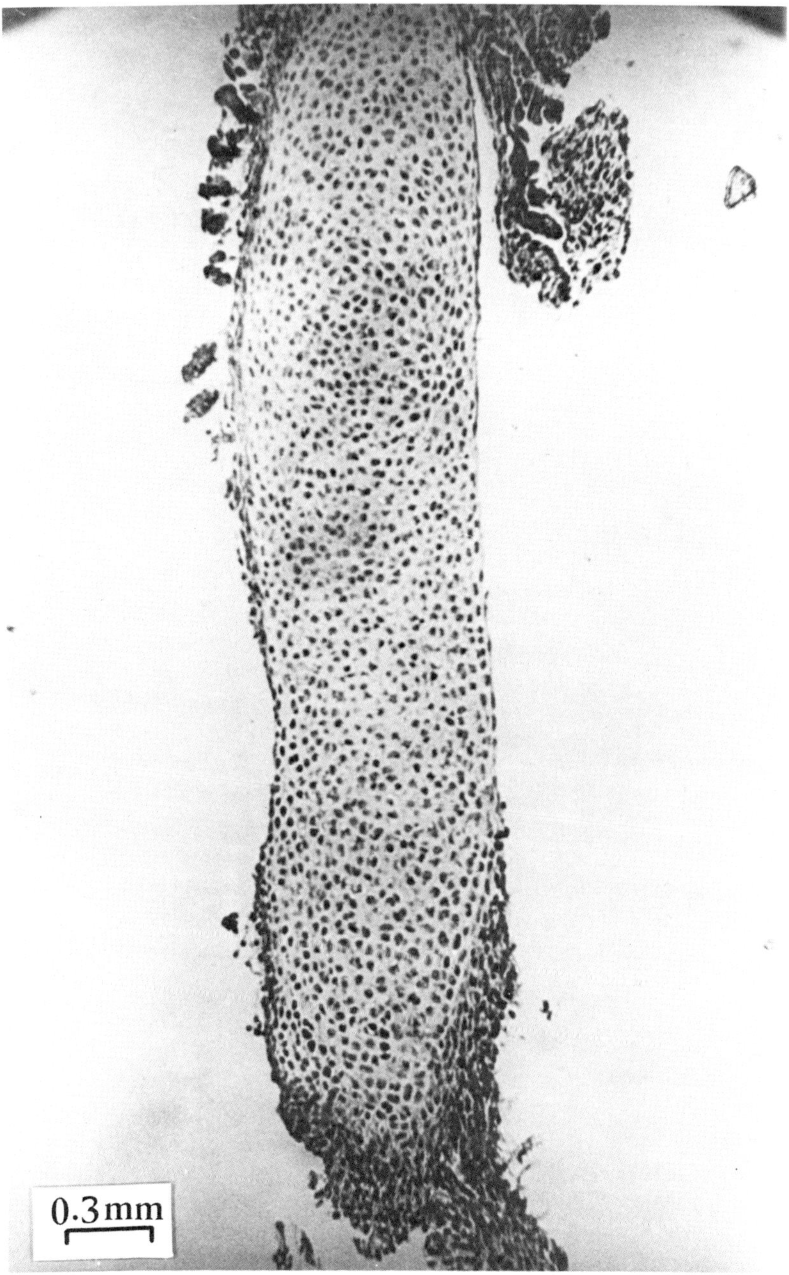
0.3 mm

TABLE XXI

Activity of Alkaline Phosphatase within the Femur, Tibia, and Fibula of Normal and Brachypod (bp^H) Mice[a,b]

		Gestational age (days)			
		14	15	16	17
	normal	0.1	0.25	0.45	1.4
Femur	bp^H	0.1	0.20	0.25	0.5
	%[c]	100%	80%	55%	36%
	normal	0.1	0.35	0.6	1.7
Tibia	bp^H	0.1	0.15	0.2	0.5
	%	100%	43%	33%	29%
	normal	—	—	—	0.4
Fibula[d]	bp^H	—	—	—	0
	%				0%

[a] Activity is expressed as μmoles phenol liberated/10 min/rudiment.

[b] Based on data in Krotoski and Elmer (1973).

[c] %, brachypod/normal × 100.

[d] Alkaline phosphatase does not appear within the fibula until the seventeenth day of gestation.

but none at all was found in brachypod fibulae (Table XXI), and none is present one day after birth (W. A. Elmer, personal communication).

Osteogenesis begins prematurely in the bp^H fibula two weeks after birth. Whether its initiation is accompanied or preceded by chondrocyte hypertrophy and by accumulation of alkaline phosphatase, and whether both subperiosteal and endochondral ossification begin at the same time are being investigated by W. A. Elmer (personal communication). Quite obviously, brachypod is a very useful model for the examination of the possible dependence of subperiosteal osteogenesis on hypertrophy of adjacent chondrocytes, and for the timing of the onset of the ability of chondrocytes to hypertrophy.

III. MAINTENANCE OF THE DIFFERENTIATED OSTEOCYTE

Once the osteoblast becomes surrounded by an extracellular matrix, whether unmineralized as in osteoid, or mineralized, as in bone proper, it becomes an osteocyte. The osteocyte is distinguished by its characteristic morphology, its surrounding calcified matrix, and by its synthesis of type I collagen.

The physiological activities of osteoblasts and osteocytes can be very high in the growing animal. In the parietal bone of the rat, bone is deposited at a rate equal to 470 μm^3 matrix/osteoblast/day (Jones, 1974), and in the American elk, osteoblastic and osteocytic activities are sufficiently high to increase the length of the antlers by up to 2.7 cm/day during the March to August growing season (Goss, 1970). The large number of texts available on the physiology of bone in comparison to the paucity of volumes on cartilage attests both to the depth of our knowledge of bone physiology and to the importance of the bony skeleton. The 25 to 30 million North Americans who are edentulous in one or both jaws because of inadequate knowledge of how to prevent excessive breakdown of alveolar bone (Klein, 1975) would no doubt argue that much more should be learned!

It is impossible to review all of the available data on environmental factors. Vitamins (A, C, and D), hormones (calcitonin, parathyroid, and growth hormones, thyroxine), biomechanical factors, bioelectrical factors, and nutrition all play their roles. For excellent reviews of the physiology and maintenance of bone and bones the reader may refer to: McLean and Urist (1968), Hancox (1972a), Little (1973), Rasmussen and Bordier (1974), Vaughan (1975), and the four-volume series edited by Bourne (1971, 1972a,b, 1976).

Unlike the chondrocyte, the osteocyte survives the mineralization of its extracellular matrix, obtaining metabolites and nutrients via the extensive canalicular system. In Chapter 1, we saw that the osteoblasts in the acellular bone of the teleosts either do not become enclosed within extracellular matrix and hence do not become osteocytes, or that they become so walled in by matrix that they no longer can survive. An understanding of why these osteocytes do not survive might provide some valuable clues as to why osteocytes of cellular bone *do* survive.

Largely from the studies of Bélanger and Jande at the University of Ottawa, it is now known that, in addition to its long recognized role in bone deposition, the osteocyte also plays a role in bone resorption. Osteocytes go through a cycle of deposition, osteocytic osteolysis, and then senescence (Bélanger *et al.*, 1963; Bélanger, 1969; Jande and Bélanger, 1969, 1971, 1973; Jande, 1971; also see Aaron, 1973; Aaron and Pautard, 1973; Schulz *et al.*, 1974; Yeager *et al.*, 1975). Since osteo*clasts* also resorb bone, there has been considerable recent rediscussion of the role of hormones in activating precursor cells and in maintaining or redirecting the physiological activity of both osteocytes and of osteoclasts (McGuire and Marks, 1974; Rasmussen and Bordier, 1974; Hall, 1975c). One example of the specificity of the responses of osteocytes and osteoclasts that still requires adequate explanation is the opposing nature of the responses of different portions of the skeleton to estrogen stimulation.

A. Estrogen and Resorption of the Innominate Bones

Estrogen normally stimulates bone formation. The best studied example is the formation of medullary bone in the laying hen (Gardner and Pfeiffer, 1938;

Haines, 1957; Bonucci and Gherardi, 1975), in which estrogen stimulates both modulation of osteoprogenitor cells into osteoblasts and bone deposition by osteocytes. Estrogen has no direct effect on bone resorption (Nutik and Creuss, 1974; Liskova, 1976). In fact, at the end of pregnancy, the bones are stronger than they were in virgin animals (Currey, 1973). The one exception is that estrogen stimulates resorption of the medial edges of the innominate (pubic) bones during pregnancy. In the presence of exogenous estrogen, the innominate bones of the mouse can be completely resorbed and can disappear entirely (Gardner, 1936). The mechanism of resorption involves estrogen-stimulated modulation of osteoprogenitor cells to osteoclasts (Corwin and Morehead, 1971). The uniqueness of this physiological response is clearly shown by three studies performed by Crelin and his colleagues.

In the first study (Crelin, 1954c), pieces of rib were transplanted into contact with the innominate bones and with the pubic symphysis of castrated male mice, which were then treated with estrogen and relaxin. An interpubic ligament formed in the symphysis (see Section II,E,3), and the median edges of the innominate bones were resorbed. The adjacent rib grafts remained intact.

Crelin and Haines (1955) gonadectomized male and female mice at three weeks of age. Two weeks later, they implanted estrone capsules in contact with the pelvic symphysis, and under the patellar ligament in contact with the patella. Thirty days later, extensive resorption of the innominate bones had ensued. However, the *formation* of extensive medullary bone had occurred in the patella, femur, and tibia, sufficient to almost completely fill the medullary cavities.

In their third study, Pinnell and Crelin (1963) placed pieces of innominate bone or rib into defects in the tibiae of host mice, into whose back musculature estradiol benzoate pellets were then placed. Six weeks later, the grafted innominate bones were undergoing resorption, while the rib grafts and host tibiae were depositing new bone. The possibilities for studying the differential behavior of osteoblastic and osteoclastic cell populations and their precursors under such conditions are considerable.

B. Isolation of Osteoblastic Cells *in Vitro*

The matrix surrounding the osteocyte obviously plays an important role in modifying the osteocyte's response to environmental perturbations. At the same time, it is difficult to study the physiology of osteocytes when they are embedded in a heavily mineralized matrix. This has led to attempts to obtain populations of osteoblastic and osteocytic cells *in vitro*. Osteocytes do not long survive release from their extracellular matrices, and any information gained *in vitro* should be cautiously applied to the understanding of their physiological activity *in vivo*. For example, *in vitro* and *in vivo* studies with growth hormone give quite disparate results, for *in vivo,* growth hormone stimulates the liver to synthesize and/or to release somatomedin, which mediates the effect of growth hormone on the skeleton (McConaghey and Sledge, 1970; Daughaday, 1971; Urist, 1972; Sledge, 1973; Ash and Francis, 1975).

A good portion of the *in vivo* response of the individual osteoblasts and osteocytes is initiated because they are in contact and communication with one another through cell-to-cell junctions (Furseth, 1973; Weinger and Holtrop, 1974). Wounding one area of a bone initiates a compensatory response not only at the wound site but also some distance from it, and for substantial lengths of time after the initial stimulus (Melcher and Accursi, 1972; Pritchard, 1972b; Hyldebrandt *et al.*, 1974). Isolated *in vitro* populations of osteoblasts and osteocytes are obviously not able to show these physiological responses, even though they may be terminally differentiated and metabolically active. Nevertheless, some valuable information on the responses of osteogenic cells has begun to accumulate from the pioneering studies on isolated osteoblasts *in vitro*.

Two main populations have been studied: (a) osteoblasts isolated from bone rudiments after osteogenesis had already begun *in vivo*, to study the maintenance of osteogenic activity, and (b) mesenchymal cells isolated before *in vivo* osteogenesis has been initiated, to determine whether osteogenesis can be initiated *in vitro*.

Peck *et al*. (1964), utilizing buffered collagenase to isolate osteoblasts from rat calvaria, were able to show cell viability after seven days in culture, although by then the cells had become fibroblastic. No osseous matrix was deposited by these cells. The method by which they isolated bone cells, or modifications of it (e.g., Yagiela and Woodbury, 1977), has provided the basis for much subsequent work. They were able to show that such isolated cells synthesized collagen and responded to ascorbic acid (Peck *et al.*, 1967a); that the isolated cells reacted to hormones such as hydrocortisone and parathormone, and also to adenosine by synthesis of adenylate cyclase (Peck *et al.*, 1967b, 1974). A recent study shows that such cells will proliferate in a serum-free medium (Burks and Peck, 1978), opening up a wide range of possible biochemical and metabolic studies on isolated bone cells.

Bard *et al*. (1972) report the isolation of viable osteoblasts from the femoral heads of various adult animals. Osteoblasts from the calvaria of newborn rats are capable of *de novo* synthesis of lipid *in vitro* (Schuster *et al.*, 1975), and D. M. Smith *et al*. (1970, 1973) have provided data on oxygen consumption and lactate production over short-term (2–4 hr) incubations. Calcium uptake by such cells was investigated by Dziak and Brandt (1974a,b) and by Dziak and Stern (1976). An uptake of 17 mmoles Ca/kg wet wt was observed and was shown to be independent of parathyroid stimulation but increased by the ionophore A 23187. All of these cultivations were relatively short term. Ongoing osteogenesis was not reported in any of these studies.

In 1974, Binderman *et al*. provided the first convincing evidence for continued osteogenesis by an isolated population of osteoblastic cells *in vitro*. Using trypsin and EDTA, these workers were able to isolate cells from calvaria and from calvarial periostea from 16- to 20-day-old fetal rats. In some isolates, they traumatized the periosteum in order to initiate cell division so as to augment the population of cells obtained. These cells were maintained for up to eight weeks in BGJ medium plus 10% fetal calf serum in an atmosphere of 10% CO_2. The cells retained the general morphological appearance of osteoblasts (see their Figures 6 and 7), and continued to produce and deposit a collagenous extracellular matrix. This extracellular matrix contained collagen with the typical banding pattern of bone collagen and mineralized the collagenous matrix, in addition to responding to calcitonin stimulation by increased calcium uptake. Similar cells, examined using scanning electron microscopy, have been shown to have both a distinctive morphology and distinctive cell-to-cell contacts (Boyde *et al.*, 1976). They can be separated from fibroblasts on the basis of their ability to bind, but not to be lysed by, anti-rat tail collagen serum (Duksin *et al.*, 1975). Retention of osteoblastic and/or osteocytic activity and differentiation from fibroblasts are well established. I must include one note of caution regarding the use of response to stimulation by calcitonin and/or parathyroid hormone as a marker for differentiated osteogenic cells that comes from the studies of Osdoby and Caplan (1976) and Oakes *et al.* (1977). High-density cultures of limb bud mesenchymal cells chondrify, and after about 13 days *in vitro* produce a calcified tissue having the features of an osteogenic matrix, i.e., of bone rather than of calcified cartilage. These cultured cells respond to parathyroid hormone stimulation after only one or two days *in vitro*. Rather than arising with terminal differentiation of osteoblasts or of osteocytes, this ability is present in the precursor cells, even when those cells first go through a cartilaginous phase.

Hormone-sensitive responses have been utilized by Wong and Cohn (1974, 1975), Wong *et al.* (1977), and Luben *et al.* (1977) in their important refinement of the isolation and separation of populations of osteogenic cells *in vitro*. Using EDTA, trypsin, and collagenase in three sequential extractions, these experimenters were able to obtain from mouse calvaria three populations of cells representing cells progressively deeper in the bone. The first population extracted (the surface and superficial cells) remained fibroblastic *in vitro*, and was unresponsive to both parathyroid hormone and to calcitonin. The third population, whose cells were stellate and contained many connective cell processes (like those described by Binderman *et al.* and by Boyde *et al.* above), showed the greatest response to hormone stimulation (measured by production of 3′,5′-cyclic AMP). The responses to parathyroid hormone and to calcitonin occurred in parallel: both peaked in cells cultured for one week, and both declined together. To test whether the same cells were responding to both hormones, Wong and Cohn (1975) used gentler enzymatic treatment to further subdivide this third population into three subpopulations. One was responsive only to parathyroid hormone, one to both parathyroid hormone and to calcitonin, and the third unresponsive to either. Evidently, different cells are involved, but these cannot yet be related to the classically recognized osteogenic cell types. Their later studies have indicated that isolated osteoclasts, as well as isolated osteoblast-like cells, respond to parathyroid hormone and to 1,25-dihydroxycholecalciferol, the active metabolite of vitamin D (Luben *et al.*, 1977; Wong *et al.*, 1977).

Rao *et al.* (1977) have used a slightly different enzymatic digestion (collagenase, elastase, and DNase, but no trypsin or EDTA) to isolate five populations of rat calvarial osteogenic cells that retain the ability to respond to parathyroid hormone and to prostaglandin E_1, either after several subculturings or after prolonged storage at −80°C. None of these populations responded to calcitonin. The morphological arrangement of calvarial osteoblasts *in vitro* is also ordered by parathyroid hormone (Jones and Boyde, 1976a).

At this stage, differentiated osteogenic cells can be maintained *in vitro,* they will respond in physiological fashion to exogenous hormone stimulation, and of perhaps greatest importance, they will deposit and mineralize an extracellular osseous matrix. The response of the isolated cells and that of the cells embedded within the extracellular matrix, can now be compared, and both can be compared with the *in vivo* environment.

There are surprisingly few studies on the ability of isolated mesenchymal cells to initiate osteogenesis *in vitro* in situations other than those in which cartilage is replaced by bone. However, many studies do indicate that osteogenesis can be initiated in organ culture from mesenchyme that retains the three-dimensional organization it had within the embryo (Glasstone, 1968, 1971). The study of Osdoby and Caplan (1976) referred to above provides a good example of *in vitro* osteogenesis subsequent to chondrogenesis.

Fitton Jackson (1965) cultured embryonic chick fibroblasts and obtained osteogenesis; however, the best example to my knowledge is the study of Marvaso and Bernard (1977). Dissociated cells from the calvarial mesenchyme of 14-day-old fetal mice, i.e., before onset of osteogenesis, were obtained using collagenase, as proposed by Peck *et al.* (1964), and were cultured in BGJ plus 20% fetal calf serum and 5% beef embryo extract. Over the first week in culture, these cells differentiated into osteoblasts and synthesized and deposited collagen into an extracellular matrix, which was then mineralized as the cells formed nodules of woven bone. I am very impressed with the quality of the bone produced in such a short time and under avascular conditions.

Osteoclasts are especially dependent on the presence of bone matrix in order to express their physiological activity. Nevertheless, there have been several reports of methods for isolating populations of osteoclasts (Walker, 1972; Nelson and Bauer, 1977). Further development of these procedures to the point where osteoclasts can be cultured with bone chips, with or without the addition of a pure population of osteoblastic cells, will enable much closer study of bone cell physiology than has been possible in the past.

8

Polarity and Morphogenesis

I. INTRODUCTION

Bones and cartilages come in all shapes and sizes. There are long bones, flat bones, curved bones, bones of irregular and geometrically undefinable shapes, large bones, and small bones. Bones exhibit bumps, ridges, grooves, holes, and depressions where they articulate with other bones, where they attach to tendons and ligaments, and where nerves and blood vessels cross or enter them. The shapes, sizes, and external forms of bones, cartilages, and teeth are the final expressions of the interaction between genotype and environment that constitutes the process of morphogenesis (Hall, 1971, 1975a; Kollar, 1975).

While environmental factors do play a role in morphogenesis, the embryonic rudiments of bone and cartilage exhibit many of their characteristic morphological features even when allowed to develop in such bizarre environments as organ culture, subcutaneously, intramuscularly, intracerebrally, or on the chorioallantoic membrane (Murray, 1926, 1928; Felts, 1961; Mawdsley and Ainsworth-Harrison, 1963). The attainment of the *fundamental* form of the skeletal element, the presence and position of the condyles, articular surfaces, tuberosities, and grooves are independent of both functional demand and biomechanical factors (Howell, 1917; Duerden, 1920; Monson and Felts, 1961; Chalmers and Ray, 1962; Grüneberg, 1963; Hall, 1971, 1975a; Yasuda, 1973; Howes, 1977b). The basic form of the endochondral bone of long bones is laid down in the three-dimensional organization of the cartilaginous model that preceded bone formation. The basic form of the subperiosteal intramembranous bone is laid down by the osteogenic precursors of the periosteum (although the initiation of their osteogenic activity depends upon stimuli from the adjacent cartilage (Chapter 7, Section F,3,b). Both features are nicely illustrated when periosteum from the tibia is grafted over defects in the tracheal cartilage of rabbits. The transplanted periosteum continues to produce bone, but the osteogenic process is intramembranous and the bone forms as flat sheets (Ritsila and Alhopuro, 1973).

The development and maintenance of the *minor* morphogenetic features of the skeleton (e.g., the ridges for attachment of muscles and ligaments) depend very much upon the functional demand to which the skeleton is subjected (Murray, 1936; Chalmers, 1965; Drachman and Sokoloff, 1966; Murray and Drachman, 1969; Hall, 1971, 1975b).

The blueprint for the skeletal form, which is contained within the genes, is a blueprint that is not rigidly expressed. The skeleton of a particular individual or a particular skeletal element may adapt to its owner's way of life. Although the four young in each litter of the armadillo result from the fission of a single zygote, all of their features are not identical. In other species in which litter size varies, the maturity of the skeleton at particular developmental ages is greater when litter size is small, and reduced when litter size increases and body size is reduced (Garrard *et al.,* 1974). The uterine environment modifies form, function, and growth. All of the Australopithecines are variations of the same animal over a range of sizes (Pilbeam and Gould, 1974).

Morphogenesis, the molding of form, should not be equated with growth, which is permanent increase in size. Although these two developmental processes can occur simultaneously, form can change without concomitant change in size, and vice versa. Both morphogenesis and growth can be analyzed with reference to axes and to polarity, for both are directional processes. D'Arcy Thompson's (1917) analysis of the polarization and maintenance of form is a classical study. He analyzed form as an adaptation to the environment but was unaware of the relationship between ontogenetic adaptation and the inheritance of form and pattern. Murray (1936), Enlow (1968a, 1973), and Lanyon (1974) provide discussions of the structural evidence within trabecular bone of the axes along which its growth and form were molded. The orientation of collagen fibers and the permanent electrical polarization of the skeleton (Athenstaedt, 1969, 1970, 1974) provide further evidence of the directional nature of these processes.

Despite the evidence indicating that morphogenesis and growth are polarized processes, little attention has been paid to analyzing how polarity and axial symmetry are expressed during differentiation, although some information is available for tooth development.

The dental lamina is unpolarized in the jaw of the larval salamander (regeneration can occur in either the anterior or posterior direction), but is polarized in the adult, in which regeneration only occurs in the anterior direction (Graver, 1973).

Presumably, the hormonal milieu associated with metamorphosis effects the initiation of this polarity.

In the X-linked tabby mutant of the mouse, the molars are reduced in size, and there is abnormal proliferation of the dental lamina anterior to the first molar (Sofaer, 1975). A preliminary report also suggests that in the mouse, the positioning and formation of the other molar teeth are determined by the first molar tooth germ (Lumsden and Osborn, 1976). Intraocular grafting of the first molar germ led to the formation of three molar teeth, whereas only one tooth developed when either molar two or molar three was grafted alone. The polarity of the murine dental lamina is evidently set by differentiation of the molar teeth. In some rodents, the odontoblasts deposit dentine on the mesial face of the incisor, while depositing a vascularized dentine (vasodentine) on the lingual face (Moss-Salentijn and Moss, 1975), presumably in response to the continuous growth of the rodent incisor.

Scanning electron microscopy has been used by Jones and colleagues (Jones and Boyde, 1976a,b; Jones *et al.*, 1975) to begin assessing the orientation and polarization of the osteoblasts within membrane bone. The orientation of osteoid fibers in relation to osteoblast orientation and ordering of osteoblasts by parathyroid hormone have been examined.

Microfilaments and/or microtubules are obvious contenders for roles in the polarization of skeletal cells. Directional transport of metabolites from the surface to the depth of bones is facilitated by the presence of 240 Å-diameter microtubules within osteoblasts and osteocytes, as well as by 50–70 Å-diameter microfilaments within osteocyte processes, and tight junctions coupling cells to one another within the bone (Holtrop and Weinger, 1972; Stanka, 1975). The high degree of polarization of odontoblasts and ameloblasts depends upon the presence of intact, polymerized microtubules and microfilaments (Karcher-Djuricic *et al.*, 1975).

Three recent studies (Ede *et al.*, 1977; Holmes and Trelstad, 1977; Trelstad, 1977) have utilized the orientation of the Golgi body to show that chondrogenic and prechondrogenic cells are polarized, and that the polarization can change during morphogenesis.

Undoubtedly, understanding morphogenesis is a formidable problem. Ultimately, however, all of the modifying factors must act by regulating the proliferation, differentiation, and/or selective activation or inhibition of osteo- and chondroprogenitor cells and their cellular and molecular products. The fact that these processes are multidimensional presents one of the major stumbling blocks. Oppenheimer (1974) provides an excellent statement of the problems and their status.

There *is*, however, considerable evidence on the way in which the major axes and polarity of the developing limb bud are determined and expressed; these data will be discussed next.

II. ESTABLISHMENT OF AXES AND POLARITY DURING LIMB DEVELOPMENT

The limb bud has three major axes of symmetry. The anterior–posterior (A–P) axis of the bud corresponds to the primary A–P axis of the embryo and is fixed very early during development, i.e., before H.H. stage 8 (ca. 28 hours of incubation). The dorsoventral (D–V) axis fixes the dorsal and ventral aspects of the limb bud and the position of paired skeletal elements when one is dorsal to the other (e.g., radius and ulna). Like the A–P axis, the D–V axis is fixed early in development—before H.H. stage 11 (ca. 42 hours of incubation). Both of these axes are fixed well before any localization or outgrowth of limb mesoderm has been initiated. The third major axis of the limb is the proximodistal (P–D) axis, corresponding to the direction of limb outgrowth away from the flank. The apical ectodermal ridge plays a role in the determination of this axis, as confirmed by ridge removal, which prevents further distal outgrowth or development of more distal limb structures (Chapter 4, Section III,A). Unlike the other two axes, the P–D axis is determined progressively and *after* development of the limb bud has been initiated.

A. Anterior–Posterior (A–P) and Dorsal–Ventral (D–V) Axes

The initial studies on these axes were part of the classical study on amphibian development by Harrison (1918). Rotation of the limb disc of the urodele, *Ambystoma,* at the tail bud stage, such that the A–P and D–V axes of the disc no longer corresponded to those of the body, resulted in the formation of a limb whose A–P axis was reversed, but whose D–V axis was normal (Fig. 31). The A–P axis had already been fixed, but the D–V axis conformed to the host D–V axis (Swett, 1937). Similar experiments with avian embryos gave similar results (Chaube, 1959). The A–P axis is fixed very early during development, perhaps at the time when the basic A–P embryonic axis is laid down.

The A–P axis is determined by certain properties of the mesoderm, while the D–V axis is determined by certain properties of the ectoderm. When pieces of limb mesoderm are wrapped in ectodermal jackets and grafted into somites, or if the ectoderm is rotated, the limb structures that subsequently form have an A–P axis corresponding to the orientation of the mesoderm and a D–V axis corresponding to that of the ectoderm (Errick and Saunders, 1974; MacCabe *et al.*, 1974; Pautou, 1977). Only one area of the limb mesoderm carries this information on limb polarity—the mesoderm that comes to lie in the so-called "zone of polarizing activity" (ZPA). Where these mesenchymal cells lay when the A–P axis was determined early in development is not known.

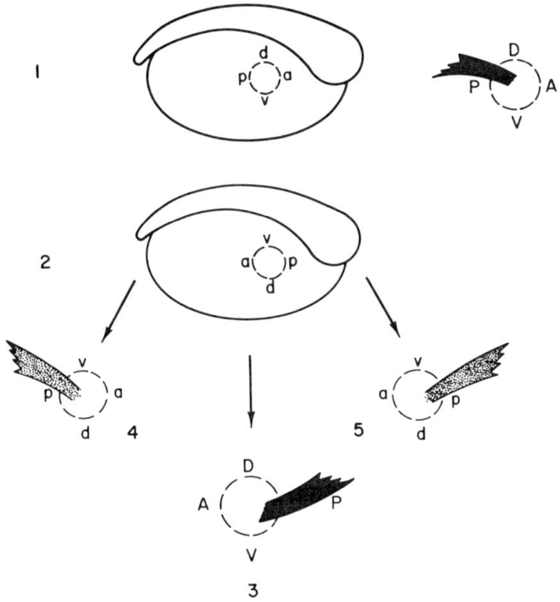

Fig. 31. (1). The major axes of the limb field of an amphibian neurula, in which a is anterior, p, posterior, d, dorsal, and v, ventral. On the right is shown the normal orientation of the limb. (2). The limb disc has been rotated 180° and reimplanted. (3). The resulting limb has its A–P but not its D–V axis reversed, indicating that the A–P axis had been fixed before rotation of the limb disc. (4). The expected result if the D–V but not the A–P axis had already been determined and (5) the expected result if both axes had been determined.

The ZPA was discovered by Gasseling and Saunders (1964) and was named by Balcuns *et al.* (1970). Its role in A–P orientation has been reviewed by Saunders and Gasseling (1968) and also by Saunders (1977). Determination of this A–P axis occurs very rapidly (Amprino and Camosso, 1959, 1963). When an area of mesoderm that corresponds in position to either the location of the posterior necrotic zone of the wing or to the equivalent but nonnecrotic area of the leg (the proximal and posterior mesoderm of the bud, see Fig. 10) was grafted beneath the ectoderm of a second limb bud, an accessory limb bud developed immediately *anterior* to the graft. The posterior face of this new limb bud always faced the grafted mesoderm. Other areas of limb mesoderm, dissociated limb mesenchyme lacking the ZPA, or dissociated ZPA, produce a distorted limb bud without recognizable A–P polarity (MacCabe *et al.*, 1973; Crosby and Fallon, 1975; MacCabe and Parker, 1975, 1976a).

Tickle *et al.* (1975) have performed some elegant experiments to test whether there is a gradient of morphogenetic information within the ZPA. The digits of the wing are produced from mesoderm within the posterior half of the limb bud, adjacent to somites 18 and 19. The ZPA lies immediately posterior to this zone, at the level of somites 19 and 20. If the ZPA represents the posterior portion of a gradient going anteriorly in the limb, then grafting a ZPA into the anterior of the limb bud should induce supernumerary digits, with the posterior digit closest to the ZPA. If grafted sufficiently posterior in the limb bud, only posterior supernumerary digits should form. Both hypotheses were confirmed when the ZPA's from H.H. stage 18–20 embryos were grafted. Tickle and co-workers calculated that the gradient covered 500–1000 μm, and formed within approximately ten hours. As cells leave the distal portion of the limb bud (the zone that specifies the P–D character of the limb; see Section B), their position within the A–P gradient would determine whether they would form an anterior or a posterior limb element (Wolpert, 1971). Summerbell (1974a) has tested and confirmed this. The cells from the distal tip can show alterations in their A–P character but not in their P–D character when they are grafted anteriorly into the limb bud.

The limb bud of the talpid[2] mutant (which will be discussed extensively in Section III,B) is symmetrical, with nine or ten syndactylous digits arranged on a broad and short limb (Abbott, 1967). The symmetrical nature of these limb buds is evident early in development (H.H. stage 18 for the wing, and H.H. stage 20 for the leg), indicating that the A–P axis is probably never established. The symmetry is expressed in the symmetrical digits and in the fusion of adjacent proximal elements, such as metacarpals and radius and ulna. The ZPA in these mutants is normal, for it has been shown to induce normal mesoderm to form duplicated limbs with normal A–P polarity (MacCabe and Abbott, 1974). The mechanism responsible for the lack of A–P polarity must, therefore, be sought elsewhere (Section III,B).

Recently, Cameron and Fallon (1977a) have shown that the hindlimb bud of *Xenopus laevis* possesses a polarizing zone similar to that of the chick, both in location and in the ability to induce a polarized supernumerary limb after the tip of the limb is rotated 180° and transplanted into an anterior *Xenopus* limb bud. Transplantation into the avian limb bud did not produce duplication; however, the equivalent zone from limb buds of the pig, ferret, mouse, partridge, guinea fowl, turtle, hamster, and human will induce duplications when implanted into the embryonic chick limb bud (MacCabe and Parker, 1976b; Tickle *et al.*, 1976; Fallon and Crosby, 1977). The existence of polarizing zones is evidently widespread throughout the vertebrates.

B. Proximodistal (P–D) Polarity

The skeletal elements of the embryonic limb bud are laid down in a proximodistal sequence. The older the limb bud, the more restricted its mesenchyme becomes in terms of its ability to form proximal structures. The role of the apical ectodermal ridge in this phenomenon has already been discussed (Chapter 4, Section III,A).

Summerbell *et al.* (1973) have postulated the existence of a *progress zone* through which mesenchymal cells pass and within which the proximodistal level of organization is specified. Immediately behind the tip of the limb bud is a labile zone in which cytodifferentiation does not take place. The length of time spent by the cells in this labile zone specifies whether they will form a proximal or a distal limb element. This model of polarity of the limb axis links pattern formation to limb bud growth. Either slowing down or speeding up the growth of the bud should alter the pattern of skeletal elements produced. This model is also time graded, for the P–D sequence is established by the time spent within the progress zone, rather than by external cues, the apical ectodermal ridge, or the proximal tissues of the bud. Is the evidence consistent with this model?

There is a zone approximately 300–500 μm from the tip of the limb bud where cells remain relatively undifferentiated until at least H.H. stage 28; here, cell division is rapid (Stark and Searls, 1973; Summerbell and Lewis, 1975), and morphogenetic ability is retained the longest (Finch and Zwilling, 1971). The apical ectodermal ridge does not control the P–D sequencing of skeletal elements—sequencing is normal when "old" ridges are placed onto "young" limb mesoderm, and vice versa (Rubin and Saunders, 1972). The apical ectodermal ridge may play a role in maintaining the lability of the cells at the tip. Stark and Searls (1974) have reported evidence for the role of dorsal and ventral limb ectoderm in the stabilization of limb pattern. Regulation does not occur after exchange of limb bud tips between embryos of various ages (H.H. stages 19 and 24), indicating that influences from the stump of the limb bud are not involved in P–D sequencing.

Summerbell *et al.* (1973) repeated Saunder's classical experiment, removing the apical ectodermal ridge from limb buds of various developmental ages. The fact that increasingly distal structures were affected the older the limb bud at the time of ridge removal was interpreted as halting the change in positional value within the progress zone. In the absence of the ectodermal ridge, the cells of the progress zone lose their lability; as a result, they differentiate a structure at the level specified at the time the ridge was removed. Until at least H.H. stage 19, both Japanese quail and chick limb buds can shift their proximodistal values either more distally or more proximally, but they cannot compensate for reversal of the proximodistal gradient (Kieny and Pautou, 1977).

Smith *et al.* (1974) and Smith and Crawley (1977) have extended this theory to limb regeneration, especially to the distal transformation and growth of the regenerate. They postulate that as the blastema is established, a progress zone is set up that initially corresponds in value to the level of the amputation. But this progress zone imposes more and more distal values onto the cells as the blastema grows and as regeneration ensues. In his [^{3}H]thymidine-labeling study, Maden (1976) maintains that neither a progress zone nor centers of proliferation exist within the blastemata; rather, level-specific properties, and not growth-related ones, specify proximal or distal differentiation. Bryant and her co-workers have established a model for pattern formation in the regenerating limb. This model is dependent on level-specific properties, viz., the shortest intercalation to complete a segment and the completion of a circle of polar coordinates to determine the polarity of the regenerating structures (Bryant, 1976, 1977; French *et al.*, 1976; Bryant *et al.*, 1977; Glass, 1977; but see Slack and Savage, 1978, for evidence against the complete circle rule). One way of specifying such properties would be through centers of mitotic activity; Globus and Vethamany-Globus (1977) have shown that neurons can specify such centers *in vitro*.

Once cytodifferentiation of the limb bud commences, polarity of its outgrowth and the morphogenesis of cartilaginous elements are maintained via the differential synthesis and deposition of extracellular matrix. This has been discussed in Chapter 7, Section II.

III. MECHANISMS OF MORPHOGENESIS

The differential deposition of extracellular matrix by central and peripheral cells within a mesodermal condensation or developing anlage is one mechanism whereby the shape of a cartilaginous skeletal element is molded. As emphasized in the introduction to this chapter, the control over morphogenesis is a mixture of tight genetic control and modification by environmental influences. In this section, I should like to discuss two mechanisms involved in morphogenesis. These are (a) programed cell death as a factor in molding the gross morphology of entire skeletal systems, such as an entire limb or hand; and (b) differential adhesion of cell membranes, as expressed in the talpid^2 and talpid^3 mutants.

A. Programmed Cell Death

Cell death has been defined as "an amplification of the currently mysterious control of catabolism, which ultimately crosses the threshold of irreversibility" (Lockshin and Beaulaton, 1975). We have already seen (Chapter 4, Section III,A) that the process of precocious cell death prevents the early limb bud from continuing its development in limbless vertebrates.

Several regions of physiological programmed cell death are observed during the development of the limb buds of limbed vertebrates. Among these are the zones at the junction of limb bud and flank (the posterior and anterior necrotic zones, Fig. 10); zones of cell death between presumptive digits (interdigital zones); the "opaque patch" in the center of the wing bud; zones in developing scleral ossicles, and areas of cell death within the mesoderm, both at the site of future articular surfaces and within sutures. For general reviews of these areas of cell death, the reader should consult Hale (1956b), Saunders (1966), Saunders and Fallon (1967), Whitten (1969), Mitrovic (1971, 1977), Lockshin and Beaulaton (1975), and Ten Cate *et al.* (1977).

1. Posterior and Anterior Necrotic Zones (PNZ, ANZ)

The existence of localized areas of cell death on the posterior face of the wing bud (the PNZ) and on the preaxial border of the hindlimb bud (the ANZ) was described by Saunders *et al.* (1962). The cell death is quite extensive (approximately 2500 cells die within the PNZ) and rapid (within an 8-hour period during H.H. stage 24, 1500–2000 cells die within the PNZ) (Saunders, 1966).

The first necrotic cells within the PNZ are visible at H.H. stage 24, corresponding in time to the first sculpturing of the paddle-shaped limb bud. From as early as H.H. stage 17, these cells are programmed to die at H.H. stage 24, but this program can be altered up until H.H. stage 22, in which case they survive (Saunders and Fallon, 1967). This can be effected by grafting these cells to other areas of the limb bud or into nonlimb mesoderm, where they not only survive but where they can differentiate into chondroblasts. When grafted into the PNZ region, cells from outside the PNZ do not show programmed cell death. Evidently, the cells of the PNZ represent a subpopulation of limb mesenchymal cells that have been set aside for this particular developmental role.

Programmed cell death is a metabolically controlled phenomenon. Time-to-die is slowed at 30°C and postponed indefinitely at 20°C (Fallon and Saunders, 1968). Programmed cell death is not suicidal, because phagocytic or lysosomal enzymes are not produced. The exact mechanism of this cell death still remains in doubt. The attainment at H.H. stage 22 of irreversible determination for cell death is accompanied by a sharp decline in DNA synthesis (Saunders and Fallon, 1967). This prompted Webster and Gross (1970) to propose that cell death resulted from increased cross-linking of DNA. When these workers injected agents to increase cross-linking, they found that necrotic cells appeared all over the embryo. However, the DNA of cells from the PNZ is no more highly cross-linked than DNA from other cells.

Three studies present evidence for biochemical degeneration preceding programmed cell death in these zones. Mottet and Hammar (1972) observed protein and ribosomal crystals within the cells of the PNZ at H.H. stages 22 to 24. Pollak and Fallon (1974), following protein synthesis autoradiographically, found that, from H.H. stage 22 onward, protein synthesis decreased significantly within the PNZ compared to that in cells at the distal tip of the limb bud. DNA and RNA synthesis declined between H.H. stages 21 and 22, approximately six hours before the decline in protein synthesis (Pollak and Fallon, 1976).

Does this programmed cell death help to sculpt the base of the limb bud as it grows out from the body (more probably, the recession of the flank from the body is as prominent as the outgrowth of the limb bud at these stages; see Chapter 9)? Like many cherished hypotheses, this one became somewhat tarnished in the face of experimental data—when cell death was blocked *in ovo,* limb morphogenesis was found to be normal (Saunders, 1966; Saunders and Gasseling, 1968). However, in at least two mutants, there is presumptive evidence for an association between programmed cell death and alteration of limb morphology (also recall the limbless vertebrates). In the wingless mutant, precocious cell death occurs in the ANZ beginning at H.H. stage 19 and extends, during H.H. stages 20 to 23, well beyond the normal bounds of the ANZ (Hinchliffe and Ede, 1973). The removal of this additional mesenchyme may rob the limb bud of an apical ectodermal maintenance factor, and may contribute to the lack of an apical ectodermal ridge. Talpid^3 has a longer than normal apical ectodermal ridge, and lacks both ANZ and PNZ. This prompted Hinchliffe and Ede (1967) to propose that these zones normally act as "end-stops" for the apical ectodermal ridge, limiting its extent. They, and Ede and Agerback (1968), feel that altered adhesiveness may be involved in the loss of these necrotic zones (see Section B).

This, then, is a curious situation: in mutants, the absence of necrotic zones is associated with abnormal limb morphogenesis, but in normal embryos, experimental interference with the action of necrotic zones does not affect limb shape. It rather looks as if the experimental procedures for analyzing the role of cell death in normal embryos need to be modified.

The tissues of the limb buds of normal embryos also have an effect on cell death in associated tissues. The ventral horns of the grey matter in the developing spinal cord are enlarged adjacent to the limb buds, in part because the limb bud slows the rate of neuroblast death (Deuchar, 1975). It is not known which limb cells have this action or how it is mediated.

2. Interdigital Cell Death

Areas of necrotic cells appear at the distal tips of the limb buds of birds (Deleanu, 1965; Saunders and Fallon, 1967), in reptiles and turtles (Fallon and Cameron, 1977), and in rodents and man (Milaire, 1963, 1965a,b; Menkes *et al.*, 1965; Ballard and Holt, 1968; Kelley, 1973), but not in amphibians (Cameron and Fallon, 1977b). Several lines of evidence indicate that interdigital necrosis is responsible for the sculpturing of the interdigital area.

In the duck, the pattern of webbing between the digits can be related to the presence or absence of interdigital cell death. Necrotic areas are prominent between the first and second toes, where no web forms, but necrosis is reduced between other digits, where webs do form (Pautou, 1974). No interdigital necrosis is observed in syndactylous mice (Milaire, 1965a); neither in Janus Green-induced syndactyly in the embryonic chick (Saunders and Fallon, 1967), nor in the syndactylous limb buds of the talpid^3 mutant (Hinchliffe and Thorogood, 1974).

In the embryonic chick, interdigital cell death occurs between H.H. stages 30 and 35. The dead cells are engulfed by phagocytes. Biochemical degeneration may precede cell death. In a histochemical analysis, Hammar and Mottet (1971) found that succinate dehydrogenase was absent from the interdigital areas at H.H. stages 26 and 27. Fallon *et al.* (1974), using biochemical methods of assay and histochemistry on frozen sections, repeated Hammar and Mottet's study, and claimed that succinate dehydrogenase was present in the interdigital areas before necrosis set in. However, Fallon's group also obtained a reaction in the whole limb in the absence of the enzyme substrate, so the question of whether metabolic activity is reduced within these cells before the onset of necrosis remains unanswered.

Some evidence suggests that the apical ectodermal ridge suppresses cell death in the adjacent mesenchyme, although it has not yet been possible to selectively test the ectoderm overlying digital and interdigital mesenchyme (Barasa, 1960; Janners and Searls, 1971; Ede and Flint, 1972; Cairns, 1975).

Cell necrosis in the interdigital mesenchyme does play a role in patterning the distal portion of the limb. For an incisive discussion of the implications of such mechanisms for the evolution of the vertebrate limb, the reader should see Maderson (1975).

3. The Opaque Patch

The opaque patch, an area of less translucent mesenchyme in the center of the wing bud of the embryonic chick, is observed at H.H. stages 24 and 25. In this area, cell death commences at H.H. stage 23 and is at a maximum at H.H. stages 24 and 25. The loss of these cells serves to separate the condensation of mesenchyme for the future ulna from that for the future radius. The initial appearance of the opaque patch at H.H. stage 23 precedes the condensation of mesenchymal cells. Before that stage, these cells were incorporating ^{35}S into glycosaminoglycans. Glycosaminoglycan synthesis is suppressed at the stage when cell death is maximal (Dawd and Hinchliffe, 1971). In talpid^3, in which the opaque patch is either absent or very reduced, these cells continue to synthesize glycosaminoglycans and are responsible for the lack of separation of the radius and ulna (Hinchliffe and Thorogood, 1974). These normally necrotic cells can then retain their chondrogenic ability in the presence of factors suppressing cell death.

In conclusion, it is evident that programmed cell death is one way of fashioning the entire limb, or of molding individual elements within the limb skeleton. We are almost completely ignorant as to the metabolic and cellular mechanisms responsible for programmed cell death.

B. Cell Adhesion and Morphogenesis in the Talpid (*ta*) Mutant

The particular morphology of a skeletal element is generated from the accumulated actions of various processes: namely, cell division, cell aggregation and/or migration, cell differentiation, and the deposition of extracellular matrix. Any alteration in cell-to-cell adhesion early in development would affect all of these processes. The role of altered cell adhesion at the time of mesenchymal cell condensation in the subsequent abnormal morphogenesis of the limb of the brachypod mouse has already been discussed (Chapter 7, Section II,F,3,c). Increased cell-to-cell adhesion has also been proposed as the mechanism underlying the generation of the abnormal morphology of the talpid limb bud in the chick.

Talpid, the polydactylous lethal, has been isolated on three previous occasions [see Abbott (1967) for details]. In all cases, severe polydactyly was produced, evidenced by eight to ten digits in a foreshortened, broad limb bud, reminiscent of the forelimbs of the mole, *Talpa*. All three mutations are homozygous recessive and lethal. Talpid^1, which is now extinct, induced death between five and thirteen days of incubation. The talpid^2 mutant, which has been studied extensively by Abbott and her colleagues in the U.S., dies at around thirteen days of incubation, and talpid^3, studied by Ede and Hinchliffe and their colleagues in the U.K., dies between five and six days of incubation.

Talpid^2 and talpid^3 both produce abnormalities in the face and beak, severe polydactyly, shortened vertebral columns, shortened proximal long bones of the appendicular skeleton, and fusion of lateral elements (radius and ulna, metacarpals). Despite these similarities between the two mutants, I shall discuss each separately. We do not know whether they are allelic, or whether their modes of action are identical.

1. Talpid^2

Abnormalities of the beak and face were described by Abbott *et al.* (1960). The extent of the apical ectodermal ridge is normal up to H.H. stage 22, when it begins to enlarge at a rate faster than normal. By H.H. stage 25, it is 50% larger than the ridge of similarly aged normal embryos (Goetinck and Abbott, 1964). The apical ectodermal maintenance factor is distributed more extensively within talpid mesoderm than within normal mesoderm (Goetinck and Abbott, 1964; MacCabe and Abbott, 1974), allowing the larger than normal ectodermal ridge to be maintained. The localization of the primary defect within the mesoderm has been convincingly shown by recombinations between talpid and normal limb bud mesoderm and ectoderm grown as flank grafts (Table XXII). Ectoderm from talpid limb buds can respond to normal limb bud mesoderm by forming a

TABLE XXII

Results of Recombinations between Talpid², Scaleless, and Normal Limb Bud Mesoderm and Ectoderm To Show That Talpid² Is a Mesodermal Deficiency[a,b]

Source of ectoderm	Source of mesoderm	Results
talpid²	++	85% ++
++[c]	++	82% ++
scaleless	talpid²	87% talpid²
scaleless	++	71% ++

[a] Recombinants were grown as flank grafts within normal hosts.
[b] Based on data in Goetinck and Abbott (1964).
[c] ++, Normal (wild-type) embryos.

normal limb. As indicated in Section III,A,1, the zone of polarizing activity is distributed normally although talpid limb buds lack anterior–posterior polarity (MacCabe and Abbott, 1974). No other information is available on the mechanism of action of talpid².

2. Talpid³

The gross abnormalities of talpid³ have been described extensively by Ede and Kelley (1964a,b) and by Ede (1971). The basic defect affects the segregation of the prechondrogenic mesenchyme into condensations. Bilateral symmetry of the head is not established because the prechordal mesenchyme fails to separate (Ede and Kelley, 1964a).

As in talpid², the condensations of the limb buds of talpid³ are laid down in the correct proximodistal sequence; however, the anterior–posterior polarity is disturbed, and as a result, some elements fuse and some undergo symmetrical duplication. Chondrogenesis is abnormal, and osteogenesis is not initiated in the primordia of the normally endochondral bones (Ede and Kelley, 1964b; Hinchliffe and Ede, 1967, 1968). Intramembranous ossification is normal.

The apical ectodermal ridge is abnormally elongated, especially at H.H. stages 27 and 28 (Hinchliffe and Ede, 1967), and the junction between the ridge and the subjacent mesoderm is unusually wide (Ede *et al.*, 1974). The areas of programed cell death normally seen in limb buds are absent (Hinchliffe and Ede, 1967; Dawd and Hinchliffe, 1971; Hinchliffe and Thorogood, 1974; also see Section III,A). Reduced cell death is one mechanism operating within the talpid³ limb bud that could provide additional cells for the broad, spatulate limb. Recombination studies between *ta*³ mesoderm and ++ ectoderm similar to those described for *ta*² in Table XXII have not been performed.

Donald Ede and his colleagues have explored the thesis that cell adhesion and movement are disturbed in talpid³ and are responsible for the abnormal morphogenesis (Ede and Agerback, 1968; Ede and Law, 1969; Ede, 1971; Ede and Flint, 1972, 1975a,b; also see Chapter 5, Section III). Ede and Agerback (1968) dissociated H.H. stage 24 and 26 normal and *ta*^{3} limb buds into single cells with trypsin digestion and observed their pattern of reaggregation in rotation culture. Because they were more adhesive and less mobile than cells from normal embryonic limb buds, *ta*^{3} cells aggregated more rapidly and formed smaller aggregates. *In vivo,* this cell behavior would be expressed as the failure of condensations to separate from one another, with fusion of adjacent skeletal elements also taking place. Distal movement of mesenchymal cells was envisaged by Ede and Agerback to be a component of limb outgrowth; accordingly, the altered cellular properties of talpid^3 cells should slow this outgrowth. A computer simulation of limb outgrowth generated a talpid^3-like limb pattern when distal movement of cells was removed from the input (Ede and Law, 1969). [It is now known that the gradient in mitotic activity in talpid^3 limb buds is reversed. It is distoproximal, and not proximodistal, as in normal limb buds (Ede *et al.,* 1975). The computer simulation incorporated a proximodistal gradient.]

Ede and Flint (1975a,b) have confirmed that the rates and durations of cell movement in talpid^3 cells increase the probability of cell-to-cell contact *in vitro.* Viscometric analysis has confirmed that talpid^3 cells are more adhesive than cells from normal embryos. Niederman and Armstrong (1972), reaggregating talpid^2 cells with normal limb mesenchymal cells, expected to obtain sorting out if talpid^2 cells are more adhesive than normal, as is the case for talpid^3. No sorting out was obtained, although the methodology used by Ede's group has not been applied to talpid^2.

Talpid^3 limb buds, then, are distinguished by increased cell adhesion, decreased cell movement, decreased or lack of cell death, decreased rates of cell proliferation, and by reversal of the normal gradient of cell division within the limb bud. The initial defects in the mesenchymal condensations produced by these cell behaviors have their later expression in abnormal chondrogenesis and in defective morphogenesis of the limb buds.

9

Initiation of Skeletal Growth

I. INTRODUCTION

In the introduction to Chapter 8 and in the context of the control of morphogenetic processes, a distinction was made between morphogenesis and growth without really defining what was meant by growth. I am attracted to the definition proposed by Moss (1972b): "In the broadest sense, growth may be defined as any temporal change in any parameter that is measurable;" however, this does not draw a sufficient distinction between the change in shape that is morphogenesis and the permanent increase in size that is growth. Moss' explanation, however, does highlight the fact that any definition of growth must be operational. To be true growth, an increase in size must involve an increase in the amount of protoplasm; such increases, however, are not often readily measured, for even dry weight is not necessarily a measure of protoplasm. I would, then, modify Moss's definition to read: [growth is] "a permanent increase in the size of any parameter that is measurable," and without defining the extent of the permanence (the fetus grows, the pregnant woman does not), I would emphasize that, for estimates of growth to be accurate, as many individual parameters as possible must be measured. The latter is necessary, of course, because growth has a multifactorial basis.

In a sense, the bulk of the discussion of the preceding eight chapters has concerned the initiation of skeletal growth. In this chapter, I would like to concentrate not on the ways of measuring rates or amounts of growth, but on the developmental processes that initiate and regulate growth as a developmental phenomenon.

II. COMPONENTS OF SKELETAL GROWTH

A. The Number of Stem Cells

The proliferation of precursor cells and the accumulation of progenitor cells into condensations represent the earliest growth of the skeleton, a growth that is controlled by the number of initial stem cells and their rates of proliferation, accumulation, and/or loss. Very little information is available on the number of stem cells that are sufficient to produce particular skeletal elements. Exceptions are the studies of Beatrice Mintz and her colleagues, who used the fusion of cells from early blastocysts of mice of different strains to obtain tetraparental mice for the study of stem cell origins within the vertebrae (Mintz, 1971, 1972; Gearhart and Mintz, 1972; Moore and Mintz, 1972). Using pure strains of mice (C3H and C57BL/6) with morphologically distinct vertebral characteristics, these authors found that individuals produced after blastocyst fusion possess individual vertebrae of the C3H form on one side and of the C57BL/6 form on the other. These vertebrae were interpreted as cellular genotype mosaics, indicating that the left and right sides of the individual vertebrae were of independent origins. These cells were clonally derived, each lateral sclerotomite providing one cell line to each vertebra. Each vertebra, therefore, was derived from four clones of cells (Moore and Mintz, 1972). The individual somites were shown not to be derived from a single clone by the presence of both forms of the dimeric enzyme glucosephosphate isomerase (5.3.1.9) within single somites, one dimer being contributed by each parent (Gearhart and Mintz, 1972).

In some of these mosaic embryos, paired bones of the skulls (notably the palatines, pterygoids, and occipital condyles) also occasionally evidenced a mosaic organization within a bilateral half, indicating that they originated from at least two cell lines (Moore and Mintz, 1972).

This technique has thus provided one of the few pieces of information on the number of stem cells that contribute the cells for a particular skeletal element. These authors have also used their data to argue that the time of establishment of these cell lines is after five days of gestation (when a sufficient number of cells are first present to form all the necessary clones for the vertebral column) and before seven days of gestation, when the somites first appear. [Cartilage and bone form when blastocysts are cultivated *in vitro* or after intratesticular grafting of ova (Stevens, 1968; Hogan and Tilly, 1977).] There are some severe criticisms of this technique being used to determine the timing of the establishment of such clones (Lewis *et al.*, 1972; McLaren, 1972), but it does nevertheless represent the first approach to a very important question.

Skeletal defects could very well arise by the elimination of one clonal line [as may happen when mesenchymal condensations are reduced in size (Grüneberg,

1954)], from excessive proliferation or initial duplication of one clonal cell line [as may happen when condensation size is abnormally large (Grüneberg and des Wickramaratne, 1974)], or from disruption of a whole body gradient (Cooke, 1975). If so, it would be necessary to look for the control of growth processes not during the "growth period," but during very early embryonic development [see Laird (1966) and Saxén (1976) for the development of a similar concept in the context of comparative growth analysis and teratogenesis].

B. Rates and Patterns of Cell Proliferation

Some information on the rate of proliferation of progenitor cells may be found in Chapter 6; a discussion of the accumulation of precursor cells into condensations is presented in Chapter 5. In this section, I want to concentrate on the initiation of cell proliferation as a factor in skeletal outgrowth. The outgrowth of the avian embryonic limb bud has received the most attention and provides perhaps the best understood system for the analysis of gradients of cell proliferation and their role in outgrowth. Some general reviews in which cell proliferation is considered are Faber (1971), Zwilling (1972), and Searls (1973a,b).

It has been known since the initial carbon-marking experiments of Saunders (1948) that the growth of the embryonic limb bud is apical. The continued outgrowth of the mesoblast depends upon the continued presence of the apical ectodermal ridge (Chapter 4, Section III), but removal of the apical ectodermal ridge does not affect mitotic activity in the adjacent mesoderm (Janners and Searls, 1971).

The growth of the limb bud does not merely consist of the production of more cells. Outgrowth involves the number of cells, their proliferation, position, movement, size, shape, packing density, and the constraint of the ectoderm. Ede and Law (1969) formulated a computer simulation model for limb bud outgrowth in an attempt to discriminate between these factors and to determine the least number of components necessary to produce a normal limb bud. A combination of cell proliferation in a proximodistal gradient with distal movement of the cells produced both a form and growth pattern remarkably similar to those of the normal limb bud.

The computer simulation devised by Ede and Law required that there be a gradient of cell division within the limb bud. Amprino (1965) was the first to produce evidence for the existence of such a gradient. Within the wing bud, mitotic activity is relatively constant from H.H. stage 16 to early stage 22. At H.H. stage 22, mitotic activity declines to a new rate, which is lower but relatively constant (Janners and Searls, 1970; Searls and Janners, 1971). The decline in mitotic rate after H.H. stage 22 is greatest in the central proximal chondrogenic mesenchyme, where the [^{3}H]thymidine-labeling index drops 75%

between H.H. stages 19 and 24. The decline in this mesenchyme can be compared with those of the dorsal and ventral proximal prospective myogenic mesenchyme and the subridge mesenchyme, where the decline over the same developmental stages is approximately 25%. At H.H. stages 23 and 24, mitotic activity is highest in the subridge ectoderm, is next to highest in the dorsal and ventral proximal mesoderm, and is at its lowest level in the central mesoderm. Removal of the apical ectoderm ridge at H.H. stage 19 does not reduce mitotic activity in the subridge mesoderm. Stark and Searls (1973) concluded that cell proliferation without cell migration was sufficient to account for outgrowth of the limb bud.

Hornbruch and Wolpert (1970) performed counts of mitotic activity on wing buds from embryos between H.H. stages 18 and 30. Throughout all stages, mitotic activity was constant within the ectoderm at around 2% of cells dividing. Overall mitotic activity within the mesoderm declined with developmental stage, averaging 12% at H.H. stage 18 and declining smoothly to a rate of 2% at H.H. stage 30. No proximodistal gradient of mitotic activity was observed until H.H. stage 24, when the mitotic index in the distal mesenchyme exceeded that in the proximal mesenchyme. Much limb bud outgrowth has already been initiated by H.H. stage 24. Hornbruch and Wolpert indicate that there exists no statistically significant gradient before H.H. stage 24. Nevertheless, at H.H. stage 20, there is a difference in mitotic activity of 6.5% distally to 9.5% proximally. The specification of the length of skeletal elements between H.H. stage 22 (4 days of incubation) and H.H. stage 36 (10 days of incubation) is accurate to within one cell in twenty (Summerbell and Wolpert, 1973); accordingly the lack of statistical significance in the gradient should not invite worry about the biological significance of the differences in mitotic activity. Even so, evidence for a proximodistal gradient in mitotic activity is not substantial until H.H. stages 24 or 25, at which time chondrogenesis is stabilized. This becomes very apparent upon reworking the data of Hornbruch and Wolpert, as Lewis (1975) has done. I have reexpressed Lewis's Table 4 in graphical form in Figure 32, using as a common base the length of the limb bud, with the distal tip as 0 and the proximal base of the limb bud *at that H.H. stage* as 10; the percent mitotic index is then plotted as a percent of the rate at the distal tip at that H.H. stage (rate at distal tip as 100%). These plots clearly show that no gradient in mitotic activity is present at H.H. stage 18 and that there is a slight decline in distoproximal rate during H.H. stages 19 and 20. At H.H. stages 21 and 22, there is a rise in mitotic index midway along the limb bud, with the slope of this increase steepening considerably between H.H. stages 21 and 23. The peak at H.H. stage 23 corresponds to the condensation of proximal skeletal elements within the limb bud. At H.H. stage 24, there is a sharp decline in this peak of mitotic activity, and by H.H. stage 25 the distoproximal gradient is firmly established. In addition to this within-stage gradient in mitotic activity, there is also an absolute decrease in mitotic activity from stage to stage (Lewis, 1975).

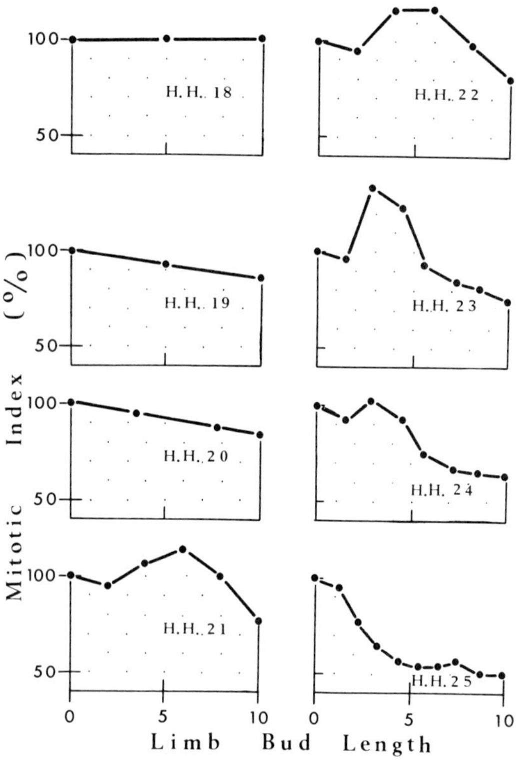

Fig. 32. The mitotic index within the central mesenchyme of the limb bud between H.H. stages 18 and 25 (as percent of the rate at the distal tip at *that* H.H. stage, with the rate at the distal tip taken as 100%), plotted against the relative length of the limb bud (where 0 is the distal tip and 10 is the proximal base of the bud at *that* H.H. stage). No distoproximal gradient is evident at H.H. stage 18. There is a slight distoproximal decline in the mitotic index at H.H. stages 19 and 20. The peak in mitotic index at H.H. stage 23 represents the central condensation, which precedes chondrification. By H.H. stage 25, a prominent gradient in mitotic index is apparent. Adapted from data in Lewis (1975).

Recent evidence indicates that between H.H. stages 19 and 26, there is a proximodistal decline in the agglutination of limb mesenchymal cells to the plant lectin, concanavalin A, that parallels the decline in morphogenetic potential, as well as the stabilization of cells for chondrogenesis and the establishment of the mitotic gradient (Paulsen and Finch, 1977). The possibility that cell surface changes and cell-to-cell adhesion play a role in the establishment of the mitotic gradient should now be considered.

Ede *et al.* (1975) have provided more detailed information on the differential distoproximal gradient between central and peripheral limb mesenchyme. They confirm that there is a gradient in mitotic rate within the central mesenchyme during H.H. stages 24 to 27. This gradient is highest distally, drops sharply in the distal one-third of the limb bud, and then more slowly toward the proximal base of the bud (cf. H.H. stage 25, Fig. 32). At H.H. stage 25, the gradient in peripheral mesenchyme was present (they did not examine H.H. stage 24), but declined much more gradually than the gradient within the central mesenchyme. No peripheral gradient was found at H.H. stage 27. The order of magnitude of the observed gradients was the distal mitotic rate (4–6%) being approximately twice the proximal rate (1–3%). As noted in Chapter 8, Section III,B, this gradient is reversed in talpid[3].

The manner in which mitotic rate is controlled may differ in the central and peripheral mesenchyme. Kieny (1975) blocked mitosis in embryos incubated for 6½ days by injecting vinblastine. She found that mitosis was blocked in the peripheral, but not in the central mesenchyme. Similar studies on earlier embryos might provide valuable information on the establishment of regional differences in mitotic activity across the limb bud.

The fact that the apical ectodermal ridge is not responsible for maintaining the high distal mitotic index has already been mentioned. However, the division of these apical mesenchymal cells may be constrained by the physical presence of the ectoderm, against whose resistance the cells must work (Amprino, 1965; Hornbruch and Wolpert, 1970; Amprino and Ambrosi, 1973; Summerbell, 1974b, 1977a). It is within this 300–500-μm zone of heightened mitotic activity that the proximodistal cell sequence is fixed (Chapter 8, Section II,B), and, since the amount of time spent within this progress zone determines positional information, any change in mitotic rate should have pronounced effects on the length of particular segments of the skeleton. Summerbell and Wolpert (1973) have calculated that, between four and ten days of incubation, each mesenchymal cell divides three times; as well, the length of such individual skeletal elements as the humerus, ulna, and digits is controlled very precisely to within ±4–5%. Accordingly, given the number of starting cells, the rate of proliferation, and the rate of accretion of extracellular matrix, then growth rates and final size should be predicted with considerable accuracy.

Since it is the limb bud which grows out from the flank, we tend to assume that the growth mechanism lies completely within limb bud cells, which divide at a higher rate than the adjacent cells of the flank. However, Searls and Janners (1971) found that the initial outgrowth of the limb bud resulted from a *suppression* of mitotic activity in the flank between H.H. stage 16 (when the limb bud is just appearing) and H.H. stage 20. During this time, the mitotic rate in the flank declined by about 25%, while the mitotic rate in the limb bud remained constant. The initial outgrowth of the limb bud, then, is really a regression of the flank away from the limb bud. Differential rates of *decline* in cell division have also been shown to be responsible for the growth and morphogenesis of the frontonasal process of the embryonic chick (Minkoff and Kuntz, 1977) and, in fact, may be more common than the differential increase in cell division in the growth of processes or buds.

Ede and Agerback (1968), Ede and Law (1969), Ede *et al.* (1974), and Wilby and Ede (1975) maintain that cell movement within the limb bud is also an important component of limb outgrowth. Ede *et al.* (1974) produced ultrastructural evidence of attachment of mesenchymal cells to, and movement toward, the ectoderm. Ede's studies on cell adhesion in the talpid^3 mutant (Chapter 8) are important in this regard. Kwasigroch and Kochhar (1975) found that the mesenchymal cells within the proximal portion of the mouse limb bud were more motile than those in the distal portion of the bud, and that applying vitamin A to the mother or to limb buds *in vitro* decreased both cell motility and the size of the developing limb. Any reduction in cell motility, or decrease in surface adhesiveness, would tend to increase the generation time for cell division and would appear as a reduced labeling index after [^{3}H]thymidine (Riley, 1974).

The initiation of growth within the limb bud, then, is a complex series of processes. These processes include: the initiation of cell division; the alteration of rates of division with time; the establishment of gradients in cell division across or along the limb bud; and, as the limb develops, the alteration of such cell surface-related processes as attachment and cell movement.

C. Epigenetic Control

The differentiation and maintenance of scleroblasts, as well as the deposition of extracellular matrices (Chapters 6 and 7), provide a second major component of skeletal growth. At this point, growth is a composite process, consisting of an increase in the size of the cell population and regulation of the rates of physiological activity of that cellular population. Controls at this stage of development may differ from those controlling earlier cellular proliferation (Cook, 1974), for the earlier the stage in development, and the more fundamental* the biological unit under study (molecule–organelle–cell–tissue–organ), the more independent of epigenetic control is the growth of that unit. (The science of epigenetics is concerned with the causal analysis of development, in particular of the mechanisms by which genes express their phenotypic effects.) Storey (1972), Moss (1972b,c), and the volume edited by Moffett (1972) provide excellent discussions of the relative contributions of intrinsic and extrinsic factors to the control of growth of cartilage and bone. In addition, Bonner (1974) has introduced a novel experimental method involving blastocyst transfer.

*For those who abhor the reductionist philosophy that our knowledge of mechanisms increases inversely with the size of the biological unit under study, the term "smaller" may be substituted for "more fundamental."

Mechanical stimulation is obviously important as a regulator of skeletal growth. The possible way in which mechanical factors control cartilage growth was discussed extensively in Chapter 6, Section VI,A.

1. Hormones—The Growth of Antlers

Among the epigenetic factors that stimulate the production of a population of skeletal cells are the hormones (Tonna, 1974). Parathyroid hormone stimulates osteoclasts and osteocytes to resorb bone. Calcitonin, while inhibiting osteoclastic resorption, stimulates osteoblastic function. Both hormones influence the differentiation of precursor cells. This balance between deposition and resorption controls both bone turnover and whether bone will accumulate (grow) or be lost (atrophy).

Sexually dimorphic elements of the skeleton are particularly dependent upon gonadal hormones for their differentiation, morphogenesis, and/or growth. The transformation of the interpubic cartilage into an interpubic ligament in female mice under the influence of estrogen and relaxin has already been discussed (Chapter 7, Section II,E). An example relevant to the control of skeletal growth is the growth of antlers on male deer in response to testosterone.

The histogenesis of antler formation and shedding was discussed in Chapter 1. The initiation of antler outgrowth is initially dependent upon the formation of a bony pedicle on the frontal bone, and, in addition, on the interaction between the pedicle and the overlying epidermis and dermis (Chapter 4).

The growth of the antler bud, and that of the antler, occur very rapidly. In fact, antlers are the fastest growing bones of the postnatal skeleton of any vertebrate. Their final size varies from species to species, but, in all species the period during which they grow is approximately six months long, throughout the spring and summer seasons. It follows that growth rates must and do vary greatly from species to species. Goss (1970) has recorded maximal growth rates as high as 2.75 cm/day in some species (Table XXIII).

The antler bud, as well as the antlers, branches as it grows. Antler growth, unlike that of horns, is apical (terminal). Surgical removal of the apex of the antler prevents any further growth of that antler (Goss, 1961).

Variations in the endogenous level of testosterone act to mediate growth of both pedicle and antler. As has been amply demonstrated by Goss (1968a,b, 1969b,c, 1970, 1976), these levels are under the control of the local exogenous photoperiod.

The bony knob on the frontal which presages the development of the pedicle is present in male but not in female fetal red deer (*Cervus elaphus*). Its appearance and development over gestation days 60 to 100 parallel testicular development and the accumulation of testosterone (Lincoln, 1973). No further development occurs until the fawn receives a photoperiodic cue. If fawn are castrated before

TABLE XXIII

Maximum Daily Growth Rates (cm/day) and Final Sizes (cm) of Antlers from Various Species of Deer[a,b]

Species	Final antler length (cm)	Maximum growth rate (cm/day)	Growth period (months)
Cervus canadensis (American elk)	129.5	2.75	March to August
Cervus dama (fallow deer)	64.0	0.68	April to September
Cervus nippon (sika deer)	42.2	0.37	April to September

[a] Antler length is variable between species, but the growth period (months) is constant.
[b] Based on data in Goss (1970).

antlerogenesis has commenced, the pedicle fails to form and the first set of antlers does not develop (Goss, 1963). If such castrated fawn are given testosterone, formation of a pedicle begins. If, in order to provide a wound stimulus, this pedicle is amputated, the development and growth of a set of antlers commence (Jaczewski and Krzywinska, 1974). Therefore, testosterone is essential for pedicle formation.

In the adult, testosterone controls the ossification of the antlers so that growth and shedding are coordinated (Goss, 1968a). In reindeer and caribou, the level of plasma testosterone is normally about 1 ng testosterone/ml plasma, but during the period of maximal antler growth, levels rise to between 30 and 60 ng testosterone/ml plasma (Whitehead and McEwan, 1973). If adults are castrated in the fall and winter, at which time the antlers have stopped growing and consist of bony tissue, the antlers are not shed but remain as permanent structures. If castration is performed in the summer, when the antlers are still growing, the velvet is retained and the antlers are not shed (Goss, 1963). The time of antler shedding is thus under hormone control.

Bubenik *et al.* (1975) administered cyproterone acetate, a testosterone inhibitor, to adult white tail deer (*O. virginianus,* 3.5 mg inhibitor/kg body weight/week), and found that mineralization of the antlers was blocked, that fewer haversian systems were produced, and that the antlers were retained. The antlers are target organs for testosterone. Immunohistochemical procedures show that testosterone accumulates in the prechondroblastic blastema as the growth season progresses (Bubenik *et al.,* 1974a). The absence of testosterone either prevents antlers from forming, or, if antlers are already present, stops their growth and prevents their normal shedding. Excess testosterone leads to increased rates of ossification, premature cessation of antler growth and shedding of the velvet, and to retention of the set of antlers.

The action of estrogen opposes the action of testosterone, while hypophysectomy prevents antler growth altogether (Goss, 1969a, 1970), a finding which implicates the pituitary hormones. Although growth hormone may be involved

(Bubenick *et al.*, 1974b), the exact role of the pituitary hormones remains to be established.

Graham *et al.* (1962) investigated some aspects of blood chemistry during the cycle of antler growth in the Virginia deer. In the female, the blood levels of calcium, phosphorus, and alkaline phosphatase were all stable over the time of the year when males were growing antlers. Serum levels of calcium and phosphorus remained stable in the males, but the level of alkaline phosphatase rose. The increase in alkaline phosphatase levels was related to antler growth by a premature cessation of antler growth by clamping off the base of the antler in order to constrict its blood flow. In these animals, the levels of alkaline phosphatase dropped. Within the antler, alkaline phosphatase accumulates within the preosseous tissue (Kuhlman *et al.*, 1963).

The control of antler growth by testosterone, coupled with transformation of the symphyseal cartilage into an interpubic ligament, and the resorption of the innominate bones under the influence of estrogen (Chapter 7), are two examples of hormonal elicitation of site-specific responses from the skeleton. A major subject for future investigation is the basis of this regional specificity within the skeleton.

2. Metabolic Regulation—Creeper (*Cp*) in the Chick

The microenvironment surrounding the skeletal cells (local shifts in pH, temperature, pO_2 and pCO_2 tensions, osmolarity, metabolic inhibitors, etc.) must modify the rates of physiological processes, and thus must alter the rates of growth. Very little is known about this microenvironment, so it is difficult to accurately determine the roles played by these factors and changes. Alterations in the extracellular matrices have been more amenable to analysis and have yielded some valuable information on the control of cartilage growth by positive feedback from matrix products to chondroblast and chondrocyte biosynthetic activity. These abnormalities in feedback and in the interactions between extracellular matrix products have been shown to retard growth as was described in the various achondroplasias (Chapter 7, Section II,D). The Creeper (*Cp*) mutant in the fowl provides another useful example of how metabolic inhibition can severely retard both growth and morphogenesis.

Creeper (*Cp*), a dominant mutation on the C chromosome of the common fowl, is a homozygous lethal. The embryos, which die at the end of the third day, exhibit retarded heads, limbs, and extra embryonic vasculature. The heterozygotes are viable but phocomelic (Landauer and Dunn, 1930; Landauer, 1932a). The long bones, especially those of the hindlimb and, in particular, the tibia, are severely shortened. The tibia is severely bent, although the fibula may be larger than normal. The dissimilarity in the action of *Cp* on the two bones is thought to reflect the faster growth rate of the tibia between seven and nine days of incubation and the fact that *Cp* acts by retarding the growth rate of appendicular skeletal elements. [For a fibula to develop at all, a critical volume of mesenchyme must be present in the early limb bud; there is competition for mesenchyme between the presumptive tibia and the presumptive fibula as they initiate their growth. Ninety-degree rotation of the presumptive fibular mesenchyme and reimplantation into the limb bud change the direction of growth of the fibula, allowing it to grow to the same length as the tibia (Hampé, 1958, 1960).]

Creeper is genetically linked with rose comb (Landauer, 1932b)—a useful linkage, since Creeper embryos are often difficult to distinguish from normal embryos at early ages. The changes in the growth rates of the long bones have been studied by Landauer and Dunn (1930), Landauer (1932a), Lerner (1936), Cock (1966), and Perlov (1968), all of whom attribute the action of Creeper to retardation of growth rate. The chondrocytes of the tibia fail to undergo hypertrophy on time. This in turn slows subperiosteal osteogenesis (Perlov, 1968).

The theory that growth retardation could lead to specific skeletal defects was tested by Fell and Landauer (1935) in what might well be the first *in vitro* test of a genetic mutant. They cultured limb buds from normal embryos of 3½ to 5 days of incubation in a medium that would restrict their growth (plasma clot–embryo extract, with dilution of the embryo extract). They found that the growth of the limb buds was retarded (perhaps not unexpectedly), and, although the smaller limb buds did produce chondrocytes, these did not undergo hypertrophy. Due to the lack of chondrocyte hypertrophy, osteogenesis was not initiated in the perichondria. Mandibular membrane bones cultured on the same medium ossified normally, even though their growth was retarded. Fell and Landauer argued that this mechanism, i.e., retardation of chondrocyte maturation, was sufficient to explain the defects produced by the *Cp* gene *in vivo*. Van Limborgh (1970, 1972) may be consulted for a more recent discussion of the differing controls that operate in the growth of cartilage and bone.

This *in vitro* study on growth retardation of skeletal rudiments from normal embryos stimulated research on the *Cp* mutant. There have probably been more studies utilizing either *in vitro* cultivation or transplantation to investigate the mode of action of *Cp* than for any other skeletal mutant. Despite this, *Cp* is one of the most poorly understood mutants—perhaps because of its being fundamentally a defect in growth, and because of our meager knowledge about how the many processes involved in control of growth are regulated and coordinated. Hamburger and Waugh (1940) showed that, when grown as transplants, limb buds from *Cp* embryos were less retarded than limb buds from normal embryos; accordingly, these workers have cautioned the use of transplantation as a method for analyzing growth and its control. Similar cautions have been voiced by Hall (1977b) in the context of chorioallantoic grafting.

Morphogenesis of *Cp* limb buds remains abnormal after transplantation into normal hosts (Hamburger, 1941; Rudnick, 1945), indicating that *Cp* acts directly on the skeleton. There is also a systemic effect of the Creeper gene on the whole embryo (Wallace *et al.*, 1969) that involves the potentiation of the action of other agents that influence skeletal development and growth. Creeper embryos show earlier and more pronounced rickets in response to vitamin D than normal embryos (Landauer, 1934, 1969a). Retardation of tibial growth in response to administration of sex hormones (testosterone, estradiol, dehydroepiandrosterone) is also more pronounced in Creeper embryos (Scopelliti, 1975). Discussions of the potentiation of agents influencing skeletal growth may be found in Hall (1972a,c, 1977b).

The Creeper embryo produces a growth inhibitor that can inhibit the growth, but not the morphogenesis, of tibiae from normal embryos when they are cultured in its presence (Wolff and Kieny, 1957, 1963; Kieny and Abbott, 1962). This provides a nice example of the independence of the control processes over morphogenesis and growth. If tibiae from 6–6½-day-old normal embryos are cultured in a medium containing embryo extract from 9-day-old Creeper embryos rather than extract from normal embryos, their growth is retarded by some 12% (Wolff and Kieny, 1963; Elmer, 1968a). Total protein in these explants is reduced because of deficient leucine metabolism (Elmer, 1968a). Coculture of normal and Creeper tibiae retards growth of the normal tibia in comparison to its growth when cocultured with a second normal tibiae (Elmer, 1968b). The tibiae of the Creeper seem, therefore, to be one source of the growth inhibitor, releasing it *in vitro* and perhaps synthesizing it as well.

Tibiae from normal embryos exhibit an accelerated accumulation of protein and an accelerated rate of growth between eight and nine days of incubation. This acceleration is slowed in tibiae from Creeper embryos (Elmer, 1968b; Perlov, 1968). Protein/μg DNA, hydroxyproline/μg DNA, and hydroxyproline/mg protein in tibiae from Creeper embryos have been examined by Dinner (1970). Protein/μg DNA and hydroxyproline/μg protein both decreased at the beginning of the eighth day relative to normal embryos, but the levels of hydroxyproline/μg DNA were normal. Dinner concluded that noncollagenous protein was reduced in Creeper. This comprises essentially all of the available data on the biochemistry of Creeper tibiae. No data are available on glycosaminoglycan metabolism or on collagen biosynthesis, although some histochemical work has been done on young embryos (Loewenthal, 1957). No ultrastructural data are available, and there are surprisingly few histological data other than the studies of chondrocyte hypertrophy by Perlov (1968).

The *Cp* strain in Japan is said to show precocious differentiation of chondrocytes coupled with decreased rates of proliferation of precursor cells (Kuroda and Shibuya, 1969; Shibuya *et al.*, 1972; Shibuya and Kuroda, 1973). However, I am unable to completely follow their evidence.

In sum, even though the mechanism of action of the Creeper gene is not completely understood, it obviously provides a good example of how growth can be controlled through regulation of metabolic processes.

3. The Growth of Long Bones

The tissues adjacent to the skeleton (its functional matrix) can either impose limits on or can accelerate skeletal growth. These adjacent tissues would include the fibrous periosteum, the tendons, the muscles, the vascular supply, and the innervation. They can control primary skeletal growth, as occurs when the tension in the periosteum of a long bones changes, or they can lead to compensatory skeletal growth, which, for example, occurs when the brain undergoes expansive growth. I should like to discuss the growth of embryonic long bones in the context of their functional matrices.

It has been well established that the increase in the length of long bones is due to the proliferation of precursor cells into chondroblasts in the cartilaginous growth plate and to the deposition of extracellular matrix by the chondroblasts and chondrocytes (Hales, 1727; Bhaskar *et al.*, 1954; Kember, 1960, 1971; Kember and Walker, 1971; Walker and Kember, 1972; Thorngren and Hansson, 1973a,b; Silbermann and Kadar, 1977). Growth in length ceases when the growth plate has used up its supply of precursor cells. Exceptions to this are the fish, amphibia, and many reptiles which possess no growth plate and in which growth is indefinite (indeterminate) (Goss, 1974). Haines and Mohuiddin (1960) have provided some evidence that the human epiphyseal plate incorporates cartilage derived from precursors within the marrow, and that these cells "replenish exhausted reserves of undifferentiated cells."

Rates of increase in bone length vary (Humphrey, 1861). The amount and the rate of growth of the proximal and distal epiphyses of long bones are often not the same. In man, 75% of the growth of the humerus takes place at the proximal growth plate, while 75% of the growth of the radius occurs at the distal growth plate. How these differential growth rates are established is not known, but it is generally thought that the proliferation of the precursor cells is controlled intrinsically, perhaps by specification of the length of the cell cycle early in fetal life (Dixon, 1970; Moss-Salentijn, 1974). However, the rates of growth at both epiphyses can be modified experimentally. While stapling the epiphysis to the metaphysis or irradiation of one growth plate stops its further growth, the opposite growth plate is stimulated to increased cell proliferation (Hall-Craggs, 1969; Dawson and Kember, 1974).

Recent studies on the interaction between the periosteum and the cartilaginous growth plate indicate that cell proliferation within the growth plate is damped by the tension exerted by the periosteum on the growth plate. Sĕbek *et al.* (1972) measured the strength of various portions of the periosteum of bovine long bones and found that periosteal strength was lowest in the diaphysis and increased toward the metaphyses. They postulated that sufficient pressure was exerted by the periosteum to influence proliferation within the growth plate. Severing the periosteum of the chicken radius across the bone's long axis, Crilly (1972) found that longitudinal growth of the bone was stimulated. Longitudinal splitting of the periosteum had little effect on bone growth. Crilly proposed that the periosteum placed the growth plate under tension, and that proliferation of the precursors of the growth plate was stimulated when this tension was released. The developing growth plate would normally be opposed by the resistance of the periosteum, whereby its growth is regulated (see also Hert, 1969). Kéry (1972) and Bex and Maltha (1976) have confirmed these results *in vivo,* and Harkness and Trotter (1975) have shown that sectioning the periosteum of subcutaneously transplanted rat tibiae similarly stimulates their growth over that of unsectioned tibiae.

Via stripping of the periosteum from the long bones, this feedback loop between periosteum and cartilaginous growth plate has been used to advantage in stimulating the bone growth of children suffering from poliomyelitis (Jenkins *et al.,* 1975). Moss (1972b) sees this feedback as a cycle. The fibrous periosteum adapts to the tension placed upon it as the growth plate elongates by initiating its own compensatory growth. This decreases the tension on the growth plate, slows proliferation of cartilage precursors, and allows the cycle to be repeated. Lutfi (1974), who studied tibiae from 4-day-old avian embryos throughout their development into 32-week-old-chicks, implicates this loop in subperiosteal osteogenesis. In this view, the periosteum is lengthened, and local mechanical stimuli enhance deposition of subperiosteal bone, enabling the shaft of the long bone to grow in width. Growth of the epiphysis in width arises from the zone where the periosteum attaches to the epiphysis, the so-called Ranvier's groove, or perichondrial or periosteal ring (Badi, 1972a; Shimomura *et al.,* 1973; Ogden *et al.,* 1975; Shapiro *et al.,* 1977). This is a site of fibrogenesis, chondrogenesis, and osteogenesis. Interstitial growth within the epiphysis also contributes to its growth in width (Hert, 1972).

The proliferative activity of the precursors within the growth plate is not intrinsically fixed; rather, it can respond to functional demands, to which it is subjected. What distinguishes these prechondrogenic cells is that they can continue to divide, and the skeletal rudiment can continue to grow, in the absence of these demands. The rate of division, and not the ability to divide, is controlled extrinsically. This is not true for other cartilages.

4. Growth of the Condylar Process

The contribution of the growth of the condylar process of the mammalian mandible to the morphogenesis and growth of the face has been a controversial topic for many years (for reviews, see Moss, 1962c, 1968d; Moffett, 1972; Moore and Lavelle, 1974; Enlow, 1975). Is the condylar cartilage a primary growth center for the mandible, or is its growth compensatory and adaptive?

As discussed in Chapter 6, Section IV,A, preferential differentiation of chondroblasts over osteoblasts on the condylar process requires movement and associated mechanical forces. A considerable body of evidence now indicates that the presence and continued growth of the condylar process are also dependent upon the presence of functional musculature, and, in particular, that the maintenance of a pool of proliferating precursor cells on the apex of the condylar process depends upon the presence of local mechanical stimuli.

The continued deposition of bone on the mandibular processes is also dependent on local mechanical stimuli. This can be illustrated by the development and growth of the angular and coronoid processes—mandibular processes on which secondary cartilage is a transient feature.

During the first six weeks of postnatal life in the rat, when growth of the angular process is rapid, the number of muscle fibers in the medial pterygoid muscle doubles (Rayne and Crawford, 1975). If either the medial pterygoid or the masseter muscle is extirpated, approximately half of the angular process is resorbed and not maintained. If both muscles are extirpated, the angular process undergoes complete resorption. Under conditions of congenital absence of these muscles, the angular process fails to form; however, if the muscles are congenitally hypertrophic, the angular process becomes unusually large. Thus, the angular process depends upon local muscle action for its development and for its maintenance, adjusting the balance between bone deposition and resorption in response to the local mechanical conditions.

Similar reductions in the size of the coronoid process follow removal of the masseter or temporal muscles (Pratt, 1943; Washburn, 1947; Moore, 1969; Soni and Mallory, 1974). Mechanoreceptors within the capsule of the temporomandibular joint play a role in coordinating the activity of these various muscles (Abe *et al.*, 1973).

The condylar process, which consists partially of membrane bone and partially of secondary cartilage, is similarly dependent upon local muscle action for its growth. This has been shown by Duterloo and Wolters (1971), who made use of the fact that the secondary cartilage of the angular process of the rat is not maintained beyond 17 days after birth. These workers condylectomized rats and transplanted the angular process into the position normally occupied by the condyle. A new condyle and temporomandibular joint formed, and the secondary cartilage of the transplanted angular process persisted in its new functional environment.

Soft diet or removal of the incisors reduces the muscular activity associated with mastication and slows the growth of the condylar process (Moore, 1965; Simon, 1977). If the condylar secondary cartilage is transplanted intracerebrally, chondrogenesis stops after about ten days (Ronning, 1966; Duterloo and Wolters, 1971), indicating that this cartilage lacks independent growth potential (Koski and Mäkinen, 1963). If cultured *in vitro* under conditions that allow epiphyseal or Meckelian cartilage to divide and grow, condylar secondary cartilage survives but exhibits neither cell division nor growth (Charlier and Petrovic, 1967).

Charlier and Petrovic continued these studies and produced a testable model for growth of condylar secondary cartilage in response to functional demand from the lateral pterygoid muscle. Charlier *et al.* (1969a) placed an orthodontic hyperpropulsive device onto the heads of rats, whereby the lower incisor was projected in front of the upper incisor for four weeks to increase the stress on the temporomandibular joint. (Normally, as the condyle grows up and back, the mandible moves down and forward.) After this four-week treatment, the size of the prechondroblast zone, as well as the thickness of the fibrous articular disc, were increased over those observed in control unstimulated rats (also see Lindsay, 1977). When a chin cup was employed to reduce the functional stimulus on the joint, the thicknesses of the prechondroblast and chondroblast zones were reduced within one week (Charlier *et al.*, 1969b). This explains the clinical value of chin cup therapy (Graber, 1977). It was proposed that the continuous proliferation of prechondroblasts on the condylar process was dependent upon mechanical stimuli; Petrovic (1970, 1972) has reviewed the early work of Charlier *et al.* on this model.

Petrovic has sought to relate this data to *in vivo* muscular activity. Hyperpropulsion was found to increase, and the chin cup to decrease, the number of sarcomeres in the lateral pterygoid muscle, as well as the amount of proliferation in the prechondroblast zone (Petrovic *et al.*, 1973). They proposed that contraction of the lateral pterygoid muscle controlled the rate of proliferation of prechondroblasts, and that this muscle also provided the common link in translating other stimuli such as hormones and tongue movement to the prechondroblast zone. Condylar cartilage growth is viewed as entirely adaptive and responsive to this muscular control. Petrovic and co-workers have extended this model to the coronoid and angular cartilages and their growth control via the temporal and masseter muscles, respectively.

Stutzman and Petrovic (1974) tested this model for condylar secondary cartilage by resecting the lateral pterygoid muscle of young growing rats, and by subsequently following the proliferation of prechondroblasts autoradiographically. Proliferation was slowed in these treated rats and, as noted in Chapter 6, Section IV,A, the prechondroblasts differentiated into osteoblasts and not into chondrocytes. Petrovic (1974) may be consulted for a review of this model and for a cybernetic model of growth of the entire mandible. This is among the most promising control systems presently available for investigating the nature of growth control through the functional matrix of a skeletal component.

Also deserving of mention is the study of Meikle (1973a,b), whose experiments led him to conclude that proliferation of prechondroblast was *independent* of functional demand. He transplanted the mandibular joint intracerebrally and three weeks later introduced [^{3}H]thymidine to label any dividing cells. Because cell proliferation was still active, Meikle concluded that it therefore must be independent of function. It is difficult to reconcile this study with Petrovic's experiments. Comparative data on cell proliferation *in situ* and in the transplants might indicate that cell proliferation within the transplants was indeed slowed. Comparative data on transplantation or *in vitro* cultivation of whole temporomandibular joints versus condylar cartilage, coupled with quantitative analysis of mitotic activity and comparison with controls, might help to resolve this rather important question. Meikle did find that the prechondroblasts of the transplants differentiated into osteoblasts rather than into chondrocytes, as Petrovic's group had also observed.

From the condylar cartilage of the temporomandibular joint, we can now turn to a discussion of the contributions of the joints to the initiation of skeletal growth.

D. Joints as Centers of Skeletal Growth

1. The Initial Formation of Joints

Whereas joint formation does not depend upon functional demand or extrinsic stimulation, the maintenance of joints does. The mesenchyme of a presumptive joint region can be placed under *in vitro* conditions, in which it will differentiate into a morphologically normal joint with articulating surfaces (Fell and Canti, 1934a,b; Glasstone, 1968, 1971, 1973; Hall, 1968e; Tyler and Hall, 1977). Either prolonged cultivation *in vitro,* transplantation to an ectopic site, or *in ovo* paralysis leads to fusion of tissues across preexisting joints and to fusions of joints with other connective or skeletal tissues (Pellegrini, 1934; Lelkes, 1958; Barnett *et al.,* 1961; Drachman and Coulombre, 1962; Drachman, 1964; Chalmers, 1965; Murray and Smiles, 1965; Drachman and Sokoloff, 1966; Hall, 1967b, 1968d,e, 1972b; Murray and Drachman, 1969; Mitrovic, 1971, 1972, 1974). These studies clearly establish that the *maintenance* of a functioning joint requires the presence of extrinsic mechanical factors. Are the initiation and continuance of the growth of a joint also dependent upon extrinsic control, i.e., is the joint growth compensatory, or is it controlled intrinsically, i.e., are joints primary centers of expansive growth?

I have already mentioned that the temporomandibular joint can form in the absence of extrinsic stimuli (Glasstone, 1971), and that the maintenance of the population of proliferating progenitor cells on the condylar process requires the presence of mechanical stimuli normally associated with joint function. The condylar process and its secondary cartilage do not constitute a primary growth center for the mandible, but rather, comprise a center of adaptive and compensatory growth (Section C,4). Two other types of joints will be discussed in this section, viz., cranial synchondroses and cranial sutures.

2. Cranial Synchondroses

The synchondroses are the joints within the skeleton whose bony surfaces are united by an intervening cartilage. Organizationally, they somewhat resemble two epiphyseal growth plates back to back, but with a common central zone of resting cells. Because of the presence of this cartilage, synchondroses are exquisitely adapted to resist pressure and might be expected to regulate their growth in response to that pressure.

There are several synchondroses in the cranium of the developing mammal. At birth, three are present in man, namely, the midsphenoidal, the septoethmoidal, and the sphenooccipital. That which has received the most attention is the sphenooccipital synchondrosis at the base of the skull. Since it is the last of the synchondroses to fuse, it has been argued that the role of the sphenooccipital in cranial growth is proportionately bigger than those of the earlier fusing synchondroses. Several excellent reviews are available on various aspects of the synchondroses and of their role in skull growth (Felts, 1961; Enlow, 1968b, 1975; van Limborgh, 1972; Moore and Lavelle, 1974; Koski, 1975; Moss, 1975).

Do the synchondroses direct the growth of the cranium or do they respond adaptively to cranial growth? At least three possible answers to this question have been argued by different workers at different times and using different experimental techniques.

The classical view is that the cartilaginous synchondroses and, indeed, cartilage in general (e.g., nasal and condylar cartilage) serve as the pacemakers for the growth of the bony portion of the skull (Scott, 1954). Synchondroses are viewed as the growth centers of the skull and as equivalent in their growth potential to epiphyseal growth cartilages. Evidence for this view is the inherent growth potential demonstrated by synchondroses when isolated either *in vitro* or as grafts (Petrovic and Charlier, 1967; Baume, 1968, 1970; Servoss, 1973). In his autoradiographic study, Servoss followed progenitor cells from the resting zone through the proliferative zone and observed some mature into hypertrophic chondrocytes.

An alternate view, compatible with some of the above *in vitro* studies, is that the growth potential of the synchondrosis is not equivalent to that of the epiphyseal growth plate. Rather, the synchondrosis is considered to have some limited potential for growth—less than the epiphysis, but more than the condylar secondary cartilage, which is totally dependent upon functional stimulation for its growth. The major difference between these two views is that the second does not envisage synchondroses as primary growth centers pushing cranial bones apart. Koski and Rönning (1970) transplanted synchondroseal cartilage from the presphenoid–basisphenoid synchondrosis of 5- to 7-day-old rats intracerebrally for periods of between 30 and 60 days and compared their growth with transplanted condylar and epiphyseal cartilage. Their growth potential was intermediate between those of the condylar and epiphyseal cartilages. When transplanted subcutaneously, the synchondroseas survived but failed to grow (Koski and Rönning, 1969); however, Harkness (1976) reports that some growth does occur at such sites, albeit less than is normally seen *in vivo*. Part of the difference between the growth of synchondroseal and condylar cartilages lies in the fact that, while growth of condylar secondary cartilage is appositional, growth of synchondroseal cartilage is interstitial (Jolly and Moore, 1975; Kvinnsland and Kvinnsland, 1975; Brewer *et al.*, 1977).

These transplant studies support the concept that synchondroses have some inherent potential for growth in the absence of functional stimuli. But is the amount of growth that is expressed after transplantation or upon cultivation *in vitro* the best criterion for evaluating the inherent growth potential of cartilage? Durkin *et al.* (1973) have addressed themselves to this question with respect to differentiating adaptive from growth plate-like cartilage. They maintain that it is possible to use histological criteria—the arrangement of the cells, the pattern of calcification, the mechanism of erosion, and whether or not subchondral bone is present. Brown and Heeley (1974) used these criteria in their study of the rat septopresphenoidal joint, which they maintained was a primary growth center, and the septoethmoidal joint, which they thought to be partially adaptive and partially growth plate-like. Kenrad and Vilmann (1977) have confirmed the usefulness of these histological criteria and also have utilized the differential distribution of alkaline and acid phosphatases to separate synchondroseal from condylar and epiphyseal cartilage. Kvinnsland (1974) administered repeated doses of papain to growing rats to disrupt the deposition of cartilaginous extracellular matrix. The reduced growth of the entire skull was attributed to the impaired growth of the synchondroseal cartilages. Agreement on the validity of test criteria will have to be reached before the contribution of synchondroseal cartilages to skull growth can finally be elaborated.

The third view of synchondroseal cartilages argues that their growth is entirely adaptive and occurs as a response to the functional matrix in which they develop. Response to the mass of neural tissue present and to its growth is seen as especially important (Moss, 1962c; Moss *et al.*, 1972). While it does appear that some cartilaginous growth is adaptive, it is equally clear that it is not entirely adaptive.

3. Cranial Sutures

Sutures are joints within the skeleton where one bone articulates with another through intervening fibrous connective tissue. Because of the presence of the connective tissue, sutures are especially well adapted to resist tension such as is found in the growing skull. The basic structure of sutures has been well documented by Pritchard *et al.* (1956), Moss (1957a), and Koskinen *et al.* (1976).

The classical view of skull growth contended that osteogenesis at the sutures made no contribution to cranial growth (Brash, 1934a,b), and that the growth rate of the skull was entirely determined by the growth of the cranial cartilages (Scott, 1954; Hoyte, 1966; also see Section 2). That deposition and resorption of bone do occur at the sutural margins has been demonstrated utilizing alizarin red and fluorescent calcium-binding markers (Isotupa *et al.*, 1965; Hoyte, 1971;

Cleall, 1974). Growth on opposite sides of a single suture may differ in rate, duration, direction, and with time (de Angelis, 1969; Moore and Lavelle, 1974; Smith and McKeown, 1974).

Do the deposition and resorption of bone and connective tissue at the suture provide a primary growth force for the dermocranium, or is such growth secondary and adaptive? Experimental excision, transplantation, and fusion of cranial sutures indicate that sutural growth is secondary and adaptive. For detailed discussions of various aspects of this literature on skull growth, the reader is referred to Hoyte (1966), van Limborgh (1970, 1972), Moss (1972b,c), Moss *et al.* (1972), Persson (1973), Storey (1973), Moore and Lavelle (1974), and Enlow (1975).

Moss (1954, 1957a,b, 1959), extirpating sutures from the cranium of the rat, found that bone was deposited over the original suture line. He therefore argued that the position of the suture was not intrinsically fixed. Growth of these skulls continued after excision of the sutures, indicating that the sutures were not involved in generating the forces necessary in separating the cranial bones from one another. When sutures are extirpated or repositioned, both the growth and the morphology of the suture depend upon the functional characteristics of the new site (Smith and McKeown, 1974).

Isolated sutures lack inherent growth potential. Stutzmann and Petrovic (1970) cultured the midpalatal sutures of young (5-day-old) rapidly growing rats for 18 days without observing any growth. [See Moore and Lavelle (1974) for additional examples and for an excellent analysis of the literature.]

The fusion of sutures creates growth disturbances in the other components of the skull (Alhopuro *et al.*, 1973a,b; Ritsila *et al.*, 1973). Premature fusion is thus to be avoided, although in some instances premature fusion has evolved as a mechanism for strengthening the face during rapid growth (Herring, 1974). Growth of sutures is therefore secondary and adaptive.

Van Limborgh (1970, 1972) has provided incisive reviews of both dermocranial and chondrocranial growth and has developed the thesis that bone growth is primarily an adaptation to local epigenetic factors, whereas cartilage growth is an expression of inherent growth potential. The growth of both sutures and synchondroses provides strong support for his view. Moss (1972b) and Moss *et al.* (1972) have discussed the adaptiveness of sutural growth in the context of the skull and its associated structures as a functional matrix. They emphasize analysis of the growth of the functional unit in order to understand the growth of the individual components. For example, as the brain expands, the calvarium is displaced outwardly, subjecting the sutures to tension. The sutures respond adaptively by depositing bone to counteract the tension.

The adaptive response of sutures to tension can be used to experimentally initiate bone deposition as a way of healing bone defects. A screw device used to place the palatomaxillary suture under tension induces sufficient bone deposition on the palatine bones such that they will approximate one another and obliterate cleft palates (Latham *et al.*, 1973). Elder and Tuenge (1974) placed facial sutures under tension in order to initiate resorption of bone to reposition the maxilla. The application of tension for appropriate periods of time could be used to establish cycles of deposition and resorption (Hinrichsen and Storey, 1968; Murray and Cleall, 1971; Ten Cate *et al.*, 1977).

A complication not often considered in the study of sutural growth is the presence of cartilage within sutures. Such secondary cartilage is a normal component of many sutures (Pritchard *et al.*, 1956; Moss, 1958; Hall, 1970; Stutzmann and Petrovic, 1970; Aaron, 1973); it appears when sutures are repositioned (Smith and McKeown, 1974) and also in cleft lip and cleft palate (Curtin and Pruzansky, 1971; Pruzansky, 1971; Friede, 1973). The contribution made by such cartilages to the growth of sutures in growing mammals has not been assessed, nor has the possibility that growth of cartilage may explain why apparently similar sutures exhibit a diversity of responses when subjected to identical treatments. It is the heterogeneity underlying apparently similar skeletal tissues that makes the developing skeleton such a fascinating and profitable organ system for study.

Appendix

The hours or days of incubation of embryonic chick development corresponding to H.H. stages 1 to 38 are listed along with the major events of skeletal development. For a complete description of all H.H. stages and developmental events, the reader should consult Hamburger and Hamilton (1951) and Freeman and Vince (1974). Table VIII may be consulted for details on limb bud development.

H.H. Stage	Age	Developmental event
1	6 hr	Major embryonic morphogenetic movements
2	7 hr	Morphogenetic movements
3	12.5 hr	Morphogenetic movements
4	18.5 hr	Primitive streak present
5	20 hr	Initiation of notochord development
6	1 day	Neurulation
7	25–26 hr	Neurulation; somite 1 appears
8	28 hr	Four somites present
9	31 hr	Seven somites present; heart formation; eye formation; cells leaving neural crest
10	35 hr	Ten somites present; brain vesicles
11	42 hr	13 somites present
12	46 hr	16 somites present; sclerotomal cells migrating (see Fig. 11)
13	2 day	19 somites present
14	51 hr	22 somites present; initiation of visceral arches; sclerotomal cells migrating; neural crest cells reach optic cup
15	53 hr	25–26 somites present; neural crest cells in mandibular arch
16	55 hr	26–28 somites present; initiation of wing bud

continued

H.H. Stage	Age	Developmental event
17	60 hr	30 somites present; initiation of hindlimb bud (see Table VIII)
18	66 hr	33–34 somites present; allantois appears
19	3 days	38 somites present; periocular mesenchyme determined for chondrogenesis
20	3 days	40–42 somites present; eye pigmentation begins
21	3½ days	44 somites present
22	3½ days	Somites extend to tip of tail
23,24	4 days	First extracellular matrix around sclerotomal vertebral primordia; membrane bone primordia now independent of epithelia
25	4½ days	Elbow and knee joints distinct
26	5 days	Demarcation of digits; embryonic movement starts; Meckel's cartilage begins to chondrify
27	5¼ days	Chondrogenesis begins in limb buds
28	5¾ days	Digits distinct; beak visible
29	6–6½ days	AER lost ability to initiate outgrowth; tibial perichondrium becoming a periosteum
30	6½–7 days	Egg tooth, feather germs appear; two scleral papillae present
31	7–7½ days	Six scleral papillae, first chondroblasts in presumptive vertebrae; first bone in mandible and quadratojugal
32	7½ days	6–8 scleral papillae
33	7½–8 days	13 scleral papillae; scleral cartilage differentiating
34	8 days	
35	8½–9 days	
36	10 days	Determination of cells for secondary chondrogenesis on membrane bones
37	11 days	Secondary cartilage present on membrane bones
38	12 days	Ossification of scleral ossicles begins

Bibliography

Aaron, J. E. (1973). *Calcif. Tissue Res.* **12,** 259–279.

Aaron, J.E., and Pautard, F. G. E. (1973). *In* "The Cell Cycle in Development and Differentiation" (M. Balls and F. S. Billett, eds.), pp. 325–330. Cambridge Univ. Press, London and New York.

Abbott, J., and Holtzer, H. (1966). *J. Cell Biol.* **28,** 473–487.

Abbott, J., and Holtzer, H. (1968). *Proc. Natl. Acad. Sci. U.S.A.* **59,** 1144–1151.

Abbott, J., Schiltz, J., Dienstman, S.R., and Holtzer, H. (1974). *Proc. Natl. Acad. Sci. U.S.A.* **71,** 1506–1510.

Abbott, U. K. (1967). *In* "Methods in Developmental Biology" (F. H. Wilt and N. K. Wessels, eds.), pp. 13–52. Crowell-Collier, New York.

Abbott, U. K. (1975). "Genetics Lectures," Vol. IV, pp. 69–84. Oregon State Univ. Press, Corvallis, Oregon.

Abbott, U. K., Taylor, L. W., and Ablanalp, H. (1960). *J. Hered.* **51,** 195–202.

Abdin, M., and Friedenstein, A. Y. (1972). *Clin. Orthop. Relat. Res.* **82,** 182–194.

Abe, K., Takata, M., and Kawamura, Y. (1973). *Arch. Oral Biol.* **18,** 301–304.

Abrahamsohn, P. A., Lash, J. W., Kosher, R. A., and Minor, R. R. (1975). *J. Exp. Zool.* **194,** 511–518.

Adelstein, R. S., and Conti, M. A. (1971). *Cold Spring Harbor Symp. Quant. Biol.* **37,** 599–606.

Adelstein, R. S., Pollard, T. D., and Kuehl, W. M. (1971). *Proc. Natl. Acad. Sci. U.S.A.* **68,** 2703–2707.

Adelstein, R. S., Conti, M. A., Johnson, G. S., Pastan, I., and Pollard, T. D. (1972). *Proc. Natl. Acad. Sci. U.S.A.* **69,** 3693–3697.

Agnish, N. D., and Kochhar, D. M. (1977). *Dev. Biol.* **56,** 174–183.

Ahmed, Y. Y. (1966). *Anat. Rec.* **155,** 133–138.

Aho, A. J. (1966). *Acta Pathol. Microbiol. Scand.* Suppl. **184,** 1–95.

Ahrens, P. B., Solursh, M., and Reiter, R. S. (1977a). *Devel. Biol.* **60,** 69–82.

Ahrens, P. B., Solursh, M. and Meier, S. (1977b). *J. Exp. Zool.* **202,** 375–388.

Åkeson, W. H., Woo, S.L.Y., Amiel, D., and Matthews, J. V. (1974). *Connect. Tissue Res.* **2,** 315–324.

Alcock, N. W. (1972). *Clin. Orthop. Relat. Res.* **86,** 287–311.

Alhopuro, S., Ranta, R., and Ritsilä, V. (1973a). *Proc. Finn. Dent. Soc.* **69,** 166–167.
Alhopuro, S., Ranta, R., and Ritsilä, V. (1973b). *Proc. Finn. Dent. Soc.* **69,** 263–265.
Ali, S. Y., and Evans, L. (1973a). *Fed. Proc., Fed. Am. Soc. Exp. Biol.* **32,** 1494–1498.
Ali, S. Y., and Evans, L. (1973b). *Biochem. J.* **134,** 647–650.
Ali, S. Y., Sajdera, S. W., and Anderson, H. C. (1970). *Proc. Natl. Acad. Sci. U.S.A.* **67,** 1513–1520.
Ali, S. Y., Elves, M. W., and Leaback, D. H., eds. (1974). "Normal and Osteoarthritic Articular Cartilage." Institute of Orthopaedics, University of London, London.
Allenspach, A. L., and Babiarz, B. S. (1975). *J. Ultrastruct. Res.* **51,** 348–353.
Allyn, P. G., and Minkin, C. (1976). *J. Dent. Res.* **55,** B219.
Ambrosi, G., Camosso, M. E., and Roncali, L. (1973). *Boll. Soc. Ital. Biol. Sper.* **49,** 135–140.
Amprino, R. (1955a). *Acta Anat.* **24,** 121–163.
Amprino, R. (1955b). *Experientia* **11,** 19–21.
Amprino, R. (1955c). *Experientia* **11,** 65–67.
Amprino, R. (1956). *Bone Struct. Metab., Ciba Found. Symp. 1955,* pp. 89–100.
Amprino, R. (1965). *In* "Organogenesis" (R. L. de Haan and H. Ursprung, eds.), pp. 255–282. Holt, New York.
Amprino, R. (1974). *Rech. Biol. Contemp.,* pp. 245–250.
Amprino, R. (1975a). *Nova Acta Leopold.* [N. S.] **41,** 235–270.
Amprino, R. (1975b). *Arch. Anat., Histol., Embryol.* **58,** 29–40.
Amprino, R. (1976). *Arch. Biol.* **87,** 1–41.
Amprino, R. (1977a). *Acta Anat.* **98,** 295–312.
Amprino, R. (1977b). *Acta Embryol. Exp.* **1,** 51–70.
Amprino, R., and Aiello-Malmberg, P. (1971). *Acta Anat.* **80,** 183–203.
Amprino, R., and Ambrosi, A. (1973). *Arch. Biol.* **84,** 35–86.
Amprino, R., and Amprino-Bonetti, D. (1967a). *Bull. Assoc. Anat.* **52,** 134–140.
Amprino, R., and Amprino-Bonetti, D. (1967b). *Nature (London)* **214,** 826–827.
Amprino, R., and Bonetti, D. A. (1964). *Nature (London)* **204,** 298.
Amprino, R., and Camosso, M. E. (1959). *Arch. Anat. Microsc. Morphol. Exp.* **48,** 261–306.
Amprino, R., and Camosso, M. E. (1963). *Acta Embryol. Morphol. Exp.* **6,** 241–259.
Amsel, S., and Dell, E. S. (1971). *Int. J. Radiat. Biol.* **20,** 119–127.
Amsel, S., and Dell, E. S. (1972). *Blood* **39,** 267–273.
Amsel, S., Maniatis, A., Tavassoli, M., and Crosby, W. H. (1969). *Anat. Rec.* **164,** 101–112.
Anderson, C. E., and Parker, J. (1966). *J. Bone Joint Surg., Am. Vol.* **48,** 899–914.
Anderson, H. C. (1966). *Electron Microsc., Proc. Int. Congr., 6th, 1966* Vol. II, pp. 573–574.
Anderson, H. C. (1967). *J. Cell Biol.* **35,** 81–102.
Anderson, H. C. (1969). *J. Cell Biol.* **41,** 59–72.
Anderson, H. C. (1976a). *Clin. Orthop. Relat. Res.* **119,** 211–224.
Anderson, H. C. (1976b). *Int. Rev. Connect. Tissue Res.* **7,** 251–322.
Anderson, H. C., and Coulter, P. R. (1967). *J. Cell Biol.* **30,** 165–177.
Anderson, H. C., and Reynolds, J. J. (1973). *Dev. Biol.* **34,** 211–227.
Anderson, H. C., Marker, P. C., and Fogh, J. (1964). *Am. J. Pathol.* **54,** 507–519.
Anderson, H. C., Matsuzawa, T., Sajdera, S. W., and Ali, S. Y. (1970). *Trans. N.Y. Acad. Sci.* [2] **32,** 619–630.
Anderson, H. C., Cecil, R., and Sajdera, S. W. (1975). *Am. J. Pathol.* **79,** 237–254.
Andrew, W. (1971). "The Anatomy of Aging in Man and Animals." Heinemann, London.
Arrambide, E., Patrone, M. G., and Calcagno, M. (1968). *Experientia* **24,** 896.
Arsenis, C., and Huang, S. M. (1977). *Differentiation* **8,** 183–190.
Ash, P., and Francis, M. J. O. (1975). *J. Endocrinol.* **66,** 71–78.
Ashley, D. J. B. (1970). *J. Pathol.* **102,** 186–188.

Asling, C. W. (1976). *J. Dent. Res.* **55,** B111.

Athenstaedt, H. (1969). *Z. Zellforsch. Mikrosk. Anat.* **97,** 537–548.

Athenstaedt, H. (1970). *Nature (London)* **228,** 830–834.

Athenstaedt, H. (1974). *Ann. N.Y. Acad. Sci.* **238,** 68–94.

Attinger, E., and Parakkal, P. (1977). "The Relevance of Biomedical Engineering to Dentistry." DHEW Publ. (NIH) 77–1198. DHEW, Washington, D.C.

Avery, G., Chow, M., and Holtzer, H. (1956). *J. Exp. Zool.* **132,** 409–426.

Badi, M. H. (1972a). *J. Anat.* **111,** 201–210.

Badi, M. H. (1972b). *J. Anat.* **112,** 415–422.

Baker, B. E., Becker, R. O., and Spadaro, J. (1974a). *Clin. Orthop. Relat. Res.* **102,** 251-267.

Baker, B. E., Spadaro, J. A., Marino, A. W., and Becker, R. O. (1974b). *Ann. N.Y. Acad. Sci.* **238,** 491–499.

Balcuns, A., Gasselling, M. T., and Saunders, J. W., Jr. (1970). *Am. Zool.* **10,** 323.

Balduini, C., Brovelli, A., deLuca, G., Galligani, L., and Castellani, A. A. (1973). *Biochem. J.* **133,** 243–254.

Balinsky, B. I. (1974). *J. Exp. Zool.* **188,** 195–202.

Ballard, K. J., and Holt, S. J. (1968). *J. Cell Sci.* **3,** 245–262.

Banks, W. J. (1974). *Calcif. Tissue Res.* **14,** 257–274.

Banks, W. J., Epling, G. P., Kainer, R. A., and Davis, R. W. (1968a). *Anat. Rec.* **162,** 387–397.

Banks, W. J., Epling, G. P., Kainer, R. A., and Davis, R. W. (1968b). *Anat. Rec.* **162,** 399–405.

Barasa, A. (1960). *Riv. Biol.* **52,** 257–292.

Barasa, A. (1962). *Atti Soc. Ital. Sci. Vet.* **16,** 416–420.

Baratz, R. S. (1974). *J. Dent. Res.* **53,** 77.

Bard, D. R., Dickens, M. J., Smith, A. U., and Zarek, J. M. (1972). *Nature (London)* **236,** 314–315.

Bard, J., and Wright, M. O. (1974). *J. Cell. Physiol.* **84,** 141–146.

Bargman, G. J., Mackler, B., and Shepard, T. H. (1972). *Arch. Biochem. Biophys.* **150,** 137–146.

Barnes, D. W. H., Loutit, J. F., and Sansom, J. M. (1975). *Proc. R. Soc. London, Ser. (B)* **188,** 501-505.

Barnett, C. H., Davies, D. V., and MacConaill, M. A. (1961). "Synovial Joints. Their Structure and Mechanics." Thomas, Springfield, Illinois.

Barnicot, N. A. (1941). *Am. J. Anat.* **68,** 497–531.

Baron, R. (1972). Thèse Doctorat Troisième Cycle. Science Odontol., pp. 1–239. Université de Paris.

Baron, R. (1973). *J. Biol. Buccale* **1,** 151–170.

Barratt, M. E. J. (1973). *J. Cell Sci.* **13,** 205–220.

Bassett, C. A. L. (1962). *J. Bone Joint Surg., Am. Vol.* **44,** 1217–1244.

Bassett, C. A. L. (1972a). *Clin. Orthop. Relat. Res.* **87,** 49–59.

Bassett, C. A. L. (1972b). *Acta Morphol. Neer. Scand.* **10,** 71–86.

Bassett, C. A. L., and Herrmann, L. (1961). *Nature (London)* **190,** 460–461.

Bassett, C. A. L., and Pawluk, R. J. (1972). *Science* **178,** 982–983.

Baume, L. J. (1961). *Science* **134,** 53–54.

Baume, L. J. (1962a). *Science* **138,** 904–905.

Baume, L. J. (1962b). *Trans. Eur. Orthod. Soc.* pp. 1–11.

Baume, L. J. (1968). *Int. Dent. J.* **18,** 489–513.

Baume, L. J. (1970). *Am. J. Orthod.* **58,** 537–551.

Baylink, D., Wergedal, J., and Thompson, E. (1972). *J. Histochem. Cytochem.* **20,** 279–292.

Bazin, S., Girard, H., and Delaunay, A. (1962). *Ann. Inst. Pasteur, Paris* **102,** 346–352.

Becker, R. O. (1972). *Clin. Orthop. Relat. Res.* **83,** 255–262.

Bélanger, L. F. (1954). *Can. J. Biochem. Physiol.* **32,** 161–169.

Bélanger, L. F. (1969). *Calcif. Tissue Res.* **4,** 1–12.
Bélanger, L. F., Robichon, J., Migicovsky, R. B., Copp, D. H., and Vincent, J. (1963). *Publ. Am. Assoc. Adv. Sci.* **75,** 531–556.
Beneke, G., Endres, O., Becker, H., and Kulka, R. (1966). *Virchows Arch. Pathol. Anat. Physiol.* **341,** 365–380.
Bennett, A. F., and Dalzell, B. (1973). *Evolution* **27,** 170–174.
Benoit, J. A. A. (1960a). *J. Embryol. Exp. Morphol.* **8,** 33–38.
Benoit, J. A. A. (1960b). *J. Embryol. Exp. Morphol.* **8,** 39–46.
Benoit, J. A. A. (1963). *Arch. Anat. Microsc. Morphol. Exp.* **52,** 573–590.
Benoit, J. A. A., and Schowing, J. (1970). *In* "Tissue Interactions during Organogenesis" (E. Wolff, ed.), pp. 105–130. Gordon & Breach, New York.
Bentley, G., And Greer, R. B. (1970). *J. Bone Joint Surg., Br. Vol.* **52,** 571–577.
Benveniste, K., Wilczek, J., and Stern, R. (1973). *Nature (London)* **246,** 303–305.
Benya, P. D., Padilla, S. R., and Nimni, M. E. (1977). *Biochemistry* **16,** 865–872.
Beresford, W. A. (1970). *Acta Orthop. Scand.* **41,** 134–149.
Beresford, W. A. (1975a). *J. Dent. Res.* **54,** 417.
Beresford, W. A. (1975b). *Am. J. Orthod.* **68,** 189–195.
Beresford, W. A., and Clayton, S. P. (1977). *J. Anat.* **123,** 297–312.
Beresford, W. A., and Hancox, N. M. (1965). *J. Anat.* **99,** 407–408.
Beresford, W. A., and Hancox, N. M. (1967). *Acta Anat.* **66,** 78–117.
Bergsma, D., ed. (1975). "Morphogenesis and Malformation of Face and Brain," Birth Defects Orig. Artic. Ser., Vol. 11, No. 7. Alan R. Liss, Inc., New York.
Bergsma, D., and Lenz, W., eds. (1977). "Morphogenesis and Malformation of the Limb," Birth Defects Orig. Artic. Ser., Vol. 13, No. 1. Alan R. Liss, Inc., New York.
Berkovitz, B. K. B., and Moore, M. H. (1974). *Arch. Oral Biol.* **19,** 1111–1120.
Bernstein, P., and Crelin, E. S. (1967). *Anat. Rec.* **157,** 517–526.
Bevelander, G., and Johnson, P. L. (1950). *Anat. Rec.* **108,** 1–21.
Bex, J. H. M., and Maltha, J. C. (1976). *J. Dent. Res.* **55,** D163.
Bhaskar, S. N., Weinmann, J. P., and Schour, I. (1954). *Anat. Rec.* **119,** 231–245.
Bhatnagar, R. S., and Prockop, D. J. (1966). *Biochim. Biophys. Acta* **130,** 382–392.
Bhatnagar, R. S., and Rapaka, S. S. R. (1971). *Nature (London), New Biol.* **234,** 92–93.
Bierly, J. A., Sottosanti, J. S., Costley, J. M., and Cherrick, H. M. (1975). *J. Periodontol.* **46,** 277–283.
Biltz, R. M., and Pellegrino, E. O. (1969). *J. Bone Joint Surg., Am. Vol.* **51,** 456–466.
Binderman, I., Duksin, D., Harell, A., Katzir (Katchalski), E., and Sachs, L. (1974). *J. Cell Biol.* **61,** 427–439.
Bingham, P. J., Barzell, I. A., and Owen, M. J. (1969). *J. Endocrinol.* **45,** 387–400.
Bischoff, R. (1971). *Exp. Cell Res.* **66,** 224–236.
Blumenkrantz, N., and Prockop, D. J. (1970). *Biochim. Biophys. Acta* **208,** 461–466.
Bollet, A. J. (1969). *Arthritis Rheum.* **12,** 152–163.
Bonner, J. J. (1974). *J. Dent. Res.* **53,** 1313–1316.
Bonner, P. H. (1975). *Dev. Biol.* **47,** 222–227.
Bonucci, E., and Gherardi, G. (1975). *Cell Tissue Res.* **163,** 81–98.
Bonucci, E., and Nardi, F. (1972). *Virchows Arch. A* **357,** 199–212.
Bonucci, E., Delmarco, A., Nicoletti, B., Petrinelli, P., and Pozzi, L. (1976). *Growth* **40,** 241–252.
Bonucci, E., Gherardi, G., Delmarco, A., Nicoletti, B., and Petrinelli, P. (1977). *J. Submicrosc. Cytol.* **9,** 299–306.
Bordier, P., Hioco, D., Rouquier, M., Hepner, G. W., and Thompson, G. R. (1969). *Calcif. Tissue Res.* **4,** 78–83.
Bornstein, P. (1974). *Ann. Rev. Biochem.* **43,** 567–604.

Bourne, G. H., ed. (1971). "The Biochemistry and Physiology of Bone," 2nd ed., Vol. 3. Academic Press, New York.
Bourne, G. H., ed. (1972a). "The Biochemistry and Physiology of Bone," 2nd ed., Vol. 1. Academic Press, New York.
Bourne, G. H., ed. (1972b). "The Biochemistry and Physiology of Bone," 2nd ed., Vol. 2. Academic Press, New York.
Bourne, G. H., ed. (1976). "The Biochemistry and Physiology of Bone," 2nd ed., Vol. 4. Academic Press, New York.
Bourret, L. A., and Rodan, G. A. (1976a). *J. Cell. Physiol.* **88,** 353–362.
Bourret, L. A., and Rodan, G. A. (1976b). *Calcif. Tissue Res.* **21,** Suppl., 431–436.
Bouvet, J. (1974). *Arch. Anat. Microsc. Morphol. Exp.* **63,** 79–96.
Bouvet, J. (1976). *Cell Tissue Res.* **170,** 367–382.
Boyde, A., Jones, S. J., Binderman, I., and Harell, A. (1976). *Cell Tissue Res.* **166,** 65–70.
Brachet, A. (1893). *Int. Monatsschr. Anat. Physiol.* **10,** 391–417.
Brandt, K. D. (1974). *Biochem. J.* **143,** 475–478.
Brash, J. C. (1934a). *Edinburgh Med. J.* **41,** 305–319.
Brash, J. C. (1934b). *Edinburgh Med. J.* **41,** 363–386.
Brehe, J. E., and Fleming, W. R. (1976). *J. Comp. Physiol. B* **110,** 159–170.
Brem, H., and Folkman, J. (1975). *J. Exp. Med.* **141,** 427–439.
Brewer, A. K., Johnson, D. R., and Moore, W. J. (1977). *J. Embryol. Exp. Morphol.* **39,** 59–70.
Bridges, J. B. (1959). *Int. Rev. Cytol.* **8,** 253–278.
Brigham, G., Scaletta, L., and Johnston, L., Jr. (1977). *J. Dent. Res.* **56,** A50.
Brighton, C. T., and Friendenberg, Z. B. (1974). *Ann. N.Y. Acad. Sci.* **238,** 314–320.
Brighton, C. T., and Heppenstall, R. B. (1971). *J. Bone Joint Surg., Am. Vol.* **53,** 719–728.
Brighton, C. T., and Hunt, R. M. (1974). *Clin. Orthop. Relat. Res.* **100,** 406–416.
Brighton, C. T., and Krebs, A. G. (1972a). *J. Bone Joint Surg., Am. Vol.* **54,** 323–332.
Brighton, C. T., and Krebs, A. G. (1972b). *Surg., Gynecol. Obstet.* **135,** 379–385.
Brighton, C. T., Sugioka, Y., and Hunt, R. M. (1973). *J. Bone Joint Surg., Am. Vol.* **55,** 771–784.
Brighton, C. T., Lane, J. M., and Koh, J. K. (1974). *Arthritis Rheum.* **17,** 245–252.
Brighton, C. T., Adler, S., Black, J., Itada, N., and Friedenberg, Z. B. (1975). *Clin. Orthop. Relat. Res.* **107,** 277–282.
Brighton, C. T., Cronkey, J. E., and Osterman, A. L. (1976). *J. Bone Joint Surg., Am. Vol.* **58,** 971–977.
Brown, T. A., and Heeley, T. D. (1974). *J. Dent. Res.* **53,** 246.
Bruns, R. R., Trelstad, R. L., and Gross, J. (1973). *Science* **181,** 269–271.
Bryant, S. V. (1976). *Nature (London)* **263,** 676–679.
Bryant, P. J., Bryant, S. V., and French, V. (1977). *Sci. Am.* **237,** 66–81.
Bryant, S. V. (1977). *In* "Vertebrate Limb and Somite Morphogenesis" (D. A. Ede, J. R. Hinchliffe, and M. Balls, eds). pp. 311–328. Cambridge Univ. Press, London and New York.
Bubenik, G. A., Brown, G. M., Bubenik, A. B., and Grota, L. J. (1974a). *Calcif. Tissue Res.* **14,** 121–130.
Bubenik, G. A., Brown, G. M., Wilson, D., Bubenik, A. B., and Trenkle, A. (1974b). *Proc. Can. Fed. Biol. Soc.* **17,** 174.
Bubenik, G. A., Bubenik, A. B., Brown, G. M., and Wilson, D. A. (1975). *J. Exp. Zool.* **194,** 349–358.
Burch, J. G. (1966). *Anat. Rec.* **156,** 433–438.
Burger, M. M., Bombik, B. M., Breckingridge,, B. M., and Sheppard, J. R. (1972). *Nature (London), New Biol.* **239,** 161–163.
Burgess, A. M. C. (1974). *In* "Neoplasia and Cell Differentiation" (G. V. Sherbet, ed.), pp. 106–152. Karger, Basel.

Buring, K. (1974). *Akad. Avh., Malmo,* pp. 1–20.

Buring, K. (1975). *Clin. Orthop. Relat. Res.* **110,** 293–302.

Burks, J. K., and Peck, W. A. (1978). *Science* **199,** 542–544.

Butler, E. G. (1933). *J. Exp. Zool.* **65,** 271–315.

Butler, P. M., and Joysey, K. A., eds. (1978). "Development, Function, and Evolution of Teeth." Academic Press, New York and London.

Butler, W. T., Miller, E. J., Finch, J. E., Jr., and Inagami, T. (1974). *Biochem. Biophys. Res. Commun.* **57,** 190–195.

Cabrini, R. L. (1961). *Int. Rev. Cytol.* **11,** 283–306.

Cahn, R. D., and Lasher, R. (1967). *Proc. Natl. Acad. Sci. U.S.A.* **58,** 1131–1138.

Cairns, J. M. (1965). *Dev. Biol.* **12,** 36–52.

Cairns, J. M. (1975). *J. Embryol. Exp. Morphol.* **34,** 155–169.

Cairns, J. M., and Saunders, J. W., Jr. (1954). *J. Exp. Zool.* **127,** 221–248.

Calandra, A. J., and MacCabe, J. A. (1978). *Devel. Biol.* **62,** 258–269.

Cameron, D. A. (1972). *In* "The Biochemistry and Physiology of Bone" (G. H. Bourne, ed.), 2nd ed., Vol. 1, pp. 191–236. Academic Press, New York.

Cameron, D. A., Paschall, H. A., and Robinson, R. A. (1964). *In* "Bone Biodynamics" (H. M. Frost, ed.), pp. 91–104. Little, Brown, Boston, Massachusetts.

Cameron, D. A., Paschall, H. A., and Robinson, R. A. (1967). *J. Cell Biol.* **33,** 1–14.

Cameron, I. L., and Jeter, J. R., Jr. (1971). *In* "Developmental Aspects of the Cell Cycle" (I. L. Cameron, G. M. Padilla, and A. M. Zimmerman, eds.), pp. 191–222. Academic Press, New York.

Cameron, J. A., and Fallon, J. F. (1977a). *Dev. Biol.* **55,** 320–330.

Cameron, J. A., and Fallon, J. F. (1977b). *Dev. Biol.* **55,** 331–337.

Campo, R. D. (1974). *Calcif. Tissue Res.* **14,** 105–119.

Campo, R. D., and Phillips, S. J. (1973). *Calcif. Tissue Res.* **13,** 83–92.

Caplan, A. I. (1970). *Exp. Cell Res.* **62,** 341–355.

Caplan, A. I. (1971). *J. Exp. Zool.* **178,** 351–358.

Caplan, A. I. (1972a). *Exp. Cell Res.* **70,** 185–195.

Caplan, A. I. (1972b). *J. Exp. Zool.* **180,** 351–362.

Caplan, A. I. (1972c). *Dev. Biol.* **28,** 71–83.

Caplan, A. I. (1972d). *Dev. Biol.* **28,** 344–351.

Caplan, A. I. (1977). *In* "Vertebrate Limb and Somite Morphogenesis" (D. A. Ede, J. R. Hinchliffe, and M. Balls, eds.), pp. 199–214. Cambridge Univ. Press, London and New York.

Caplan, A. I., and Koutroupas, S. (1973). *J. Embryol. Exp. Morphol.* **29,** 571–584.

Caplan, A. I., and Rosenberg, M. J. (1975). *Proc. Natl. Acad. Sci. U.S.A.* **72,** 1852–1857.

Caplan, A. I., and Stoolmiller, A. C. (1973). *Proc. Natl. Acad. Sci. U.S.A.* **70,** 1713–1717.

Caplan, A. I., Zwilling, E., and Kaplan, N. O. (1968). *Science* **160,** 1009–1010.

Carlson, E. C., and Upson, R. H. (1974). *Am. J. Anat.* **141,** 441–446.

Carlson, E. C., Upson, R. H., and Evans, D. K. (1974). *Anat. Rec.* **179,** 361–374.

Carriker, M. R., Person, P., Libben, R. M., and van Zandt, D. (1972). *Biol. Bull.* **143,** 317–331.

Cassin, C. (1977). *Rev. Stomato-Odontol.* **118,** 149–162.

Cassin, C., and Capuron, A. (1972). *C.R. Hedb. Sciences Acad. Sci.* **275,** 2953–2956.

Cassin, C., and Capuron, A. (1977). *Wilhelm Roux Arch. Entwicklungsmech. Org.* **181,** 107–112.

Chalmers, J. (1965). *Proc. Eur. Symp. Calcif. Tissues, Congr. Colloq. Univ. Liège,* 1964, 2nd Vol. **31,** pp. 177–184.

Chalmers, J., and Ray, R. D. (1962). *J. Bone Joint Surg., Br. Vol.* **44,** 149–164.

Chamay, A., and Tschantz, P. (1972). *J. Biomech.* **5,** 173–180.

Charlier, J.-P., and Petrovic, A. (1967). *Orthod. Fr.* **38,** 1–11.

Charlier, J.-P., Petrovic, A., and Herrmann-Stutzmann, J. (1969a). *Am. J. Orthod.* **55,** 71–74.

Charlier, J.-P., Petrovic, A., and Linck, G. (1969b). *Orthod. Fr.* **40,** 99–113.
Chaube, S. (1959). *J. Exp. Zool.* **140,** 29–78.
Chevallier, A., Kieny, M., and Mauger, A. (1977). *J. Embryol. Exp. Morphol.* **41,** 245–258.
Chiakulas, J. J. (1957). *J. Exp. Zool.* **136,** 287–300.
Christ, B., Jacob, H. J., and Jacob, M. (1977). *Anat. Embryol.* **150,** 171–186.
Cleall, J. F. (1974). *J. Dent. Res.* **53,** 1226–1234.
Cock, A. G. (1966). *Q. Rev. Biol.* **41,** 131–190.
Coffin, P. A., and Hall, B. K. (1974). *J. Embryol. Exp. Morphol.* **31,** 169–181.
Cohen, A. M., and Hay, E. D. (1971). *Dev. Biol.* **26,** 578–605.
Cohen, A. M., and Konigsberg, I. R. (1975). *Dev. Biol.* **46,** 262–280.
Conrad, G. W. (1970). *Dev. Biol.* **21,** 611–635.
Conrad, G. W., Hart, G. W., and Chen, Y. (1977). *J. Cell Sci.* **26,** 119–138.
Cook, P. R. (1974). *Biol. Rev. Cambridge. Philos. Soc.* **49,** 51–84.
Cooke, J. (1975). *Nature (London)* **254,** 196–199.
Coon, H. G. (1966). *Proc. Natl. Acad. Sci. U.S.A.* **55,** 66–73.
Cooper, G. W. (1965). *Dev. Biol.* **12,** 185–212.
Cooper, R. R., and Ponseti, I. V. (1973). *J. Bone Joint Surg., Am. Vol.* **55,** 485–495.
Cooper, R. R., Ponseti, I. V., and Maynard, J. A. (1973). *J. Bone Joint Surg., Am. Vol.* **55,** 475–484.
Copher, G. H. (1935). *Ann. Surg.* **102,** 927–940.
Corsin, J. (1974). *Arch. Anat. Microsc. Morphol. Exp.* **63,** 231–238.
Corsin, J. (1975a). *J. Embryol. Exp. Morphol.* **33,** 335–342.
Corsin, J. (1975b). *Acta Embryol. Exp.* **1,** 15–22.
Corsin, J. (1977). *J. Embryol. Exp. Morphol.* **38,** 139–150.
Corwin, J. A., and Morehead, J. R. (1971). *Anat. Rec.* **171,** 509–516.
Coulombre, A. J. (1965). *In* "Organogenesis" (R. L. de Haan and H. Ursprung, eds.), pp. 219–251. Holt, New York.
Coulombre, A. J., and Coulombre, J. L. (1973). *Dev. Biol.* **33,** 257–267.
Coulombre, A. J., Coulombre, J. L., and Mehta, H. (1962). *Dev. Biol.* **5,** 382–401.
Coulombre, J. L., and Coulombre, A. J. (1975). *Dev. Biol.* **45,** 291–303.
Cowden, R. R. (1967). *Histochemie* **9,** 149–163.
Cowden, R. R., and Fitzharris, T. P. (1975). *Histochemie* **43,** 1–10.
Craft, P. D., and Mani, M. M., Pazel, J., and Masters, F. W. (1974). *Plastic Reconstruct. Surg.* **53,** 321–325.
Crelin, E. S. (1954a). *Am. J. Anat.* **95,** 47–73.
Crelin, E. S. (1954b). *Proc. Soc. Exp. Biol. Med.* **86,** 22–24.
Crelin, E. S. (1954c). *Anat. Rec.* **120,** 23–32.
Crelin, E. S. (1956). *Anat. Rec.* **124,** 279.
Crelin, E. S. (1957a). *Science* **125,** 650.
Crelin, E. S. (1957b). *Anat. Rec.* **127,** 282–283.
Crelin, E. S. (1958a). *Anat. Rec.* **130,** 290.
Crelin, E. S. (1958b). *Anat. Rec.* **130,** 400–401.
Crelin, E. S. (1958c). *Anat. Rec.* **130,** 400.
Crelin, E. S. (1958d). *Anat. Rec.* **130,** 400.
Crelin, E. S. (1959). *Anat. Rec.* **133,** 369.
Crelin, E. S. (1960a). *Ann. N.Y. Acad. Sci.* **84,** 479–512.
Crelin, E. S. (1960b). *Anat. Rec.* **136,** 179.
Crelin, E. S. (1963). *Anat. Rec.* **146,** 149–163.
Crelin, E. S. (1969). *Trans. N.Y. Acad. Sci.* [2] **31,** 1049–1058.
Crelin, E. S., and Brightman, M. W. (1957). *Anat. Rec.* **128,** 467–484.

Crelin, E. S., and Grillo, M. A. (1957). *Anat. Rec.* **127,** 407.

Crelin, E. S., and Haines, A. L. (1955). *Endocrinology* **56,** 461–470.

Crelin, E. S., and Koch, W. E. (1965). *Anat. Rec.* **153,** 161–172.

Crelin, E. S., and Koch, W. E. (1967). *Anat. Rec.* **158,** 473–484.

Crelin, E. S., and Levin, J. (1955). *Endocrinology* **57,** 730–747.

Crelin, E. S., and Southwick, W. D. (1960). *Yale J. Biol. Med.* **33,** 243–244.

Crelin, E. S., and Southwick, W. D. (1964). *Anat. Rec.* **149,** 113–134.

Crilly, R. G. (1972). *J. Anat.* **112,** 11–18.

Crosby, G. M., and Fallon, J. F. (1975). *Dev. Biol.* **46,** 28–39.

Currey, J. D. (1973). *Calcif. Tissue Res.* **13,** 99–112.

Curtin, J. W., and Pruzansky, S. (1971). *World Congr. Plast. Reconstr. Surg., 5th, 1970,* pp. 181–184.

Dahlberg, A. A., ed. (1971). "Dental Morphology and Evolution." Univ. of Chicago Press, Chicago, Illinois.

Daimon, T. (1973a). *Acta Histochem. Cytochem.* **6,** 280–293.

Daimon, T. (1973b). *Acta Histochem. Cytochem.* **6,** 294–311.

Damjanov, I., and Urbanke, A. (1969). *J. Urol.* **101,** 863–865.

Daneo, L. S., and Filogamo, G. (1973). *J. Submicrosc. Cytol.* **5,** 219–225.

Daniel, J. C. (1976). *Cell Differ.* **5,** 247–254.

Daniel, J. C., Kosher, R. A., Lash, J. W., and Hertz, J. (1973). *Cell Differ.* **2,** 285–298.

Daughaday, W. H. (1971). *Am. J. Med.* **50,** 277–280.

Davidovitch, Z. (1973). *Am. J. Orthod.* **64,** 314–315.

Davidovitch, Z., and Shanfeld, J. L. (1974). *J. Dent. Res.* **53,** 132.

Davidovitch, Z., and Shanfeld, J. L. (1975). *Arch. Oral Biol.* **20,** 567–574.

Davidovitch, Z., Shanfeld, J. L., and Batastini, P. J. (1974). *Am. J. Orthod.* **65,** 320–321.

Davidovitch, Z., Montgomery, P. C., and Eckerdal, O. (1976a). *Calcif. Tissue Res.* **19,** 305–316.

Davidovitch, Z., Montgomery, P. C., Eckerdal, O., and Gustafson, G. T. (1976b). *Calcif. Tissue Res.* **19,** 317–330.

Davidovitch, Z., Montgomery, P. C., and Shanfeld, J. L. (1977a). *Calc. Tissue Res.* **24,** 81–91.

Davidovitch, Z., Montgomery, P. C., and Shanfeld, J. L. (1977b). *Calc. Tissue Res.* **24,** 73–79.

Davison, P. F., and Berman, M. (1973). *Connect. Tissue Res.* **2,** 57–64.

Dawd, D. S., and Hinchliffe, J. R. (1971). *J. Embryol. Exp. Morphol.* **26,** 401–424.

Dawson, A., and Kember, N. F. (1974). *Cell Tissue Kinet.* **7,** 285–292.

deAngelis, V. (1969). *Am. J. Anat.* **123,** 359–368.

de Beer, G. R. (1937). "Development of the Vertebrate Skull." Oxford Univ. Press, London.

de Beer, G. R. (1947). *Proc. R. Soc. London, Ser. B* **134,** 377–398.

de Beer, G. R. (1958). "Embryos and Ancestors," 3rd ed. Oxford Univ. Press, London.

DeHaan, R. L. (1956). *J. Exp. Zool.* **133,** 73–85.

Delbrück, A. (1970). *Enzymol. Biol. Clin.* **11,** 130–153.

Delcher, H. K., Eisenbarth, G. S., and Lebovitz, H. E. (1973). *J. Biol. Chem.* **248,** 1901–1905.

Deleanu, M. (1965). *Rev. Roum. Embryol. Cytol. Ser. Embryol.* **2,** 45–56.

Delides, G. S. (1972). *Br. J. Urol.* **44,** 571–581.

DeLuca, G., Speziale, P., Balduini, C., and Castellani, A. A. (1975). *Connect. Tissue Res.* **3,** 39–48.

Denison, R. H. (1963). *Clin. Orthop. Relat. Res.* **31,** 141–152.

Deporter, D. A., and Ten Cate, A. R. (1973). *J. Anat.* **114,** 457–461.

de Ricqlès, A. (1968). *Ann. Paleontol.* **54,** 133–145.

de Ricqlès, A. (1969). *Ann. Paleontol.* **55,** 3–52.

de Ricqlès, A. (1972). *Ann. Paleontol.* **58,** 17–60.

de Ricqlès, A. (1973). Thèse de Doctorat d'Etat, Université de Paris VII.

de Ricqlès, A. (1974a). *Evol. Theory* **1,** 51–80.

de Ricqlès, A. (1974b). *Ann. Paleontol.* **60,** 1–39.
de Ricqlès, A. (1974c). *Ann. Paleontol.* **60,** 171–216.
de Ricqlès, A. (1975). *Colloq. Int. C.N.R.S.* **218,** 351–363.
Deshmukh, K. (1974). *Proc. Soc. Exp. Biol. Med.* **147,** 726–729.
Deshmukh, K., and Kline, W. G. (1976). *Eur. J. Biochem.* **69,** 117–124.
Deshmukh, K., and Sawyer, B. D. (1977). *Proc. Natl. Acad. Sci. U.S.A.* **74,** 3864–3868.
Deshmukh, K., Kline, K. D., and Sawyer, B. D. (1977). *Biochim. Biophys. Acta* **499,** 28–35.
Desmond, A. J. (1976). "The Hot-Blooded Dinosaurs: A Revolution in Palaeontology." Blond & Briggs, London.
Detwiler, S. R., and van Dyke, R. H. (1934). *J. Exp. Zool.* **68,** 321–346.
Deuchar, E. M. (1975). "Cellular Interactions in Animal Development." Chapman & Hall, London.
Dhouailly, D., and Kieny, M. (1972). *Dev. Biol.* **28,** 162–175.
Diegelmann, R. F., and Peterkofsky, B. (1972). *Dev. Biol.* **28,** 443–453.
Dienstman, S. R., Biehl, J., Holtzer, S., and Holtzer, H. (1974). *Dev. Biol.* **39,** 83–95.
Diethelm, F., and Schowing, J. (1973). *Bull. Assoc. Anat.* **57,** 1–8.
Diethelm, F., and Schowing, J. (1974). *Wilhelm Roux Arch. Entwickluhgsmech. Org.* **175,** 163–172.
Dingle, J. T., Horsfield, P., Fell, H. B., and Barratt, M. E. J. (1975). *Ann. Rheum. Dis.* **34,** 303–311.
Dinner, B. J. (1970). M.Sc. Thesis, Dept. Orthodontics, University of Nebraska, Lincoln, Nebraska.
Dittmer, J. E., and Goss, R. J. (1972). *Am. Zool.* **12,** 730–731.
Dittmer, J. E., and Goss, R. J. (1974). *Cardiology* **58,** 355–363.
Dittmer, J. E., Goss, R. J., and Dinsmore, C. E. (1974). *Am. J. Anat.* **141,** 155–160.
Dixon, B. (1970). *Rev. Eur. Etud. Clin. Biol.* **15,** 875–878.
Dodds, G. S., and Cameron, H. C. (1939). *Am. J. Pathol.* **15,** 723–740.
Dondi, P., and Muir, H. (1976). *Biochem. J.* **160,** 117–120.
Drachman, D. B. (1964). *Science* **145,** 719–721.
Drachman, D. B., and Coulombre, A. J. (1962). *Lancet* **2,** 523–526.
Drachman, D. B., and Sokoloff, L. (1966). *Dev. Biol.* **4,** 401–420.
Drews, Ulrich, and Drews, Ute (1972). *Wilhelm Roux Arch. Entwicklungsmech. Org.* **169,** 70–86.
Drews, Ulrich, and Drews, Ute (1973). *Wilhelm Roux Arch. Entwicklungsmech. Org.* **173,** 208–227.
Drews, Ulrich, Kocher-Becker, U., and Drews, Ute (1972). *Wilhelm Roux Arch. Entwicklungsmech. Org.* **171,** 17–37.
Duerden, J. E. (1920). *Am. Nat.* **54,** 289–312.
Duhamel de Monceau, H. (1742). *Hist. Mem. Acad. Inscriptions Belles Lettres* **2,** 481.
Duke, J., and Elmer, W. A. (1977). *J. Embryol. Exp. Morphol.* **42,** 209–217.
Duksin, D., Maoz, A., and Fuchs, S. (1975). *Cell* **5,** 83–86.
Durkin, J. F. (1972). *Am. J. Orthod.* **62,** 15–41.
Durkin, J. F., Heeley, J., and Irving, J. T. (1973). *Oral Sci. Rev.* **2,** 29–99.
Duterloo, H. S., and Wolters, J. M. (1971). *Trans. Eur. Orthod. Soc.* pp. 103–116.
Dziak, R., and Brand, J. S. (1974a). *J. Cell. Physiol.* **84,** 75–84.
Dziak, R., and Brand, J. S. (1974b). *J. Cell. Physiol.* **84,** 85–96.
Dziak, R., and Stern, P. H. (1976). *Calcif. Tissue Res.* **22,** 137–147.
Ede, D. A. (1971). *Sym. Soc. Exp. Biol.* **25,** 235–254.
Ede, D. A. (1976). *Cell Surf. Rev.* **1,** 495–544.
Ede, D. A., and Agerback, G. S. (1968). *J. Embryol. Exp. Morphol.* **20,** 81–100.
Ede, D. A., and Flint, O. P. (1972). *J. Embryol. Exp. Morphol.* **27,** 245–260.
Ede, D. A., and Flint, O. P. (1975a). *J. Cell Sci.* **18,** 97–112.
Ede, D. A., and Flint, O. P. (1975b). *J. Cell Sci.* **18,** 301–314.
Ede, D. A., and Kelly, W. A. (1964a). *J. Embryol. Exp. Morphol.* **12,** 161–182.
Ede, D. A., and Kelly, W. A. (1964b). *J. Embryol. Exp. Morphol.* **12,** 339–356.
Ede, D. A., and Law, J. T. (1969). *Nature (London)* **221,** 244–248.

Ede, D. A., Bellairs, R., and Bancroft, M. (1974). *J. Embryol. Exp. Morphol.* **31,** 761–786.

Ede, D. A., Flint, O. P., and Teague, P. (1975). *J. Embryol. Exp. Morphol.* **34,** 589–607.

Ede, D. A., Flint, O. P., Wilby, O. K., and Colquhoun, P. (1977). *In* "Vertebrate Limb and Somite Morphogenesis" (D. A. Ede, J. R. Hinchliffe, and M. Balls, eds.) pp. 161–180. Cambridge Univ. Press, London and New York.

Edelman, G. M. (1976). *Science* **192,** 218–226.

Eggert, R. C. (1966). *J. Exp. Zool.* **161,** 369–390.

Eguro, H., and Goldner, J. L. (1974). *J. Bone Joint Surg., Am. Vol.* **56,** 129–141.

Ehrlich, M. G., Mankin, H. J., Treadwell, B. V., and Jones, H. (1974). *J. Bone Joint Surg., Am. Vol.* **56,** 1239–1245.

Ehrlich, M. G., Mankin, H. J., Jones, H., Wright, R., Crispen, C., and Vigliani, G. (1977). *J. Clin. Invest.* **59,** 226–233.

Ehrlich, P. H., and Bornstein, P. (1972). *Nature (London) New Biol.* **238,** 257–260.

Eilberg, R. G., Person, P., and Zuckerberg, D. (1974). *J. Dent. Res.* **53,** 246.

Eilberg, R. G., Zuckerberg, D., and Person, P. (1975). *Calcif. Tissue Res.* **19,** 85–90.

Eisenstein, R., Arsenis, C., and Kuettner, K. E. (1971a). *Isr. J. Med. Sci.* **7,** 415–416.

Eisenstein, R., Sorgente, N., and Kuettner, K. E. (1971b). *Am. J. Pathol.* **65,** 515–534.

Eisenstein, R., Sorgente, N., Soble, L. W., Miller, A. W., and Kuettner, K. E. (1973). *Am. J. Pathol.* **73,** 765–774.

Eisenstein, R., Kuettner, K. E., Neapolitan, C., Soble, L. W., and Sorgente, N. (1975). *Am. J. Pathol.* **81,** 337–348.

Eisenstein, R., Kuettner, K. E., Soble, L. W., and Sorgente, N. (1976). *Protides Biol. Fluids, Proc. Colloq.*, pp. 217–219.

Elder, J. R., and Tuenge, R. (1974). *Am. J. Orthod.* **66,** 599–617.

Elliott, H. C. (1936). *Am. J. Anat.* **58,** 127–145.

Ellison, M. L., Ambrose, E. J., and Easty, G. C. (1969). *J. Embryol. Exp. Morphol.* **21,** 331–340.

Elmer, W. A. (1968a). *Dev. Biol.* **18,** 76–92.

Elmer, W. A. (1968b). *J. Exp. Zool.* **169,** 381–390.

Elmer, W. A. (1977). *Colloq. Int. C.N.R.S.* **266,** 235–241.

Elmer, W. A., and Selleck, D. K. (1975). *J. Embryol. Exp. Morphol.* **33,** 371–386.

Elves, M. W. (1974). *J. Bone Joint Surg., Br. Vol.* **56,** 178–185.

Elves, M. W. (1976). *Clin. Orthop. Relat. Res.* **120,** 232–259.

Elves, M. W., and Pratt, L. M. (1975). *Acta Orthop. Scand.* **46,** 549–560.

Engfeldt, B. (1969). *Acta Pathol. Microbiol. Scand.* **75,** 201–219.

Enlow, D. H. (1966a). *J. Dent. Res.* **45,** 213.

Enlow, D. H. (1966b). *In* "Studies on the Anatomy and Function of Bone and Joints" (F. G. Evans, ed.), pp. 93–112. Springer-Verlag, Berlin and New York.

Enlow, D. H. (1968a). *Am. J. Orthod.* **54,** 803–822.

Enlow, D. H. (1968b). "The Human Face. An Account of the Postnatal Growth and Development of the Craniofacial Skeleton." Harper (Hoeber), New York.

Enlow, D. H. (1969). *In* "Biology of the Reptilia" (C. Gans, ed.), Vol. 1, pp. 45–80. Academic Press, New York.

Enlow, D. H. (1973). *Am. J. Anat.* **136,** 403–406.

Enlow, D. H. (1975). "Handbook of Facial Growth." Saunders, Philadelphia, Pennsylvania.

Enlow, D. H., and Brown, S. O. (1956). *Tex. J. Sci.* **8,** 405–443.

Enlow, D. H., and Brown, S. O. (1957). *Tex. J. Sci.* **9,** 186–214.

Enlow, D. H., and Brown, S. O. (1958). *Tex. J. Sci.* **10,** 187–230.

Epperlein, H. H. (1974). *Differentiation* **2,** 151–168.

Epperlein, H. H., and Lehmann, R. (1975). *Differentiation* **4,** 159–174.

Errick, J. E., and Saunders, J. W., Jr. (1974). *Dev. Biol.* **41,** 338–351.

Errick, J. E., and Saunders, J. W., Jr. (1976). *Dev. Biol.* **50,** 26–34.
Espie, J. H. (1975). *J. Biol. Buccale* **3,** 157–166.
Estabrooks, L., and Schiff, R. (1972). *J. Histochem. Cytochem.* **20,** 211–219.
Faber, J. (1971). *Adv. Morphog.* **9,** 127–148.
Fallon, J. F., and Cameron, J. A. (1977). *J. Embryol. Exp. Morphol.* **40,** 285–289.
Fallon, J. F., and Crosby, G. M. (1977). *In* "Vertebrate Limb and Somite Morphogenesis" (D. A. Ede, J. R. Hinchliffe, and M. Balls., eds.). pp. 55–70. Cambridge Univ. Press, London and New York.
Fallon, J. F., and Kelley, R. O. (1977). *J. Embryol. Exp. Morphol.* **41,** 223–232.
Fallon, J. F., and Saunders, J. W., Jr. (1968). *Dev. Biol.* **18,** 553–570.
Fallon, J. F., Brucker, R. F., and Harris, C. M. (1974). *J. Cell Sci.* **15,** 17–29.
Faulk, W. P., Conochie, L. B., Temple, A., and Papamichail, M. (1975). *Nature (London)* **256,** 123–125.
Fell, H. B. (1925). *J. Morphol. Physiol.* **40,** 417–459.
Fell, H. B. (1928). *Arch. Exp. Zellforsch. Besonders Gewebezuecht.* **7,** 390–412.
Fell, H. B. (1932). *J. Anat.* **66,** 157–180.
Fell, H. B. (1969). *Embryologia* **10,** 181–205.
Fell, H. B. (1975). *Philos. Trans. R. Soc. London, Ser. B* **271,** 325–342.
Fell, H. B. (1976). *In* "Organ Culture in Biomedical Research" (M. Balls and M. A. Monnickendam, eds.), pp. 1–13. Cambridge Univ. Press, London and New York.
Fell, H. B., and Canti, G. R. (1934a). *Arch. Exp. Zellforsch. Besonders Gewebezuecht.* **15,** 311.
Fell, H. B., and Canti, G. R. (1934b). *Proc. R. Soc. London, Ser. B* **116,** 316–351.
Fell, H. B., and Landauer, W. (1935). *Proc. R. Soc. London, Ser. B* **118,** 133–154.
Fell, H. B., and Robison, R. (1929). *Biochem. J.* **23,** 767–783.
Fell, H. B., and Robison, R. (1930). *Biochem. J.* **24,** 1905–1921.
Fell, H. B., Dingle, J. T., Coombs, R. R. A., and Lachmann, P. J. (1968). *Symp. Int. Soc. Cell Biol.* **7,** 49–68.
Felts, W. J. L. (1961). *Int. Rev. Cytol.* **12,** 243–302.
Felts, W. J. L., and Spurrell, F. A. (1966). *Am. J. Anat.* **118,** 103–134.
Fenwick, J. C. (1974). *Can. J. Zool.* **52,** 755–764.
Fessler, L. I., Burgeson, R. E., Morris, N. P., and Fessler, J. H. (1973). *Proc. Natl. Acad. Sci. U.S.A.* **70,** 2993–2996.
Fietzek, P. P., Allmann, H., Rauterberg, J., and Wachter, E. (1977). *Proc. Natl. Acad. Sci. U.S.A.* **74,** 84–86.
Finch, R. A., and Zwilling, E. (1971). *J. Exp. Zool.* **176,** 397–408.
Finkle, H. I., and Goldman, R. L. (1973). *Arch. Pathol.* **95,** 48–49.
Fischman, D. A., and Hay, E. D. (1962). *Anat. Rec.* **143,** 329–338.
Fisher, M., and Solursh, M. (1977). *J. Embryol. Exp Morphol.* **42,** 195–207.
Fitch, N. (1957). *J. Exp. Zool.* **136,** 329–362.
Fitton Jackson, S. (1965). *In* "Structure and Function of Connective and Skeletal Tissues" (G. R. Tristram and S. Fitton Jackson, eds.), pp. 277–281. Butterworth, London.
Fitton Jackson, S. (1970). *Proc. R. Soc. London, Ser. B* **175,** 405–453.
Fitton Jackson, S. (1976). *In* "Organ Culture in Biomedical Research" (M. Balls and M. A. Monnickendam, eds.), pp. 165–178. Cambridge Univ. Press, London and New York.
Fitzharris, T. P. (1976). *Am. Zool.* **16,** 593–616.
Flower, M. J. (1972). *Dev. Biol.* **28,** 583–602.
Flowers, M., and Grobstein, C. (1967). *Dev. Biol.* **15,** 193–205.
Folkman, J. (1976). *Sci. Am.* **234,** 59–73.
Foret, J. E. (1970). *J. Exp. Zool.* **175,** 297–322.
Fornasier, V. L. (1977). *Hum. Pathol.* **8,** 243–254.

Fouvet, B. (1973). *Arch. Anat. Microsc. Morphol. Exp.* **62,** 269–280.
Fraser, R. A., and Abbott, U. K. (1971a). *J. Exp. Zool.* **176,** 219–236.
Fraser, R. A., and Abbott, U. K. (1971b). *J. Exp. Zool.* **176,** 237–248.
Fraser, R. A., and Goetinck, P. F. (1971). *Biochem. Biophys. Res. Comm.* **43,** 494–503.
Frasier, M. B., Banks, W. J., and Newbrey, J. W. (1975). *Calcif. Tissue Res.* **17,** 273–288.
Freeman, B. M., and Vince, M. A. (1974). "Development of the Avian Embryo. A Behavioural and Physiological Study." Chapman & Hall, London.
Freeman, E., Ten Cate, A. R., and Dickinson, J. B. (1975). *Arch. Oral Biol.* **20,** 139–140.
French, V., Bryant, P. J., and Bryant, S. V. (1976). *Science* **193,** 969–981.
Friant, M. (1964). *Acta. Anat.* **57,** 66–71.
Friant, M. (1968). *Ann. Fac. Med. Vet. Pisa, Univ. Studi Pisa* **21,** 1–17.
Friede, H. (1973). *Cleft Palate J.* **10,** 14–22.
Friedenstein, A. Y. (1962). *Byull. Eksp. Biol. Med.* **12,** 82–83.
Friedenstein, A. Y. (1973). *Hard Tissue Growth, Repair, Remineralization, Ciba Found. 1972 Symp.* No. 11, pp. 170–185.
Friedenstein, A. Y. (1976). *Int. Rev. Cytol.* **47,** 327–359.
Friedenstein, A. Y., and Kuralesova, A. I. (1971). *Transplantation* **12,** 99–108.
Friedenstein, A. Y., and Lalykina, K. S. (1970). *Calcif. Tissue Res.* **4,** Suppl., 105–106.
Friedenstein, A. Y., and Lalykina, K. S. (1972). *Eur. J. Immunol.* **2,** 602–603.
Friedenstein, A. Y., and Lalykina, K. S. (1973). "Bone Induction and Osteogenic Precursor Cells." Med. Publ. House, Moscow (in Russian).
Friedenstein, A. Y., Piatetzky-Shapiro, I. I., and Petrakova, K. V. (1966). *J. Embryol. Exp. Morphol.* **16,** 381–390.
Friedenstein, A. Y., Lalykina, K. S., and Tolmacheva, A. A. (1967). *Acta Anat.* **68,** 532–549.
Friedenstein, A. Y., Petrakova, K. V., Kurolesova, A. I., and Frolova, G. P. (1968). *Transplantation* **6,** 230–247.
Friedenstein, A. Y., Chailakhjan, R. K., and Lalykina, K. S. (1970). *Cell Tissue Kinet.* **3,** 393–403.
Friedenstein, A. Y., Gorskaja, U. F., and Kulagina, N. N. (1976). *Exp. Hematol.* **4,** 267–274.
Friedman, D. L. (1976). *Physiol. Rev.* **56,** 652–708.
Fromme, H. G., Hohling, H. J., and Riedel, H. (1971). *Calcif. Tissue Res.* **7,** 257–258.
Frommer, J., and Margolies, M. R. (1971). *J. Dent. Res.* **50,** 1260–1267.
Fullmer, H. M. (1965). *Int. Rev. Connect. Tissue Res.* **3,** 1–76.
Furseth, R. (1973). *Scand. J. Dent. Res.* **81,** 339–341.
Furthmayr, H., and Timpl, R. (1976). *Int. Rev. Connect. Tissue Res.* **7,** 61–101.
Fyfe, D. McG. (1977). B. Sc. (Hons.) Thesis, Dalhousie University, Halifax, Nova Scotia, Canada.
Gans, C. (1974). "Biomechanics: Approach to Vertebrate Biology." Lippincott, Philadelphia, Pennsylvania.
Gans, C. (1975). *Am. Zool.* **15,** 455–467.
Gardner, E. (1971). *In* "The Biochemistry and Physiology of Bone" (G. H. Bourne, ed.), 2nd ed., Vol. 3, pp. 77–118. Academic Press, New York.
Gardner, W. U. (1936). *Am. J. Anat.* **59,** 459–483.
Gardner, W. U., and Pfeiffer, C. A. (1938). *Proc. Soc. Exp. Biol. Med.* **37,** 678–679.
Garrard, G., Harrison, G. A., and Weiner, J. S. (1974). *J. Anat.* **117,** 531–540.
Gasseling, M. T., and Saunders, J. W., Jr. (1964). *Am. Zool.* **4,** 303–304.
Gay, S., Martin, G. R., Müller, P. K., Timpl, R., and Kühn, K. (1976a). *Proc. Natl. Acad. Sci. U.S.A.* **73,** 4037–4040.
Gay, S., Müller, P. K., Lemmen, C., Remberger, K., Matzen, K., and Kühn, K. (1976b). *Klin. Wochenschr.* **54,** 969–976.
Gearhart, J. D., and Mintz, B. (1972). *Dev. Biol.* **29,** 27–37.
Gelman, R. A., and Blackwell, J. (1973). *Connect. Tissue Res.* **2,** 31–36.

Gilbert, G. H., and Gorman, H. A. (1971). *J. Am. Vet. Med. Assoc.* **158,** 77–81.
Gingerich, P. D. (1972). *Anat. Rec.* **172,** 589–596.
Girgis, F. G., and Pritchard, J. J. (1958). *J. Bone Joint Surg., Br. Vol.* **40,** 274–281.
Glass, L. (1977). *Science* **198,** 321–322.
Glasstone, S. (1968). *Nature (London)* **220,** 705–706.
Glasstone, S. (1971). *Arch. Oral Biol.* **16,** 723–730.
Glasstone, S. (1973). *J. Dent. Assoc. S. Afr.* **28,** 328–334.
Glimcher, M. J., Daniel, E. J., Travis, D. F., and Kahmis, S. (1965). *J. Ultrastruct. Res.* **7** Suppl., 14–77.
Globus, M., and Vethamany-Globus, S. (1976). *Differentiation* **6,** 91–96.
Globus, M., and Vethamany-Globus, S. (1977). *Dev. Biol.* **56,** 316–328.
Godman, G. C., and Porter, K. R. (1960. *J. Biophys. Biochem. Cytol.* **8,** 719–760.
Goedbloed, J. F. (1964). *Arch. Biol.* **75,** 207–244.
Goel, S. C., and Jacob, J. (1976). *Experientia* **32,** 216–217.
Goel, S. C., and Jurand, A. (1975). *J. Embryol. Exp. Morphol.* **34,** 327–337.
Goetinck, P. F. (1964). *Dev. Biol.* **10,** 71–91.
Goetinck, P. F. (1966). *Curr. Top. Dev. Biol.* **1,** 253–283.
Goetinck, P. F., and Abbott, U. K. (1964). *J. Exp. Zool.* **155,** 161–170.
Goetinck, P. F., and Pennypacker, J. P. (1977). *In* "Vertebrate Limb and Somite Morphogenesis" (D. A. Ede, J. R. Hinchliffe, and M. Balls, eds.). pp. 139–160. Cambridge Univ. Press, London and New York.
Goetinck, P. F., and Sekellick, M. J. (1970). *Dev. Biol.* **21,** 349–363.
Goetinck, P. F., Pennypacker, J. P., and Royal, P. D. (1974). *Exp. Cell Res.* **87,** 241–248.
Goggins, J. F., and Billups, L. C. (1972). *J. Dent. Res.* **51,** 303–307.
Goggins, J. F., Johnson, G. S., and Pastan, I. (1972). *J. Biol. Chem.* **247,** 5759–5764.
Goldhaber, P., Rabadjija, L., Beyer, W. R., and Kornhauser, A. (1973). *J. Am. Dent. Assoc.* **87,** 1027–1033.
Gordon, J. S., and Lash, J. W. (1974). *Dev. Biol.* **36,** 88–104.
Goss, R. J. (1958). *J. Embryol. Exp. Morphol.* **6,** 638–644.
Goss, R. J. (1961). *J. Embryol. Exp. Morphol.* **9,** 342–354.
Goss, R. J. (1963). *In* "Mechanisms of Hard Tissue Destruction," Publ. No. 75, pp. 339–369. Am. Assoc. Adv. Sci., Washington, D.C.
Goss, R. J. (1964). *Adv. Biol. Skin* **5,** 194–207.
Goss, R. J. (1968a). *Nature (London)* **220,** 83–85.
Goss, R. J. (1968b). *Am. Zool.* **8,** 786.
Goss, R. J. (1969a). "Principles of Regeneration." Academic Press, New York.
Goss, R. J. (1969b). *J. Exp. Zool.* **170,** 311–324.
Goss, R. J. (1969c). *J. Exp. Zool.* **171,** 223–234
Goss, R. J. (1970). *Clin. Orthop. Relat. Res.* **69,** 227–238.
Goss, R. J. (1974). *Perspect. Biol. Med.* **17,** 485–494.
Goss, R. J. (1976). *J. Exp. Zool.* **197,** 307–312.
Goss, R. J., Severinghaus, C. W., and Free, S. (1964). *J. Mammal.* **45,** 61–68.
Göthlin, G. (1973). *Acta Pathol. Microbiol. Scand., Ser. A* **81,** 507–522.
Göthlin, G., and Ericsson, J. L. E. (1973). *Virchows Arch. B* **12,** 318–329.
Göthlin, G., and Ericsson, J. L. E. (1976). *Clin. Orthop. Relat. Res.* **120,** 201–231.
Gould, R. P., Day, A., and Wolpert, L. (1972). *Exp. Cell Res.* **72,** 325–336.
Gould, R. P., Selwood, L., Day, A., and Wolpert, L. (1974). *Exp. Cell Res.* **83,** 287–296.
Graber, L. W. (1977). *Am. J. Orthod.* **72,** 23–41.
Graham, E. A., Rainey, R., Kuhlman, R. E., Houghton, E. H., and Moyer, C. A. (1962). *J. Bone Joint Surg., Am. Vol.* **44,** 482–488.

Graver, H. T. (1973). *J. Embryol. Exp. Morphol.* **30,** 635–646.

Green, H., and Goldberg, B. (1968). *Symp. Int. Soc. Cell. Biol.* **7,** 123–134.

Green, W. T., Jr., (1971). *Clin. Orthop. Relat. Res.* **75,** 248–260.

Green, W. T., Jr., and Dickens, R. (1972). *Surg. Forum* **23,** 453–455.

Green, W. T., Jr., and Ferguson, R. J. (1975). *Arthritis Rheum.* **18,** 273–280.

Greenberg, J. H., and Pratt, R. M. (1977). *Cell Differ.* **6,** 119–132.

Gritzka, T. L., Fry, L. R., Cheesman, R. L., and LaVigne, A. (1973). *J. Bone Joint Surg., Am. Vol.* **55,** 1698–1720.

Grobstein, C., and Holtzer, H. (1955). *J. Exp. Zool.* **128,** 333–357.

Grüneberg, H. (1954). *J. Genet.* **52,** 441–455.

Grüneberg, H. (1963). "The Pathology of Development. A Study of Inherited Skeletal Disorders in Animals." Blackwell, Oxford.

Grüneberg, H., and Lee, A. J. (1973). *J. Embryol. Exp. Morphol.* **30,** 119–141.

Grüneberg, H., and des Wickramaratne, G. A. (1974). *J. Embryol. Exp. Morphol.* **31,** 207–222.

Gurdon, J. B. (1970). *Proc. R. Soc. London, Ser. B* **176,** 303–314.

Gustafson, G. T., Eckerdal, O., Leever, D. L., Shanfield, J. L., Montgomery, P. C., and Davidovitch, Z. (1977). *J. Dent. Res.* **56,** 407–415.

Hadházy, Cs., Conti, G., Spreca, A., Musy, J. P., Cappelli-Gotzos, B., and Gotzos, V. (1971). *Acta Anat.* **78,** 362–382.

Hadházy, Cs., Glant, T., Mandi, B., and Miltényi, L. (1972). *Acta Biol. Acad. Sci. Hung.* **23,** 277–288.

Hadházy, Cs., Glant, T., Mandi, B., Harmati, S., Bordán, L., and Balogh, K. (1975). *Acta Morphol. Acad. Sci. Hung.* **23,** 183–193.

Hahn, E., Timpl, R., and Miller, E. J. (1975). *Immunology* **28,** 561–568.

Haines, A. L. (1957). *Yale J. Biol. Med.* **30,** 121–136.

Haines, R. W. (1976). *J. Anat.* **122,** 331–334.

Haines, R. W., and Mohiuddin, A. (1960). *J. Fac. Med., Baghdad* [N.S.] **2,** 16–22.

Haines, R. W., and Mohiuddin, A. (1968). *J. Anat.* **103,** 527–538.

Hajek, A. S., and Solursh, M. (1975). *Gen. Comp. Endocrinol.* **25,** 432–446.

Hale, L. J. (1956a). *Q. J. Microsc. Sci.* [N.S.] **97,** 333–353.

Hale, L. J. (1956b). *Q. J. Microsc. Sci.* [N.S.] **97,** 355–368.

Hales, S. (1727). "Vegetable Staticks." W. & J. Innys, London.

Hall, B. K. (1967a). *Life. Sci.* **6,** 663–667.

Hall, B. K. (1967b). *Aust. J. Zool.* **15,** 685–698.

Hall, B. K. (1968a). *Histochemie* **16,** 206–220.

Hall, B. K. (1968b). *J. Exp. Zool.* **168,** 283–306.

Hall, B. K. (1968c). Ph.D. Thesis, University of New England, Armidale, New South Wales, Australia.

Hall, B. K. (1968d). *Aust. J. Zool.* **16,** 795–806.

Hall, B. K. (1968e). *Aust. J. Zool.* **16,** 815–821.

Hall, B. K. (1969). *Life Sci.* **8,** 553–558.

Hall, B. K. (1970). *Biol. Rev. Cambridge Philos. Soc.* **45,** 455–484.

Hall, B. K. (1971). *Clin. Orthop. Relat. Res.* **74,** 249–268.

Hall, B. K. (1972a). *Dev. Biol.* **28,** 47–60.

Hall, B. K. (1972b). *Anat. Rec.* **173,** 391–404.

Hall, B. K. (1972c). *Can. J. Zool.* **50,** 1527–1536.

Hall, B. K. (1973). *Can. J. Zool.* **51,** 771–776.

Hall, B. K. (1975a). *Am. Zool.* **15,** 329–350.

Hall, B. K. (1975b). *Anat. Rec.* **181,** 767–778.

Hall, B. K. (1975c). *Anat. Rec.* **183,** 1–11.
Hall, B. K. (1977a). *Adv. Anat. Embryol. Cell Biol.* **53,** Part IV, 1–50.
Hall, B. K. (1977b). *Teratology* **15,** 1–16.
Hall, B. K. (1978a). *Tissue, Cult. Assoc. Manual* (in press).
Hall, B. K. (1978b). *Anat. Rec.* **190,** 243–256.
Hall, B. K. (1978c). *Anat. Rec.* (in press).
Hall, B. K. (1978d). *Arch. Oral Biol.* (in press).
Hall, B. K., and Jacobson, H. N. (1975). *Anat. Rec.* **181,** 55–70.
Hall, B. K., and Shorey, C. D. (1968). *Aust. J. Zool.* **16,** 821–840.
Hall, K. (1947). *J. Endocrinol.* **5,** 174–182.
Hall-Craggs, E. C. B. (1969). *Nature (London)* **221,** 1245.
Halstead, L. B. (1969a). "The Pattern of Vertebrate Evolution." Oliver & Boyd, Edinburgh.
Halstead, L. B. (1969b). *Proc. Malacol. Soc. London* **38,** 552–553.
Halstead, L. B. (1969c). *Calcif. Tissue Res.* **3,** 107–124.
Halstead, L. B. (1973). *Biol. Rev. Cambridge Philos. Soc.* **48,** 279–332.
Ham, A. W. (1930). *J. Bone Joint Surg., Am. Vol.* **12,** 827–844.
Ham, A. W. (1974). "Histology," 7th ed. Lippincott, Philadelphia, Pennsylvania.
Ham, A. W., and Harris, W. R. (1971). *In* "The Biochemistry and Physiology of Bone" (G. H. Bourne, ed.), 2nd ed., Vol. 3, pp. 337–399. Academic Press, New York.
Ham, R. G., and Sattler, G. L. (1968). *J. Cell. Physiol.* **72,** 109–114.
Ham, R. G., Murray, L. W., and Sattler, G. L. (1970). *J. Cell. Physiol.* **75,** 353–360.
Hamburger, V. (1941). *Physiol. Zool.* **14,** 355–364.
Hamburger, V., and Hamilton, H. L. (1951). *J. Morphol.* **88,** 49–92.
Hamburger, V., and Waugh, M. (1940). *Physiol. Zool.* **13,** 367–382.
Hammar, S. P., and Mottet, N. K. (1971). *J. Cell Sci.* **8,** 229–252.
Hammond, W. S., and Yntema, C. L. (1953). *Anat. Rec.* **115,** 393–394.
Hammond, W. S., and Yntema, C. L. (1964). *Acta Anat.* **56,** 21–34.
Hampé, A. (1958). *J. Embryol. Exp. Morphol.* **6,** 215–222.
Hampé, A. (1959). *Arch. Anat. Microsc. Morphol. Exp.* **48,** 345–478.
Hampé, A. (1960). *J. Embryol. Exp. Morphol.* **8,** 241–245.
Hanaoka, H. (1976). *J. Bone Joint Surg., Am. Vol.* **58,** 226–229.
Hancox, N. M. (1972a). "Biology of Bone." Cambridge Univ. Press, London and New York.
Hancox, N. M. (1972b). *In* "The Biochemistry and Physiology of Bone" (G. H. Bourne, ed.), 2nd ed., Vol. 1, pp. 45–69. Academic Press, New York.
Hancox, N. M., and Wlodarski, K. (1972). *Calcif. Tissue Res.* **8,** 258–261.
Hardingham, T. E., Fitton Jackson, S., and Muir, H. (1972). *Biochem. J.* **129,** 101–112.
Harkey, M. E., and Crelin, E. S. (1963a). *Anat. Rec.* **145,** 322–323.
Harkey, M. E., and Crelin, E. S. (1963b). *Anat. Rec.* **145,** 323.
Harkness, E. M., and Trotter, W. D. (1975). *Proc. Conf. Anat. Soc. Aust. & N.Z., 13th, 1975,* p. 15.
Harkness, E. M. (1976). *J. Dent. Res.* **55,** 1134.
Harris, E. D., Jr., and McCroskery, P. A. (1974). *N. Engl. J. Med.* **290,** 1–6.
Harrison, R. G. (1918). *J. Exp. Zool.* **25,** 413–462.
Hartwig, H. (1968). *Z. Saeugetierk.* **33,** 246–248.
Hartwig, H. (1972). *Z. Jagdwiss.* **18,** 166–168.
Hartwig, H., and Schrudde, J. (1974). *Z. Jagdwiss.* **20,** 1–13.
Hassell, J. R., Greenberg J. H., and Johnston, M. C. (1977). *J. Embryol. Exp. Morphol.* **39,** 267–271.
Havdrup, T., Hulth, A., and Telhag, H. (1975). *Clin. Orthop. Relat. Res.* **113,** 246–248.

Hay, E. D. (1965). *In* "Organogenesis" (R. L. de Haan and H. Ursprung, eds.), pp. 315–336. Holt, New York.

Hay, E. D., and Dodson, J. W. (1973). *J. Cell Biol.* **57,** 190–213.

Hay, E. D., and Meier, S. (1974). *J. Cell Biol.* **62,** 889–898.

Heersche, J. N. M. (1974). *Endocrinol. Proc. Int. Congr., 4th, 1972* Int. Congr. Ser. No. 273, pp. 359–364.

Heersche, J. N. M., Fedak, S. A., and Aurbach, G. D. (1971). *J. Biol. Chem.* **246,** 6770–6775.

Heersche, J. N. M., Marcus, M. R., and Aurbach, G. D. (1974). *Endocrinology* **94,** 241–247.

Heller, M., McLean, F. C., and Bloom, W. (1950). *Am. J. Anat.* **87,** 315–348.

Heppenstall, R. B., Grislis, G., and Hunt, T. K. (1975). *Clin. Orthop. Relat. Res.* **106,** 357–365.

Herington, A. C., Adamson, L. F., and Bornstein, J. (1972). *Biochim. Biophys. Acta* **286,** 164–174.

Herring, S. W. (1974). *Anat. Embryol.* **146,** 167–180.

Hert, J. (1969). *Folia Morphol. (Prague)* **17,** 194–203.

Hert, J. (1972). *Acta Anat.* **82,** 420–436.

Hert, J., and Zalud, V. (1971). *Acta Chir. Orthop. Traum. Cech.* **5,** 280–288.

Hert, J., Liskova, M., and Landrgot, B. (1969). *Folia Morphol. (Prague)* **17,** 389–399.

Hert, J., Liskova, M., and Landa, J. (1971a). *Folia Morphol. (Prague)* **19,** 290–300.

Hert, J., Sklenská, A., and Liskova, M. (1971b). *Folia Morphol. (Prague)* **19,** 378–387.

Hert, J., Koudela, K., and Hert, J. (1972a). *Anat. Anz.* **131,** 105–123.

Hert, J., Pribylová, E., and Liskova, M. (1972b). *Acta Anat.* **82,** 218–230.

Hewitt, A. T., and Elmer, W. A. (1976). *Nature (London)* **264,** 177–178.

Hewitt, A. T., and Elmer, W. A. (1978). *Differentiation* **10,** 31–38.

Heyner, S. (1969). *Transplantation* **8,** 666–677.

Hilfer, S. R., Searls, R. L., and Fonte, V. G. (1973). *Dev. Biol.* **30,** 374–391.

Hillman, J. R., Davis, R. W., and Abdelbaki, Y. Z. (1973). *Calcif. Tissue Res.* **12,** 323–330.

Hinchliffe, J. R. (1974). *Teratology*, **9,** 263–298.

Hinchliffe, J. R., and Ede, D. A. (1967). *J. Embryol. Exp. Morphol.* **17,** 385–404.

Hinchliffe, J. R., and Ede, D. A. (1968). *J. Embryol. Exp. Morphol.* **19,** 327–339.

Hinchliffe, J. R., and Ede, D. A. (1973). *J. Embryol. Exp. Morphol.* **30,** 753–772.

Hinchliffe, J. R., and Thorogood, P. V. (1974). *J. Embryol. Exp. Morphol.* **31,** 747–760.

Hinrichsen, G. J., and Storey, E. (1968). *Angle Orthod.* **38,** 155–165.

Hirschmann, P. N., and Shuttleworth, C. A. (1976). *Arch. Oral Biol.* **21,** 771–774.

Ho, P. L., Levitt, D., and Dorfman, A. (1977). *Dev. Biol.* **55,** 233–243.

Ho, T. Y. (1967). *Biochim. Biophys. Acta* **133,** 568–573.

Hogan, B., and Tilly, R. (1977). *Nature (London)* **265,** 626–629.

Holliday, R., and Pugh, J. E. (1975). *Science* **187,** 226–232.

Holmes, L. B., and Trelstad, R. L. (1977). *Dev. Biol.* **59,** 164–173.

Holtfreter, J. (1968). *In* "Epithelial-Mesenchymal Interactions" (R. Fleischmajer and R. E. Billingham, eds.), pp. 1–30. Williams & Wilkins, Baltimore, Maryland.

Holtrop, M. E. (1966). *Calcif. Tissues 1965, Proc. Eur. Symp., 3rd, 1965,* pp. 32–36.

Holtrop, M. E. (1967). *Proc. K. Ned. Akad. Wet. Ser. C* **70,** 21–28.

Holtrop, M. E. (1971). *Isr. J. Med. Sci.* **7,** 473–476.

Holtrop, M. E. (1972a). *Calcif. Tissue Res.* **9,** 131–139.

Holtrop, M. E. (1972b). *Calcif. Tissue Res.* **9,** 140–151.

Holtrop, M. E., and King, G. J. (1977). *Clin. Orthop. Relat. Res.* **123,** 177–196.

Holtrop, M. E., and Weinger, J. M. (1971). *Proc. Parathyroid Conf., 4th, 1971* Int. Congr. Ser. No. 243, pp. 365–374.

Holtzer, H. (1951). *Anat. Rec.* **109,** 373–374.

Holtzer, H. (1952a). *J. Exp. Zool.* **121,** 121–148.

Holtzer, H. (1952b). *J. Exp. Zool.* **121,** 573–591.

Holtzer, H. (1964a). *Biophys. J.* **4,** 239–250.
Holtzer, H. (1964b). *Proc. Int. Congr. Congenital Malformations, 2nd, 1961* pp. 233–239.
Holtzer, H., and Abbott, J. (1968). *In* "Results and Problems in Cell Differentiation" (H. Ursprung, ed.), Vol. 1, pp. 1–16. Springer-Verlag, Berlin and New York.
Holtzer, H., and Detwiler, S. R. (1953). *J. Exp. Zool.* **123,** 335–366.
Holtzer, H., and Mayne, R. (1973). *In* "Pathobiology of Development—or Ontogeny Revisited" (E. V. D. Perrin and M. J. Finegold, eds.), pp. 52–65. Williams & Wilkins, Baltimore, Maryland.
Holtzer, H., Abbott, J., Lash, J. W., and Holtzer, S. (1960). *Proc. Natl. Acad. Sci. U.S.A.* **46,** 1533–1542.
Holtzer, H., Rubinstein, N., Fellini, S., Yeoh, G., Chi, J., Birnbaum, J., and Okayama, M. (1975). *Q. Rev. Biophys.* **8,** 523–557.
Horn, E. H. (1958). *Endocrinology* **63,** 481–486.
Hornbruch, A., and Wolpert, L. (1970). *Nature (London)* **226,** 764–766.
Hörstadius, S. (1944). *Acta Zool. (Stockholm)* **25,** 75–87.
Hörstadius, S. (1950). "The Neural Crest: Its Properties and Derivatives in the Light of Experimental Research." Oxford Univ. Press, London and New York.
Houben, J. J. G. (1976). *Arch. Biol.* **87,** 345–365.
Howell, J. A. (1917). *Anat. Rec.* **13,** 233–252.
Howes, R. I., Jr. (1977a). *Acta Anat.* **97,** 151–165.
Howes, R. I., Jr. (1977b). *J. Dent. Res.* **56,** A70.
Hoyte, D. A. N. (1966). *Int. Rev. Gen. Exp. Zool.* **2,** 345–407.
Hoyte, D. A. N. (1971). *J. Dent. Res.* **50,** 1447–1461.
Hsie, A. W., Jones, C., and Puck, T. T. (1971). *Proc. Natl. Acad. Sci. U.S.A.* **68,** 1648–1652.
Huang, D. (1974). *J. Cell Biol.* **62,** 881–886.
Huang, D. (1977). *Clin. Orthop. Relat. Res.* **123,** 169–176.
Huggins, C. B. (1931a). *Arch. Surg. (Chicago)* **22,** 377–408.
Huggins, C. B. (1931b). *Biochem. J.* **25,** 728–732.
Huggins, C. B., and Sammett, J. F. (1933). *J. Exp. Med.* **58,** 393–400.
Huggins, C. B., McCarroll, H. R., and Blocksom, B. H., Jr. (1936). *Arch. Surg. (Chicago)* **32,** 915–931.
Hulth, A., Lindberg, L., and Telhag, H. (1972). *Clin. Orthop. Relat. Res.* **84,** 197–199.
Humphrey, G. M. (1861). *Med.-Chir. Trans.* **44,** 117.
Hyldebrandt, N., Damholt, W., and Nordentoft, E. L. (1974). *Acta Orthop. Scand.* **45,** 175–181.
Igarashi, S., Trelstad, R. L., and Kang, A. H. (1973). *Biochim. Biophys. Acta* **P24,** 514–519.
Ioseliani, D. G. (1972). *Clin. Orthop. Relat. Res.* **88,** 183–196.
Irving, J. T. (1973). *Clin. Orthop. Relat. Res.* **97,** 225–236.
Irving, J. T., and Wuthier, R. E. (1968). *Clin. Orthop. Relat. Res.* **56,** 237–260.
Isotupa, K., Koski, K., and Makinen, L. (1965). *Am. J. Phys. Anthrop.* **23,** 19–22.
Iwata, H., and Urist, M. R. (1973). *Clin. Orthop. Relat. Res.* **90,** 236–245.
Jabaily, J., Rall, T. W., and Singer, M. (1975). *J. Morphol.* **147,** 379–384.
Jacobson, W., and Fell, H. B. (1941). *Q. J. Microsc. Sci.* [N.S.] **82,** 563–586.
Jacoby, R. K., and Jayson, M. I. U. (1975). *J. Rheumatol.* **2,** 270–279.
Jaczewski, Z., and Krzywinska, K. (1974). *Bull. Acad. Pol. Sci.* **22,** 67–72.
Jaffe, H. L. (1968). "Tumors and Tumorous Conditions of the Bones and Joints." Lea & Febiger, Philadelphia, Pennsylvania.
Jaffe, H. L. (1972). "Metabolic, Degenerative, and Inflammatory Diseases of Bones and Joints." Lea & Febiger, Philadelphia, Pennsylvania.
Jain, P., and Sabet, T. Y. (1974). *J. Dent. Res.* **53,** 282.
James, R. B., Alexander, R. W., and Traver, J. G., Jr. (1974). *Oral Surg., Oral Med. Oral Pathol.* **37,** 189–195.

Jande, S. S. (1971). *J. Ultrastruct. Res.* **37,** 279–300.
Jande, S. S., and Bélanger, L. F. (1969). *Anat. Rec.* **163,** 204.
Jande, S. S., and Bélanger, L. F. (1971). *Calcif. Tissue Res.* **6,** 280–289.
Jande, S. S., and Bélanger, L. F. (1973). *Clin. Orthop. Relat. Res.* **94,** 281–305.
Janis, R., Sandson, J., Smith, C., and Hamerman, D. (1967). *Science* **158,** 1464–1467.
Janners, M. Y., and Searls, R. L. (1970). *Dev. Biol.* **23,** 136–165.
Janners, M. Y., and Searls, R. L. (1971). *Dev. Biol.* **24,** 465–476.
Jarry, L., and Uhthoff, H. K. (1971). *Can. J. Surg.* **14,** 127–135.
Jeanloz, R. W. (1960). *Arthritis Rheum.* **3,** 233–234.
Jenkins, D. H. R., Cheng, D. H. F., and Hodgson, A. R. (1975). *J. Bone Joint Surg., Br. Vol.* **57,** 482–484.
Johnson, D. R. (1974). *J. Embryol. Exp. Morphol.* **31,** 313–318.
Johnson, D. R. (1977). *J. Embryol. Exp. Morphol.* **39,** 279–284.
Johnson, D. R. (1978). *J. Anat.* **125,** 267–276.
Johnson, D. R., and Hunt, D. M. (1974). *J. Embryol. Exp. Morphol.* **31,** 319–328.
Johnson, D. R., and Wise, J. M. (1971). *J. Embryol. Exp. Morphol.* **25,** 21–31.
Johnson, F. R., and McMinn, R. M. H. (1956). *J. Anat.* **90,** 106–116.
Johnson, L. G. (1973). *Dev. Biol.* **30,** 223–227.
Johnston, M. C. (1964). *J. Dent. Res.* **43,** 822.
Johnston, M. C. (1966). *Anat. Rec.* **156,** 143–156.
Johnston, M. C. (1975). *Birth Defects, Orig. Artic. Ser. 11.* 1–18.
Johnston, M. C. (1977). *Birth Defects, Orig. Artic. Ser. 13.* 1–6.
Johnston, M. C., and Listgarten, M. A. (1972). *In* "Developmental Aspects of Oral Biology" (H. C. Slavkin and L. A. Bavetta, eds.), pp. 55–80. Academic Press, New York.
Johnston, P. M., and Comar, C. L. (1957). *J. Biophys. Biochem. Cytol.* **3,** 231–238.
Jollie, M. (1971). *J. Morphol.* **133,** 17–40.
Jolly, R. J., and Moore, W. J. (1975). *J. Embryol. Exp. Morphol.* **33,** 1013–1022.
Jones, K. L., and Addison, J. (1975). *Endocrinology* **97,** 359–365.
Jones, S. J. (1974). *Calcif. Tissue Res.* **14,** 309–316.
Jones, S. J., and Boyde, A. (1976a). *Cell Tissue Res.* **166,** 101–108.
Jones, S. J., and Boyde, A. (1976b). *Cell Tissue Res.* **169,** 449–466.
Jones, S. J., Boyde, A., and Pawley, J. B. (1975). *Cell Tissue Res.* **159,** 73–80.
Jurand, A. (1965). *Proc. R. Soc. London, Ser. B* **162,** 387–405.
Kahn, A. J., and Simmons, D. J. (1975). *Nature (London).* **258,** 325–327.
Kahn, A. J., Stewart, C. C., and Teitelbaum, S. L. (1978). *Science* **199,** 988–990.
Kalayjian, D. B., and Cooper, R. R. (1972). *Clin. Orthop. Relat. Res.* **85,** 242–256.
Kaminski, M., Kaminski, G., Jakobisiak, M., and Brzezinski, W. (1977). *Nature (London)* **268,** 238–240.
Kantorova, V. I. (1972). *Ontogenesis,* **3,** 448–455.
Kaprio, E. A. (1977). *Wilhelm Roux. Arch. Entwicklungsmech. Org.* **182,** 213–226.
Karcher-Djuricic, V., Ruch, J.-V., Stäubli, A., and Fabre, M. (1975). *Folia Anat. Jugosl.* **4,** 53–58.
Katchburian, E. (1973). *J. Anat.* **116,** 285–302.
Katchburian, E., and Burgess, A. M. C. (1977). *Arch. Oral Biol.* **22,** 551–553.
Katzman, R. L., and Jeanloz, R. G. (1969). *Science,* **166,** 758–759.
Kay, S., and Harrison, J. M. (1969). *Cancer* **23,** 1158–1162.
Kebabian, J. W., Bloom, F. E., Steiner, A. L., and Greengard, P. (1975). *Science* **190,** 157–159.
Kelley, R. O. (1973). *J. Embryol. Exp. Morphol.* **29,** 117–131.
Kelley, R. O., and Bluemink, J. G. (1974). *Dev. Biol.* **37,** 1–17.
Kelley, R. O., and Fallon, J. F. (1976). *Dev. Biol.* **51,** 241–256.

Kelsall, M. A., and Visci, M. (1970). *Anat. Rec.* **166,** 627–634.
Kember, N. F. (1960). *J. Bone Joint. Surg., Br. Vol.* **42,** 824–839.
Kember, N. F. (1971). *Clin. Orthop. Relat. Res.* **76,** 213–230.
Kember, N. F. (1973). *J. Anat.* **116,** 445–452.
Kember, N. F., and Walker, K. V. R. (1971). *Nature (London)* **229,** 428–429.
Kemp, N. E. (1977). *Am. Zool.* **17,** 932.
Kenrad, B., and Vilmann, H. (1977). *Calcif. Tissue Res.* **181,** 349–359.
Kernek, C. B., and Wray, J. B. (1973). *Clin. Orthop. Relat. Res.* **91,** 197–209.
Kéry, L. (1972). *Acta Chir. Acad. Sci. Hung.* **13,** 133–140.
Kieny, M. (1960). *J. Embryol. Exp. Morphol.* **8,** 457–467.
Kieny, M. (1971). *Ann. Embryol. Morphog.* **4,** 281–298.
Kieny, M. (1975). *J. Embryol. Exp. Morphol.* **34,** 609–632.
Kieny, M. (1977). *In* "Vertebrate Limb and Somite Morphogenesis" D. A. Ede, (J. R. Hinchliffe, and M. Balls, eds.), pp. 87–104. Cambridge Univ. Press, London and New York.
Kieny, M., and Abbott, U. K. (1962). *Dev. Biol.* **4,** 473–488.
Kieny, M., and Pautou, M. P. (1976). *Wilhelm Roux Arch. Entwicklungsmech. Org.* **179,** 327–338.
Kieny, M., and Pautou, M. P. (1977). *Wilhelm Roux Arch. Entwicklungsmech. Org.* **183,** 177–192.
Kieny, M. Mauger, A., and Sengel, P. (1972). *Dev. Biol.* **28,** 142–161.
Kim, J. J., and Conrad, H. E. (1974). *J. Biol. Chem.* **249,** 3091–3097.
King, G. J., and Holtrop, M. E. (1974). *J. Dent. Res.* **53,** 245.
King, G. J., and Holtrop, M. E. (1975). *J. Cell Biol.* **66,** 445–450.
King, M. C., and Wilson, A. C. (1975). *Science* **188,** 107–116.
Kingsbury, J. E., Allen, C. G., and Rotheram, B. A. (1953). *Anat. Rec.* **116,** 95–111.
Kitchin, I. C. (1949). *J. Exp. Zool.* **112,** 393–415.
Klein, I. E. (1975). *J. Prosthet. Dent.* **33,** 365–379.
Klein, L., and Zika, J. M. (1976). *Calcif. Tissue Res.* **20,** 217–228.
Kleine, T. D. (1972). *Naturwissenschaften* **59,** 64–71.
Kleinman, H. K., Pennypacker, J. P., and Brown, K. S. (1977). *Growth* **41,** 171–178.
Knese, K. H. (1972). *Acta Anat.* **83,** 275–288.
Knese, K. H., and Knoop, A. M. (1961a). *Z. Zellforsch. Mikrosk. Anat.* **53,** 201–258.
Knese, K. H., and Knoop, A. M. (1961b). *Z. Zellforsch. Mikrosk. Anat.* **54,** 1–38.
Kobayashi, S. (1971). *Int. Rev. Cytol.* **30,** 257–371.
Koch, W. E. (1967). *J. Exp. Zool.* **165,** 155–170.
Koch, W. E. (1972). *In* "Developmental Aspects of Oral Biology" (H. C. Slavkin and L. A. Bavetta, eds.), pp. 151–164. Academic Press, New York.
Kochhar, D. M. (1973). *Teratology* **7,** 289–295.
Kochhar, D. M., Aydelotte, M. B., and Vest, T. K. (1976). *Exp. Cell Res.* **102,** 213–222.
Kollar, E. J. (1972). *Am. Zool.* **12,** 125–135.
Kollar, E. J. (1975). *Dent. Clin. North Am.* **19,** 141–150.
Kollar, E. J., and Baird, G. R. (1969). *J. Embryol. Exp. Morphol.* **21,** 131–148.
Kollar, E. J., and Baird, G. R. (1970). *J. Embryol Exp. Morphol.* **24,** 173–186.
Kölliker, A. (1859). *Proc. R. Soc. London* **9,** 656–668.
Konyukhov, B., and Ginter, E. (1966). *Folia Biol. (Prague)* **12,** 199–206.
Konyukhov, B., and Paschin, Y. V. (1967). *Acta Biol. Acad. Sci. Hung.* **18,** 285–294.
Konyukhov, B., and Paschin, Y. V. (1970). *Proc. Int. Congr. Anat., 9th 1970,* p. 69.
Kosher, R. A. (1976). *Dev. Biol.* **53,** 265–276.
Kosher, R. A., and Church, R. L. (1975). *Nature (London)* **258,** 327–329.
Kosher, R. A., and Lash, J. W. (1975). *Dev. Biol.* **42,** 362–378.
Kosher, R. A., and Searls, R. L. (1973). *Dev. Biol.* **32,** 50–68.
Kosher, R. A., Lash, J. W., and Minor, R. R. (1973). *Dev. Biol.* **35,** 210–220.

Koski, K. (1975). *Birth Defects, Orig. Artic. Ser.* **11,** 231–254.

Koski, K., and Mäkinen, L. (1963). *Fin. Tandlak. Sallsk. Forh.* **59,** 296–308.

Koski, K., and Rönning, O. (1969). *Acta Odontol. Scand.* **27,** 343–357.

Koski, K., and Rönning, O. (1970). *Arch. Oral Biol.* **15,** 1107–1108.

Koskinen, E. V. S., Ryöppy, S. A., and Lindholm, T. S. (1972). *Clin. Orthop. Relat. Res.* **87,** 116–131.

Koskinen, L., Isotupa, K., and Koski, K. (1976). *Am. J. Phys. Anthropol.* **45,** 511–516.

Krotoski, D. M., and Elmer, W. A. (1973). *Teratology* **7,** 99–106.

Kuettner, K. E., Wezeman, F. H., Simmons, D. J., Lisk, P. Y., Croxen, R. L., Soble, L. W., and Eisenstein, R. (1972). *Lab. Invest.* **27,** 324–330.

Kuettner, K. E., Hiti, J., Eisenstein, R., and Harper, E. (1976a). *Biochem. Biophys. Res. Commun.* **72,** 40–46.

Kuettner, K. E., Soble, L. W., Sorgente, N., and Eisenstein, R. (1976b). *Protides Biol. Fluids, Proc. Colloq.* pp. 221–225.

Kuettner, K. E., Harper, E., and Eisenstein, R. (1977a). *Arthritis Rheum.* **20,** 5124–5129.

Kuettner, K. E., Soble, L., Croxon, R. L., Marczynska, B., Hiti, J., and Harper, E. (1977b). *Science* **196,** 653–654.

Kuhlman, R. E., and McNamee, M. J. (1970). *J. Bone Joint Surg., Am. Vol.* **52,** 1025–1032.

Kuhlman, R. E., Rainey, R., and O'Neill, R. (1963). *J. Bone Joint Surg., Am. Vol.* **45,** 345–350.

Kulonen, E., and Pikkarairen, J. eds. (1973). "Biology of Fibroblast." Academic Press, New York.

Kuralesova, A. I. (1971). *Byull. Eksp. Biol. Med.* **71,** 92–95.

Kuroda, Y., and Shibuya, T. (1969). *Annu. Rep. Natl. Inst. Gen.* **20,** 29–30.

Kvinnsland, S. (1974). *Growth* **38,** 381–388.

Kvinnsland, S., and Kvinnsland, S. (1975). *Growth* **39,** 305–314.

Kvist, T. N., and Finnegan, C. V. (1970). *J. Exp. Zool.* **175,** 241–257.

Kwasigroch, T. E., and Kochhar, D. M. (1975). *Exp. Cell Res.* **95,** 269–278.

Lacroix, P. (1951). "The Organisation of Bones" (Transl. by S. Gilder). Churchill, London.

Laird, A. K. (1966). *Growth* **30,** 263–275.

Lalykina, K. S., and Friedenstein, A. Y. (1969). *Byull. Eksp. Biol. Med.* **67,** 105–108.

Lamberg, S. I., and Stoolmiller, A. C. (1974). *J. Invest. Dermatol.* **63,** 433–449.

Landauer, W. (1932a). *J. Genet.* **25,** 367–394.

Landauer, W. (1932b). *J. Genet.* **26,** 285–290.

Landauer, W. (1934). *Am. J. Anat.* **55,** 229–252.

Landauer, W. (1957). *J. Exp. Zool.* **136,** 509–530.

Landauer, W. (1965). *J. Hered.* **56,** 131–138.

Landauer, W. (1969a). *In* "Limb Development and Deformity. Problems of Evaluation and Rehabilitation" (C. A. Swinyard, ed.), pp. 120–135. Thomas, Springfield, Illinois.

Landauer, W. (1969b). *In* "Limb Development and Deformity. Problems of Evaluation and Rehabilitation" (C. A. Swinyard, ed.), pp. 540–621. Thomas, Springfield, Illinois.

Landauer, W. (1976). *Teratology* **13,** 41–46.

Landauer, W., and Dunn, L. C. (1930). *J. Genet.* **23,** 397–413.

Landis, W. J., Paine, M. C., and Glimcher, M. J. (1977a). *J. Ultrastruct. Res.* **59,** 1–30.

Landis, W. J., Hauschka, B. T., Rogerson, C. A., and Glimcher, M. J. (1977b). *J. Ultrastruct. Res.* **59,** 185–206.

Lane, J. M., and Brighton, C. T. (1974). *Arthritis Rheum.* **17,** 235–244.

Lane, J. M., and Weiss, C. (1975). *Arthritis Rheum.* **18,** 553–562.

Lane, P. W., and Dickie, M. M. (1968). *J. Hered.* **59,** 300–308.

Langer, F., and Gross, A. E. (1974). *J. Bone Joint Surg., Am. Vol.* **56,** 297–304.

Langer, R., Brem, H., Falterman, K., Klein, M., and Folkman, J. (1976). *Science* **193,** 70–72.

Langness, U., and Udenfriend, S. (1974). *Proc. Natl. Acad. Sci. U.S.A.* **71,** 50–51.

Lanyon, L. E. (1972). *Vet. Annu.* **13,** 126–129.
Lanyon, L. E. (1974). *J. Bone Joint Surg., Br. Vol.* **56,** 160–166.
Lanyon, L. E., Hampson, W. G. J., Goodship, A. E., and Shah, J. S. (1975). *Acta Orthop. Scand.* **46,** 256–268.
Larsson, S.-E., and Kuettner, K. E. (1974). *Calcif. Tissue Res.* **14,** 49–58.
Lash, J. W. (1959). *Science* **130,** 334.
Lash, J. W. (1964). *Am. J. Obstet. Gynecol.* **90,** 1193–1207.
Lash, J. W. (1967). *J. Exp. Zool.* **165,** 47–56.
Lash, J. W. (1968). *J. Cell. Physiol.* **72** Suppl. 1, 35–46.
Lash, J. W., and Whitehouse, M. W. (1960). *Biochem. J.* **74,** 351–355.
Lash, J. W., Holtzer, S., and Holtzer, H. (1957). *Exp. Cell Res.* **13,** 292–303.
Lash, J. W., Holtzer, H., and Whitehouse, M. W. (1960). *Dev. Biol.* **2,** 76–89.
Lash, J. W., Glick, M. C., and Madden, J. W. (1964). *Natl. Cancer Inst., Monogr.* **13,** 39–49.
Lasher, R. (1971). *In* "Developmental Aspects of the Cell Cycle" (I. L. Cameron, G. M. Padilla, and A. M. Zimmerman, eds.), pp. 223–241. Academic Press, New York.
Latham, R. A., Smiley, G. R., and Gregg, J. M. (1973). *Br. J. Plast. Surg.* **26,** 252–260.
Laufer, H. (1959). *J. Embryol. Exp. Morphol.* **7,** 431–458.
Lauthier, M. (1974). *Wilhelm Roux Arch. Entwicklungsmech. Org.* **175,** 185–198.
Lauthier, M. (1977). *J. Embryol. Exp. Morphol.* **38,** 1–18.
Lavietes, B. B. (1971). *Exp. Cell Res.* **68,** 43–48.
Layman, D. L., Sokoloff, L., and Miller, E. J. (1972). *Exp. Cell Res.* **73,** 107–112.
LeDouarin, N. (1973). *Dev. Biol.* **30,** 217–222.
LeDouarin, N. (1974). *Med. Biol.* **52,** 281–319.
LeDouarin, N. (1975). *Birth Defects, Orig. Artic. Ser. 11,* 19–50.
LeDouarin, N., and Teillet, M.-A. (1973a). *C.R. Hebd. Seances Acad. Sci.* **277,** 1929–1932.
LeDouarin, N., and Teillet, M.-A. (1973b). *J. Embryol. Exp. Morphol.* **30,** 31–48.
LeDouarin, N., and Teillet, M.-A. (1974). *Dev. Biol.* **41,** 162–184.
LeDouarin, N., Fontaine, J., and LeLièvre, C. (1974). *Histochemie* **38,** 297–306.
Lee-Own, V., and Anderson, J. C. (1975). *Biochem. J.* **149,** 57–64.
Leibovich, S. J., and Weiss, J. B. (1973). *Connect. Tissue Res.* **2,** 11–20.
LeLièvre, C. (1974). *J. Embryol. Exp. Morphol.* **31,** 453–477.
LeLièvre, C., and LeDouarin, N. (1975). *J. Embryol. Exp. Morphol.* **34,** 125–154.
Lelkes, G. (1958). *J. Embryol. Exp. Morphol.* **6,** 183–186.
Lemperg, R. K., Bergenholtz, A., and Smith, T. W. D. (1975). *In Vitro* **11,** 291–301.
Leonard, L., and Sharawy, M. (1974). *J. Dent. Res.* **53,** 157.
Lerner, I. M. (1936). *Am. Nat.* **70,** 595–598.
Leung, D. Y. M., Glagov, S., and Mathews, M. B. (1976). *Science* **191,** 475–477.
Leung, D. Y. M., Glagov, S., and Mathews, M. B. (1977). *Exp. Cell Res.* **109,** 285–298.
Leutenegger, W. (1973). *Am. J. Phys. Anthropol.* **38,** 251–254.
Levak-Svajger, B., and Moscona, A. A. (1964). *Exp. Cell Res.* **36,** 692–695.
Levene, C. I., and Bates, C. J. (1970). *J. Cell Sci.* **7,** 671–682.
Levenson, G. E. (1969). *Exp. Cell Res.* **55,** 225–228.
Levenson, G. E. (1970). *Exp. Cell Res.* **62,** 271–285.
Levitt, D., and Dorfman, A. (1972). *Proc. Natl. Acad. Sci. U.S.A.* **69,** 1253–1257.
Levitt, D., and Dorfman, A. (1973). *Proc. Natl. Acad. Sci. U.S.A.* **70,** 2201–2205.
Levitt, D., and Dorfman, A. (1974). *Curr. Top. Dev. Biol.* **8,** 103–149.
Levitt, D., Ho, P. L., and Dorfman, A. (1974). *In* "The Cell Surface in Development" (A. A. Moscona, ed.), pp. 101–126. Wiley, New York.
Lewis, E. A., and Irving, J. T. (1970). *Arch. Oral Biol.* **15,** 769–776.
Lewis, J. H. (1975). *J. Embryol. Exp. Morphol.* **33,** 419–434.

Lewis, J. H., Summerbell, D., and Wolpert, L. (1972). *Nature (London)* **239,** 276–279.
Libben, R. M., Ozer, R., Person, P., and Hirschman, A. (1976). *Calcif. Tissue Res.* **22,** 67–76.
Liboff, A. R., and Rinaldi, R. A., eds. (1974). *Ann. N.Y. Acad. Sci.* **238,** 1–593.
Linck, G., Oudet, C., and Petrovic, A. (1975). *Bull. Assoc. Anat.* **59,** 467–478.
Lincoln, G. A. (1973). *J. Embryol. Exp. Morphol.* **29,** 431–437.
Lincoln, G. A., and Fletcher, T. J. (1976). *J. Exp. Zool.* **195,** 247–252.
Lindsay, K. N. (1977). *Arch. Oral Biol.* **22,** 711–714.
Linsenmayer, T. F. (1974). *Dev. Biol.* **40,** 372–377.
Linsenmayer, T. F., and Smith, G. N., Jr. (1976). *Dev. Biol.* **52,** 19–30.
Linsenmayer, T. F., Toole, B. P., and Trelstad, R. L. (1973a). *Dev. Biol.* **35,** 232–239.
Linsenmayer, T. F., Trelstad, R. L., Toole, B. P., and Gross, J. (1973b). *Biochem. Biophys. Res. Commun.* **52,** 870–876.
Liskova, M. (1976). *Calcif. Tissue Res.* **22,** 207–218.
Liskova, M., and Hert, J. (1971). *Folia Morphol. (Prague)* **19,** 301–317.
Listgarten, M. A., and Shapiro, I. M. (1974). *J. Dent. Res.* **53,** 260.
Little, K. (1973). "Bone Behaviour." Academic Press, London and New York.
Liversage, R. A., Rathbone, M. P., and McLaughlin, H. M. G. (1977). *J. Exp. Zool.* **200,** 169–175.
Lockshin, R. A., and Beaulaton, J. (1975). *Life Sci.* **15,** 1549–1565.
Loewenthal, L. A. (1957). *Anat. Rec.* **128,** 201–210.
Lopez, E. (1970a). *Z. Zellforsch. Mikrosk. Anat.* **109,** 552–565.
Lopez, E. (1970b). *Z. Zellforsch. Mikrosk. Anat.* **109,** 566–572.
Lopez, E., and Deville, J. (1973). *Proc. Eur. Symp. Calcif. Tissue, 9th, 1972* pp. 169–174.
Lopez, E., and Martelly-Bagot, E. (1971). *Z. Zellforsch. Mikrosk. Anat.* **117,** 176–190.
Loutit, J. F., and Sansom, J. M. (1976). *Calcif. Tissue Res.* **20,** 251–259.
Løvtrup, S. (1974). "Epigenetics." Wiley, New York.
Løvtrup, S. (1977). "The Phylogeny of Vertebrate." Wiley, New York.
Lowther, D. A., and Natarajan, M. (1972). *Biochem. J.* **127,** 607–608.
Luben, R. A., Wong, G. L., and Cohn, D. V. (1977). *Nature (London)* **265,** 629–630.
Luk, S. C., Nopagaroonsri, C., and Simon, G. T. (1974). *J. Ultrastruct. Res.* **46,** 165–183.
Lumsden, A., and Osborn, J. (1976). *J. Dent. Res.* **55,** D136.
Lutfi, A. M. (1971). *Acta Anat.* **79,** 27–35.
Lutfi, A. M. (1974). *J. Anat.* **117,** 413–418.
MacCabe, J. A., and Abbott, U. K. (1974). *J. Embryol. Exp. Morphol.* **31,** 735–746.
MacCabe, J. A., and Parker, B. W. (1975). *Dev. Biol.* **45,** 349–357.
MacCabe, J. A., and Parker, B. W. (1976a). *Dev. Biol.* **54,** 297–303.
MacCabe, J. A., and Parker, B. W. (1976b). *J. Exp. Zool.* **195,** 311–317.
MacCabe, J. A., Saunders, J. W., Jr., and Pickett, M. (1973). *Dev. Biol.* **31,** 323–335.
MacCabe, J. A., Errick, J. E., and Saunders, J. W., Jr. (1974). *Dev. Biol.* **39,** 69–82.
MacCabe, J. A., MacCabe, A. B., Abbott, U. K., and McCarrey, J. R. (1975). *J. Exp. Zool.* **191,** 383–394.
McConaghey, P., and Sledge, C. B. (1970). *Nature (London).* **225,** 1249–1250.
McConnell, T. H. (1970). *Cancer* **25,** 611–617.
MacEwen, W. (1912). "The Growth of Bone. Observations on Osteogenesis. An Experimental Enquiry into the Development and Reproduction of Diaphyseal Bone." James Maclehose & Sons, Glasgow.
McGuire, J. L., and Marks, S. C., Jr. (1974). *Clin. Orthop. Relat. Res.* **100,** 392–405.
McHenry, F. A., Hoffman, P. N., and Salpeter, M. M. (1974). *Dev. Biol.* **39,** 96–104.
McKusick, V. A. (1972). "Heritable Disorders of Connective Tissue." Mosby, St. Louis, Missouri.
McLachlan, J., Bateman, M., and Wolpert, L. (1976). *Nature (London)* **264,** 267–269.
McLaren, A. (1972). *Nature (London)* **239,** 274–276.

McLean, F. C., and Urist, M. R. (1968). "Bone. Fundamentals of the Physiology of Skeletal Tissues," 3rd ed. Univ. of Chicago Press, Chicago, Illinois.

McLoughlin, C. B. (1961). *J. Embryol. Exp. Morphol.* **9,** 385–409.

McMahon, D. (1974). *Science* **185,** 1012–1021.

Maden, M. (1976). *J. Embryol. Exp. Morphol.* **36,** 561–574.

Maden, M., and Wallace, H. (1975). *Acta Embryol. Exp.* **2,** 77–86.

Maderson, P. F. A. (1975). *Am. Zool.* **15,** 315–328.

Manasek, F. J. (1973). *In* "Developmental Regulation: Aspects of Cell Differentiation" (S. J. Coward, ed.), pp. 193–218. Academic Press, New York.

Manasek, F. J. (1975). *Curr. Top. Dev. Biol.* **10,** 35–102.

Manasek, F. J., and Cohen, A. M. (1977). *Proc. Natl. Acad. Sci. U.S.A.* **74,** 1057–1061.

Mankin, H. J. (1962a). *J. Bone Joint Surg., Am. Vol.* **44,** 682–688.

Mankin, H. J. (1962b). *J. Bone Joint Surg., Am. Vol.* **44,** 689–698.

Mankin, H. J. (1964). *Clin. Orthop. Relat. Res.* **34,** 170–183.

Mankin, H. J. (1973). *Fed. Proc., Fed. Am. Soc. Exp. Biol.* **32,** 1478–1480.

Mankin, H. J., and Lippiello, L. (1969a). *J. Bone Joint Surg., Am. Vol.* **51,** 862–874.

Mankin, H. J., and Lippiello, L. (1969b). *J. Bone Joint Surg., Am. Vol.* **51,** 1591–1600.

Manning, W. K., and Bonner, W. M., Jr. (1967). *Arthritis Rheum.* **10,** 235–239.

Marcus, R. E. (1973). *Arthritis Rheum.* **16,** 646–656.

Marcus, R. E., and Srivastava, V. M. L. (1973). *Proc. Soc. Exp. Biol. Med.* **143,** 488–491.

Mareel, M. (1967). *Arch. Biol.* **78,** 145–166.

Marino, A. W., and Becker, R. O. (1977). *Clin. Orthop. Relat. Res.* **123,** 280–282.

Markostamou, K., and Baron, R. (1973). *Orthod. Fr.* **44,** 245–256.

Marks, S. C., Jr. (1976). *Am. J. Anat.* **146,** 331–338.

Marks, S. C., Jr. (1978a). *Am. J. Anat.* **151,** 119–130.

Marks, S. C., Jr. (1978b). *Am. J. Anat.* **151,** 131–137.

Marvaso, V., and Bernard, G. W. (1977). *Am. J. Anat.* **149,** 453–468.

Marzullo, G., and Desiderio, E. (1972). *Dev. Biol.* **27,** 13–19.

Matalon, R., and Dorfman, A. (1966). *Proc. Natl. Acad. Sci. U.S.A.* **56,** 1310–1316.

Mathews, M. B. (1965). *Biochem. J.* **96,** 710–716.

Mathews, M. B. (1967a). *Biol. Rev. Cambridge Philos. Soc.* **42,** 499–551.

Mathews, M. B. (1967b). *Nature (London)* **213,** 1255–1256.

Mathews, M. B. (1968). *In* "Biology of the Mouth" (P. Person, ed.), Publ. No. 89, pp. 199–236. Am. Assoc. Adv. Sci., Washington, D.C.

Mathews, M. B., ed. (1975). "Molecular Biology, Biochemistry and Biophysics," Vol. 19. Springer-Verlag, Berlin and New York.

Mathews, M. B., Duh, J., and Person, P. (1962). *Nature (London)* **193,** 378–379.

Mattson, P., and Foret, J. E. (1973). *Wilhelm Roux Arch. Entwicklungsmech. Org.* **173,** 169–182.

Maurer, P. H., and Hudack, S. S. (1952). *Arch. Biochem.* **38,** 49–53.

Mawdsley, R., and Ainsworth-Harrison, G. (1963). *J. Embryol. Exp. Morphol.* **11,** 537–547.

Maxwell, G. D. (1976). *Cell Tissue Res.* **172,** 325–330.

Mayne, R., Sanger, J. W., and Holtzer, H. (1971). *Dev. Biol.* **25,** 547–567.

Mayne, R., Abbott, J., and Schiltz, J. (1972). *J. Cell Biol.* **55,** 168a.

Mayne, R., Abbott, J., and Holtzer, H. (1973). *Exp. Cell Res.* **77,** 255–263.

Mayor, M. B., and Moskowitz, R. W. (1974). *J. Rheumatol.* **1,** 17–23.

Medoff, J. (1967). *Dev. Biol.* **16,** 118–143.

Medoff, J., and Zwilling, E. (1972). *Dev. Biol.* **28,** 138–141.

Meier, S., and Hay, E. D. (1973). *Dev. Biol.* **35,** 318–331.

Meier, S., and Hay, E. D. (1974a). *Proc. Natl. Acad. Sci. U.S.A.* **71,** 2310–2313.

Meier, S., and Hay, E. D. (1974b). *Dev. Biol.* **38,** 249–270.

Meikle, M. C. (1973a). *Am. J. Orthod.* **64,** 50–62.
Meikle, M. C. (1973b). *Arch. Oral Biol.* **18,** 1011–1020.
Meikle, M. C. (1975). *J. Anat.* **119,** 85–96.
Melcher, A. H. (1971a). *Isr. J. Med. Sci.* **7,** 374–376.
Melcher, A. H. (1971b). *Arch. Oral Biol.* **16,** 1379–1391.
Melcher, A. H. (1972). *Anat. Rec.* **172,** 21–36.
Melcher, A. H., and Accursi, G. E. (1972). *Anat. Rec.* **173,** 265–276.
Melcher, A. H., and Turnbull, R. S. (1974). *J. Dent. Res.* **53,** 223.
Melcher, A. H., and Turnbull, R. S. (1976). *In* "Organ Culture in Biomedical Research" (M. Balls and M. A. Monnickendam, eds.), pp. 149–164. Cambridge Univ. Press, London and New York.
Menkes, B., Deleanu, M., and Ilies, A. (1965). *Rev. Roum. Embryol. Cytol., Ser. Embryol.* **2,** 162–171.
Meredith Smith, M., and Miles, A. E. W. (1971). *Z. Zellforsch. Mikrosk. Anat.* **121,** 470–498.
Merker, H. J., Pospisil, M., and Mewes, P. (1975). *Teratology* **11,** 199–218.
Mescher, A. L. (1976). *J. Exp. Zool.* **195,** 117–128.
Mescher, A. L., and Tassava, R. A. (1975). *Dev. Biol.* **44,** 187–197.
Meunier, F., and Boivin, G. (1974). *Bull. Soc. Zool. Fr.* **99,** 495–504.
Michael, M. I., and Niazi, I. A. (1972). *Acta Embryol. Exp.* **14,** 349–363.
Milaire, J. (1956). *Arch. Biol.* **67,** 297–394.
Milaire, J. (1957). *Arch. Biol.* **68,** 429–512.
Milaire, J. (1963). *Arch. Biol.* **74,** 129–317.
Milaire, J. (1965a). *In* "Organogenesis" (R. L. deHaan and H. Ursprung, eds.), pp. 283–300. Holt, New York.
Milaire, J. (1965b). *Arch. Biol.* **76,** 457.
Milaire, J. (1967). *Arch. Biol.* **78,** 223–288.
Milaire, J. (1974). *Handb. Histochemie,* **8,** Part 3, Suppl., 1–135.
Milaire, J. (1976). *Arch. Biol.* **87,** 315–343.
Milaire, J., and Mulnard, J. (1968). *J. Embryol. Exp. Morphol.* **20,** 215–236.
Miller, E. J. (1972). *In* "Developmental Aspects of Oral Biology" (H. C. Slavkin and L. A. Bavetta, eds.), pp. 275–291. Academic Press, New York.
Miller, E. J. (1973). *Clin. Orthop. Relat. Res.* **92,** 260–280.
Miller, E. J. (1976). *Mol. Cell. Biochem.* **13,** 165–192.
Miner, R. W. (1950). *Ann. N.Y. Acad. Sci.* **52,** 943–1196.
Minkin, C., Blackman, L., Newbrey, J., Pokress, S., Posek, R., and Walling, M. (1977). *Biochem. Biophys. Res.* **76,** 875–881.
Minkoff, R., and Kuntz, A. J. (1977). *J. Embryol. Exp. Morphol.* **40,** 101–113.
Mintz, B. (1971). *In* "Methods in Mammalian Embryology" (J. Daniel, ed.), pp. 186–214. Freeman, San Francisco, California.
Mintz, B. (1972). *In* "Cell Differentiation" (R. Harris, P. Allin, and D. Viza, eds.), pp. 267–271. Munksgaard, Copenhagen.
Mitrovic, D. (1971). *C.R. Hebd. Seances Acad. Sci.* **273,** 642–645.
Mitrovic, D. (1972). *C.R. Hebd. Seances Acad. Sci.* **274,** 288–291.
Mitrovic, D. (1974). *C.R. Hebd. Seances Acad. Sci.* **278,** 1629–1632.
Mitrovic, D. (1977). *Am. J. Anat.* **150,** 333–348.
Modell, W. (1969). *Sci. Am.* **220,** 114–122.
Modell, W., and Noback, C. V. (1931). *Am. J. Anat.* **49,** 65–95.
Moffett, B. C., ed. (1972). "Mechanisms and Regulation of Craniofacial Morphogenesis." Swets & Zeitlander, Amsterdam.

Monson, J. W., and Felts, W. J. L. (1961). *Am. J. Phys. Anthropol.* **19,** 63–77.

Moore, W. J. (1965). *J. Zool.* **146,** 123–131.

Moore, W. J. (1969). *J. Zool.* **152,** 287–296.

Moore, W. J., and Lavelle, C. L. B. (1974). "Growth of the Facial Skeleton in the Hominoidea." Academic Press, New York.

Moore, W. J., and Mintz, B. (1972). *Dev. Biol.* **27,** 55–70.

Morriss, G. A., and Thorogood, P. V. (1978). *In* "Development of Mammals," (M. Johnson, ed.), North Holland, Amsterdam, **3** (in press).

Moscatelli, D., and Rubin, H. (1975). *Nature (London)* **254,** 65–66.

Moscona, A. A. (1957). *Proc. Natl. Acad. Sci. U.S.A.* **43,** 184–194.

Moscona, A. A. (1961). *Exp. Cell Res.* **22,** 455–475.

Moss, M. L. (1954). *Am. J. Anat.* **94,** 333–362.

Moss, M. L. (1957a). *Anat. Rec.* **127,** 569–590.

Moss, M. L. (1957b). *Plast. Reconstr. Surg.* **20,** 199–205.

Moss, M. L. (1958). *Am. J. Anat.* **102,** 141–166.

Moss, M. L. (1959). *Acta Anat.* **37,** 351–370.

Moss, M. L. (1961a). *Acta Anat.* **46,** 343–362.

Moss, M. L. (1961b). *Am. J. Anat.* **108,** 99–110.

Moss, M. L. (1962a). *Acta Anat.* **48,** 46–60.

Moss, M. L. (1962b). *Acta Anat.* **49,** 266–280.

Moss, M. L. (1962c). *In* "Vistas in Orthodontics" (B. S. Kraus and R. A. Riedel, eds.), pp. 85–98. Lea & Febiger, Philadelphia, Pennsylvania.

Moss, M. L. (1963). *Ann. N.Y. Acad. Sci.* **109.** 337–350.

Moss, M. L. (1964a). *Int. Rev. Gen. Exp. Zool.* **1,** 297–331.

Moss, M. L. (1964b). *Acta Anat.* **58,** 333–354.

Moss, M. L. (1965). *Acta Anat.* **60,** 262–276.

Moss, M. L. (1968a). *In* "Biology of the Mouth" (P. Person, ed.), Publ. No. 89, pp. 37–65. Am. Assoc. Adv. Sci., Washington, D.C.

Moss, M. L. (1968b). *Acta Anat.* **71,** 178–208.

Moss, M. L. (1968c). *In* "Current Problems of Lower Vertebrate Phylogeny" (T. Ørvig, ed.), pp. 359–371. Wiley (Interscience), New York.

Moss, M. L. (1968d). *Acta Anat.* **71,** 423–447.

Moss, M. L. (1970). *Acta Anat.* **77,** 161–187.

Moss, M. L. (1972a). *Am. Zool.* **12,** 27–34.

Moss, M. L. (1972b). *In* "Regulation of Organ and Tissue Growth" (R. J. Goss, ed.), pp. 127–142. Academic Press, New York.

Moss, M. L. (1972c). *Acta Morphol. Neerl. Scand.* **10,** 103–110.

Moss, M. L. (1975). *Birth Defects, Orig. Artic. Ser.* **11,** 283–295.

Moss, M. L. (1977). *Am. Zool.* **17,** 335–342.

Moss, M. L., Meehan, M. A., and Salentijn, L. (1972). *Acta Anat.* **81,** 161–182.

Moss-Salentijn, L. (1974). *Acta Anat.* **90,** 145–160.

Moss-Salentijn, L. (1975). *Acta Anat.* **92,** 595–606.

Moss-Salentijn, L., and Moss, M. L. (1975). *Acta Anat.* **91,** 386–404.

Mottet, N. K., and Hammar, S. P. (1972). *J. Cell Sci,* **11,** 403–414.

Muirden, K. D., Deutschmann, P., and Phillips, M. (1974). *J. Rheumatol.* **1,** 24–33.

Müller, P. K., and Kühn, K. (1977). *Arzneim. Forsch.* **27,** 199–202.

Mundy, G. R., Altman, A. J., Gonder, M. D., and Bandelin, J. G. (1977). *Science* **196,** 1109–1111.

Murray, J. McG., and Cleall, J. F. (1971). *J. Dent. Res.* **50,** 1654–1660.

Murray, P. D. F. (1926). *Proc. Linn. Soc. N.S.W.* **51,** 187–263.

Murray, P. D. F. (1928). *Aust. J. Exp. Biol. Med. Sci.* **5,** 237–256.
Murray, P. D. F. (1936). "Bones. A Study of the Development and Structure of the Vertebrate Skeleton." Cambridge Univ. Press, London and New York.
Murray, P. D. F. (1941). *Nature (London)* **148,** 471.
Murray, P. D. F. (1943). *J. Anat.* **77,** 225–240.
Murray, P. D. F. (1954). *Aust. J. Zool.* **2,** 364–380.
Murray, P. D. F. (1963). *Aust. J. Zool.* **11,** 368–430.
Murray, P. D. F., and Drachman, D. B. (1969). *J. Embryol. Exp. Morphol.* **22,** 349–371.
Murray, P. D. F., and Smiles, M. (1965). *Aust. J. Zool.* **13,** 351–381.
M-Z Rahmani, T. (1974). *J. Embryol. Exp. Morphol.* **32,** 431–443.
Namenwirth, M. (1974). *Dev. Biol.* **41,** 42–56.
Nardi, F., Gerlini, G., and Bonucci, E. (1974). *Virchows Arch. A* **363,** 311–322.
Nelson, R. L., and Bauer, G. E. (1977). *Calcif. Tissue Res.* **22,** 303–314.
Nelson, S. W. (1975). *J. Dent. Res.* **54,** 265–268.
Neméth-Csóka, M. (1974). *Exp. Pathol.* **9,** 256–262.
Nesbitt, R. (1736). "Human Osteogeny Explained in Two Lectures." London.
Neuman, W. F. (1969). *Fed. Proc., Fed. Am. Soc. Exp. Biol.* **28,** 1846–1850.
Nevo, Z., and Dorfman, A. (1972). *Proc. Natl. Acad. Sci. U.S.A.* **69,** 2069–2072.
Nevo, Z., Horwitz, A. L., and Dorfman, A. (1972). *Dev. Biol.* **28,** 219–228.
Newbrey, J. W., and Banks, W. J. (1975). *Calcif. Tissue Res.* **17,** 289–302.
Newman, S. A. (1977). *In* "Vertebrate Limb and Somite Morphogenesis" (D. A. Ede, J. R. Hinchliffe, and M. Balls, eds.). pp. 181–198. Cambridge Univ. Press, London and New York.
Newsome, D. A. (1972). *Dev. Biol.* **27,** 575–579.
Newsome, D. A. (1976). *Dev. Biol.* **49,** 496–507.
Newsome, D. A., and Kenyon, K. R. (1973). *Dev. Biol.* **32,** 387–400.
Niederman, R., and Armstrong, P. B. (1972). *J. Exp. Zool.* **181,** 17–32.
Nimni, M. E. (1973). *J. Oral Pathol.* **2,** 175–202.
Nimni, M. E., and Deshmukh, K. (1973). *Science* **181,** 751–752.
Niven, J. S. F. (1933). *Wilhelm Roux Arch. Entwicklungsmech. Org.* **128,** 480–501.
Noden, D. M. (1975). *Dev. Biol.* **42,** 106–130.
Norby, D. P., Malemud, C. J., and Sokoloff, L. (1977). *Arthritis Rheum.* **20,** 709–716.
Norton, L. A., and Moore, R. R. (1972). *J. Dent. Res.* **51,** 1492–1499.
Norton, L. A., Rodan, G. A., and Bourret, L. A. (1976). *J. Dent. Res.* **55,** B215.
Norton, L. A., Bourret, L. A., and Rodan, G. A. (1977). *J. Dent. Res.* **56,** A105.
Novak, V. (1964). *Plzen. Lek. Sb., Suppl.* **12,** 133–213.
Nutik, G., and Creuss, R. L. (1974). *Proc. Soc. Exp. Biol. Med.* **146,** 265–268.
Oakes, B. W., Handley, C. J., Lisner, F., and Lowther, D. A. (1977). *J. Embryol. Exp. Morphol.* **38,** 239–263,
Oberpriller, J. (1967). *Growth* **31,** 251–296.
Öbrink, B., and Wasteson, A. (1971). *Biochem. J.* **121,** 227–234.
Ogden, J. A., Hempton, R. F., and Southwick, W. O. (1975). *Anat. Rec.* **182,** 431–446.
Ohta, Y., and Matsunaga, H. (1974). *J. Bone Joint. Surg., Br. Vol.* **56,** 3–16.
Okayama, M., Pacifici, M., and Holtzer, H. (1976). *Proc. Natl. Acad. Sci. U.S.A.* **73,** 3224–3228.
Oohira, A., Kimata, K., Suzuki, S., Takata, K., and Suzuki, I. (1974). *J. Biol. Chem.* **249,** 1637–1645.
Oppenheimer, J. M. (1974). *Am. Zool.* **14,** 867–879.
O'Rahilly, R. (1962). *Acta Anat.* **48,** 335–346.
O'Rahilly, R., Gardner, E., and Gray, D. J. (1956). *J. Embryol. Exp. Morphol.* **4,** 254–264.
Ordahl, C. P., and Caplan, A. I. (1976). *Dev. Biol.* **54,** 61–72.
Orkin, R. W., Pollard, T. D., and Hay, E. D. (1973). *Dev. Biol.* **35,** 388–394.

Orkin, R. W., Pratt, R. M., and Martin, G. R. (1976). *Dev. Biol.* **50,** 82–94.
Orkin, R. W., Williams, B. R., Cranley, R. E., Poppke, D. C., and Brown, K. S. (1977). *J. Cell Biol.* **73,** 287–299.
Ørvig, T. (1951). *Ark. Zool.* **2,** 321–456.
Ørvig, T. (1957). *Ark. Zool.* **10,** 481–490.
Ørvig, T. (1965). *Ark. Zool.* **16,** 551–556.
Ørvig, T. (1967). *In* "Structural and Chemical Organization of Teeth" (A. E. W. Miles, ed.), Vol. 1, pp. 45–110. Academic Press, New York.
Ørvig, T. ed. (1968). "Current Problems of Lower Vertebrate Phylogeny," Nobel Symp. No. 4. Almqvist & Wiksell, Stockholm.
Osdoby, P., and Caplan, A. I. (1976). *Dev. Biol.* **52,** 283–299.
Osman, A., and Ruch, J. V. (1975). *J. Biol. Buccale* **3,** 117–132.
Ossenberg, N. S. (1974). *J. Dent. Res.* **53,** 77–82.
Ostrowski, K., and Wlodarski, K. (1971). *In* "The Biochemistry and Physiology of Bone" (G. H. Bourne, ed.), 2nd ed., Vol. 3, pp. 299–337. Academic Press, New York.
Overman, D. O., Seegmiller, R. E., and Runner, M. N. (1972). *Dev. Biol.* **28,** 573–582.
Overman, D. O., Graham, M. N., and Roy, W. A. (1976). *Teratology* **13,** 85–94.
Overton, J. (1977). *Exp. Cell Res.* **105,** 313–324.
Owen, M. J. (1963). *J. Cell Biol.* **19,** 19–32.
Owen, M. J. (1970). *Int. Rev. Cytol.* **28,** 213–238.
Owen, M. J. (1971). *In* "The Biochemistry and Physiology of Bone" (G. H. Bourne, ed.), 2nd ed., Vol. 3, pp. 271–298. Academic Press, New York.
Owen, M. J., and MacPherson, S. (1963). *J. Cell Biol.* **19,** 33–44.
Palmoski, M., and Goetinck, P. F. (1969). *Am. Zool.* **9,** 1132.
Palmoski, M., and Goetinck, P. F. (1970). *J. Exp. Zool.* **174,** 157–164.
Palmoski, M., and Goetinck, P. F. (1972). *Proc. Natl. Acad. Sci. U.S.A.* **69,** 3385–3388.
Partridge, S. M. (1948). *Biochem. J.* **43,** 387–397.
Patterson, C. (1977). *In* "Problems in Vertebrate Evolution." (S. M. Andrews, R. S. Miles, and A. D. Walker, eds.), Linnean Soc. Symposium Series No. 4, pp. 77–122. Academic Press, London.
Paulsen, D. F., and Finch, R. A. (1977). *Nature (London),* **268,** 639–641.
Pautou, M-P. (1974). *C.R. Hebd. Seances Acad. Sci.* **278,** 2209–2212.
Pautou, M-P. (1977). *J. Embryol. Exp. Morphol.* **42,** 177–194.
Pawelek, J. M. (1969). *Dev. Biol.* **19,** 52–72.
Pawlicki, R. (1977). *Acta Histochem.* **58,** 75–78.
Pearson, A. A. (1977). *J. Anat.* **123,** 563–578.
Peck, W. A., Birge, S. J., and Fedak, S. A. (1964). *Science* **146,** 1476–1477.
Peck, W. A., Birge, S. J., and Brandt, J. (1967a). *Biochim. Biophys. Acta* **142,** 512–525.
Peck, W. A., Brandt, J., and Miller, I. D. A. (1967b). *Proc. Natl. Acad. Sci. U.S.A.* **57,** 1599–1606.
Peck, W. A., Carpenter, A. J., and Messinger, K. (1974). *Endocrinology* **94,** 148–154.
Pedrini-Mille, A., and Pedrini, V. (1971). *Calcif. Tissue Res.* **8,** 106–113.
Pellegrini, O. (1934). *Atti. Soc. Med.-Chir. Padova Fac. Med. Chir. Univ. Padova* **11,** 927–941.
Pennypacker, J. P., and Goetinck, P. F. (1976). *Dev. Biol.* **50,** 35–47.
Perlov, F. A. (1968). M.Sc. in Dentistry Thesis, University of Nebraska, Lincoln, Nebraska.
Person, P. (1960). *Ann. N.Y. Acad. Sci.* **85,** 9–16.
Person, P. (1969). *Science* **164,** 1404–1405.
Person, P., and Fine, A. S. (1959). *Arch. Biochem. Biophys.* **84,** 123–133.
Person, P., and Mathews, M. B. (1967). *Biol. Bull. (Woods Hole, Mass.)* **132,** 244–252.
Person, P., and Philpott, D. E. (1963). *Ann. N.Y. Acad. Sci.* **109,** 113–126.

Person, P., and Philpott, D. E. (1969a). *Biol. Rev. Cambridge Philos. Soc.* **44,** 1–16.
Person, P., and Philpott, D. E. (1969b). *J. Morphol.* **128,** 67–94.
Person, P., Lash, J. W., and Fine, A. S. (1959). *Biol. Bull. (Woods Hole, Mass.)* **117,** 504–510.
Person, P., Papierman, S., Zipper, H., and Libbin, R. M. (1977). *Calcif. Tissue Res.* **24,** 37–39.
Persson, M. (1973). *Odontol. Revy* **24,** Suppl. 26, 1–146.
Pessac, B., and Defendi, V. (1972). *Science* **175,** 898–900.
Petrakova, K. V., and Friedenstein, A. Y. (1965). *Byull. Eksp. Biol. Med.* **2,** 98–101.
Petrovic, A. (1970). *Annee Biol.* **9,** 304–311.
Petrovic, A. (1972). *Acta Morphol. Neerl. Scand.* **10,** 25–34.
Petrovic, A. (1974). *Trans. Eur. Orthod. Soc.* pp. 1–7.
Petrovic, A., and Charlier, J.-P. (1967). *C.R. Hebd. Seances Acad. Sci.* **265,** 1511–1513.
Petrovic, A., Oudet, C., and Gasson, N. (1973). *Orthod. Fr.* **44,** 191–210.
Peyer, B. (1968). "Comparative Odontology." Univ. of Chicago Press, Chicago, Illinois.
Philpott, D. E., and Person, P. (1966). *Electron Microsc. Proc. Int. Congr., 6th, 1966* pp. 565–566.
Philpott, D. E., and Person, P. (1970). *J. Morphol.* **131,** 417–430.
Pictet, R. L., Rall, L. B., Phelps, P., and Rutter, W. J. (1976). *Science* **191,** 191–192.
Pieau, C., and Raynaud, A. (1976). *C.R. Hebd. Seances Acad. Sci.* **282,** 1797–1800.
Piekarski, K., and Munro, M. (1977). *Nature (London)* **269,** 80–82.
Pilbeam, D., and Gould, S. J. (1974). *Science* **186,** 892–901.
Pinnell, S. R., and Crelin, E. S. (1963). *Anat. Rec.* **145,** 345.
Pinot, M. (1969). *Arch. Anat. Microsc. Morphol. Exp.* **58,** 123–144.
Pinot, M. (1970). *J. Embryol. Exp. Morphol.* **23,** 109–151.
Pleskova, M. V., Rodionov, V. M., Bugrilova, R. S., and Konyukhov, B. (1974). *Dev. Biol.* **37,** 417–421.
Pollak, R. D., and Fallon, J. F. (1974). *Exp. Cell Res.* **86,** 9–14.
Pollak, R. D., and Fallon, J. F. (1976). *Exp. Cell Res.* **100,** 15–22.
Poswillo, D. (1975). *Br. Med. Bull.* **31,** 101–106.
Poswillo, D. (1976). *Br. Med. Bull.* **32,** 59–64.
Pratt, L. M. (1943). *J. Mammal.* **24,** 204–211.
Pratt, R. M., Larsen, M. A., and Johnston, M. C. (1975). *Dev. Biol.* **44,** 298–305.
Pritchard, J. J. (1946). *J. Anat.* **80,** 55–60.
Pritchard, J. J. (1952). *J. Anat.* **86,** 259–277.
Pritchard, J. J. (1956). *In* "The Biochemistry and Physiology of Bone" (G. H. Bourne, ed.), 1st ed., pp. 179–212. Academic Press, New York.
Pritchard, J. J. (1965). *In* "Studies in Physiology" (D. R. Curtis, and A. K. McIntyre, eds.), pp. 232–238. Springer-Verlag, Berlin and New York.
Pritchard, J. J. (1972a). *In* "The Biochemistry and Physiology of Bone" (G. H. Bourne, ed.), 2nd ed., Vol. 1, pp. 1–20. Academic Press, New York.
Pritchard, J. J. (1972b). *Acta Morphol. Neerl. Scand.* **10,** 63–69.
Pritchard, J. J. (1974). *In* "Differentiation and Growth of Cells in Vertebrate Tissues" (G. Goldspink, ed.), pp. 101–128. Chapman & Hall, London.
Pritchard, J. J., and Ruzicka, A. J. (1950). *J. Anat.* **84,** 236–261.
Pritchard, J. J., Scott, J. H., and Girgis, F. G. (1956). *J. Anat.* **90,** 73–86.
Prockop, D. J., Pettengill, O., and Holtzer, H. (1964). *Biochim. Biophys. Acta* **83,** 189–196.
Pruzansky, S. (1971). *Tandlaegebladet* **75,** 1157–1169.
Pugh, J. W., Radin, E. L., and Rose, R. M. (1974). *J. Bone Joint Surg., Am. Vol.* **56,** 313–321.
Pugin, E. (1972). *C.R. Hebd. Seances Acad. Sci.* **275,** 2543–2546.
Putchkov, V. F. (1964). *Arkh. Anat., Gistol. Embriol.* **46,** 16–24.
Rabinowitz, J. L., Tavares, C. J., Lipson, R., and Person, P. (1976). *Biol. Bull. (Woods Hole, Mass.)* **150,** 69–79.

Rajtova, V. (1971). *Anat. Anz.* **128,** 392–401.
Ramp, W. K. (1975). *Clin. Orthop. Relat. Res.* **106,** 311–322.
Rao, L. G., Ng, B., Brunette, D. M., and Heersche, J. N. M. (1977). *Endocrinology* **100,** 1233–1241.
Rasmussen, H., and Bordier, P. (1974). "The Physiological and Cellular Basis of Metabolic Bone Disease." Williams & Wilkins, Baltimore, Maryland.
Rasmussen, H., and Goodman, D. B. P. (1977). *Physiol. Rev.* **57,** 421–509.
Rawlins, R. G. (1975). *Am. J. Phys. Anthropol.* **42,** 477–488.
Raynaud, A. (1972). *Bull. Soc. Zool. Fr.* **97,** 469–485.
Raynaud, A. (1974). *Bull. Mus. Natl. Hist. Nat.* **225,** 705–720.
Raynaud, A., and Adrian, M. (1975a). *Arch. Anat. Microsc. Morphol. Exp.* **64,** 287–316.
Raynaud, A., and Adrian, M. (1975b). *C.R. Hebd. Seances Acad. Sci.* **280,** 2591–2594.
Raynaud, A., and van den Elzen, P. (1976). *Arch. Anat. Microsc. Morphol. Exp.* **65,** 17–36.
Raynaud, A., Adrian, M., and Kouprach, S. (1974a). *Ann. Embryol. Morphol.* **7,** 243–264.
Raynaud, A., Okuzumi, H., and Kouprach, S. (1974b). *Bull. Soc. Zool. Fr.* **99,** 149–153.
Raynaud, A., Gasc, J.-P., Vassé, J., Renous, S., and Pieau, C. (1974c). *Bull. Soc. Zool. Fr.* **99,** 165–173.
Raynaud, A., Jeanny, J.-C., and Gontcharoff, M. (1975). *C.R. Hebd. Seances Acad. Sci.* **280,** 2693–2696.
Rayne, J., and Crawford, G. N. C. (1975). *J. Anat.* **119,** 347–357.
Rebhun, L. I. (1977). *Int. Rev. Cytol.* **49,** 1–54.
Redler, I. (1974). *Clin. Orthop. Relat. Res.* **103,** 262–268.
Reinbold, R. (1968). *J. Embryol. Exp. Morphol.* **19,** 43–47.
Reuss, C., and Saunders, J. W., Jr. (1965). *Am. Zool.* **5,** 214.
Rhodes, R. K., and Elmer, W. A. (1972). *Am. Zool.* **12,** 709.
Rhodes, R. K., and Elmer, W. A. (1975). *Dev. Biol.* **46,** 14–27.
Richany, S. F., Bast, T. H., and Anson, B. J. (1956). *Q. Bull. Northwest. Univ. Med. Sch.* **30,** 331–355.
Riede, U. N., Zinkernagel, R., Remagen, W., and Villiger, W. (1971). *Beitr. Pathol.* **143,** 271–282.
Riley, P. A. (1974). *Cell Diff.* **3,** 233–238.
Rimoin, D. L. (1975). *Adv. Hum. Genet.* **5,** 1–118.
Rimoin, D. L., Hughes, G. N., Kaufman, R. L., Rosenthal, R. E., McAlister, W. H., and Silberberg, R. (1970). *N. Engl. J. Med.* **283,** 728–735.
Rimoin, D. L., Silberberg, R., and Hollister, D. W. (1976). *Clin. Orthop. Relat. Res.* **114,** 137–152.
Rinaldi, L., Caronna, E., and Cinquetti, R. (1974). *Rend., Ist. Lomb. Accad. Sci. Lett. B* **108,** 52–62.
Ritsila, V., and Alhopuro, S. (1973). *Scand. J. Plast. Reconstr. Surg.* **7,** 116–119.
Ritsila, V., Alhopuro, S., and Ranta, R. (1973). *Proc. Finn. Dent. Soc.* **69,** 164–165.
Robadey, M., and Schowing, J. (1972). *Boll. Zool. Ital., Atti 16th Conv. Triesti,* **39,** 656.
Roberts, W. E., and Jee, W. S. S. (1974). *Arch. Oral Biol.* **19,** 17–22.
Robertson, P. B., and Miller, E. J. (1972). *Biochim. Biophys. Acta* **289,** 247–250.
Rodan, G. A., Mensi, T., and Harvey, A. (1975a). *Calcif. Tissue Res.* **18,** 125–132.
Rodan, G. A., Bourret, L. A., Harvey, A., and Mensi, T. (1975b). *Science* **189,** 467–469.
Rodan, G. A., Bourret, L. A., and Cutler, L. S. (1977). *J. Cell Biol.* **72,** 493–501.
Rodan, G. A., Bourret, L. A., and Norton, L. A. (1978). *Science* **199,** 690–692.
Rodan, S. B., and Rodan, G. A. (1974). *J. Biol. Chem.* **249,** 3068–3074.
Rodbard, S. (1970). *Perspect. Biol. Med.* **13,** 507–527.
Romer, A. S. (1963). *Ann. N.Y. Acad. Sci.* **109,** 168–176.
Rönning, O. (1966). *Acta Odontol. Scand.* **24,** 443–457.

Rosenberg, M. J., and Caplan, A. I. (1974). *Dev. Biol.* **38,** 157–164.

Ross, G. T. (1976). Ph.D. Thesis, Dalhousie University, Halifax, Nova Scotia.

Ross, G. T., and Hall, B. K. (1978). Submitted for publication.

Ross, R. (1975). *Philos. Trans. R. Soc. London, Ser. B* **271,** 247–260.

Roth, M. (1973). *Morphol. Jahrb.* **119,** 250–274.

Royal, P. D., and Goetinck, P. F. (1977). *J. Embryol. Exp. Morphol.* **39,** 79–95.

Rubin, L., and Saunders, J. W., Jr. (1972). *Dev. Biol.* **28,** 94–112.

Rudnick D. (1945). *J. Exp. Zool.* **100,** 1–14.

Ruth, E. B. (1932). *Anat. Rec.* **53,** 207–225.

Ruth, E. B. (1935). *Anat. Rec.* **64,** 1–5.

Ruth, E. B. (1936a). *Anat. Rec.* **67,** 69–79.

Ruth, E. B. (1936b). *Anat. Rec.* **67,** 409–419.

Ryan, G. B., Cliff, W. J., Gabbiani, G., Irle, C., Statkou, P. R., and Majno, G. (1973). *Lab. Invest.* **29,** 197–206.

Salomon, D. S., and Pratt, R. M. (1976). *Nature (London)* **264,** 174–177.

Saunders, J. W., Jr. (1948). *J. Exp. Zool.* **108,** 363–404.

Saunders, J. W., Jr. (1966). *Science* **154,** 604–612.

Saunders, J. W., Jr. (1977). *In* "Vertebrate Limb and Somite Morphogenesis" (D. A. Ede, J. R. Hinchliffe, and M. Balls, eds.), pp. 1–24. Cambridge Univ. Press, London and New York.

Saunders, J. W., Jr., and Fallon, J. F. (1967). *Symp. Soc. Dev. Biol.* **25,** 289–314.

Saunders, J. W., Jr., and Gasseling, M. T. (1963). *Dev. Biol.* **7,** 64–78.

Saunders, J. W., Jr., and Gasseling, M. T. (1968). *In* "Epithelial-Mesenchymal Interactions" (R. Fleischmajer and R. E. Billingham, eds.), pp. 78–97. Williams & Wilkins, Baltimore, Maryland.

Saunders, J. W., Jr., and Reuss, C. (1974). *Dev. Biol.* **38,** 41–50.

Saunders, J. W., Jr., Gasseling, M. T., and Cairns, J. M. (1955). *Nature (London)* **175,** 673.

Saunders, J. W., Jr., Cairns, J. M., and Gasseling, M. T. (1957). *J. Morphol.* **101,** 57–87.

Saunders, J. W., Jr., Gasseling, M. T., and Gfeller,, M. D. (1958). *J. Exp. Zool.* **137,** 39–74.

Saunders, J. W., Jr., Gasseling, M. T., and Cairns, J. M. (1959). *Dev. Biol.* **1,** 281–301.

Saunders, J. W., Jr., Gasseling, M. T., and Saunders, L. C. (1962). *Dev. Biol.* **5,** 147–178.

Saunders, J. W., Jr., Gasseling, M. T., and Errick, J. E. (1976). *Dev. Biol.* **50,** 16–25.

Savostin-Asling, I., and Asling, C. W. (1973). *Anat. Rec.* **176,** 345–360.

Savostin-Asling, I., and Asling, C. W. (1975). *Anat. Rec.* **183,** 373–392.

Saxén, L. (1976). *J. Embryol. Exp. Morphol.* **36,** 1–12.

Sayegh, F. S., Solomon, G. C., and Davis, R. W. (1974). *Clin. Orthop. Relat. Res.* **99,** 267–284.

Schaeffer, B. (1961). *Trans. N.Y. Acad. Sci.* [2] **23,** 501–505.

Schaeffer, B. (1977). *In* "Problems in Vertebrate Evolution" (S. M. Andrews, R. S. Miles, and A. D. Walker, eds.), Linn. Soc. Symp. No. 4, pp. 25–52. Academic Press, London.

Schaffer, J. (1930). *In* "Handbuch der mikroskopischen Anatomie des Menschen" (W. von Mallendorff, ed.). Vol. II, Part 2, pp. 338–350. Springer-Verlag, Berlin and New York.

Schenk, R. K. (1973). *Proc. Eur. Symp. Calcif. Tissues, 9th, 1972* pp. 13–22.

Schenk, R. K., Spiro, D., and Wiener, J. (1967). *J. Cell Biol.* **34,** 275–291.

Schenk, R. K., Wiener, J., and Spiro, D. (1968). *Acta Anat.* **69,** 1–17.

Schindler, F. H., Ose, M. A., and Solursh, M. (1975). *In Vitro* **12,** 44–47.

Schowing, J. (1961). *J. Embryol. Exp. Morphol.* **9,** 326–334.

Schowing, J. (1968a). *J. Embryol. Exp. Morphol.* **19,** 9–22.

Schowing, J. (1968b). *J. Embryol. Exp. Morphol.* **19,** 23–32.

Schowing, J. (1968c). *J. Embryol. Exp. Morphol.* **19,** 83–94.

Schowing, J. (1974). *Annee Biol.* **13,** 69–76.

Schowing, J., and Robadey, M. (1971). *C.R. Hebd. Seances Acad. Sci.* **272,** 2382–2384.

Schrarer, H. (1970). "Biological Calcification: Cellular and Molecular Aspects." Appleton, New York.
Schubert, D., and Lacorbiere, M. (1976). *Proc. Natl. Acad. Sci. U.S.A.* **73,** 1989–1993.
Schulz, A., Donath, K., and Delling, G. (1974). *Virchows, Arch. A* **364,** 347–356.
Schuster, G. S., Dirksen, T. R., and Harms, W. S. (1975). *J. Dent. Res.* **54,** 131–139.
Schwartz, E. R., and Adamy, L. (1976). *Connect. Tissue Res.* **4,** 211–218.
Schwartz, N. B., and Dorfman, A. (1975). *Connect. Tissue Res.* **3,** 115–122.
Scopelliti, R. (1975). *C.R. Hebd. Seances Acad. Sci.* **280,** 901–904.
Scott, B. L. (1965). *J. Ultrastruct. Res.* **13,** 560.
Scott, B. L. (1967). *J. Cell Biol.* **35,** 115–126.
Scott, B. L. (1969). *J. Dent. Res.* **48,** Suppl., 753–760.
Scott, B. L., and Glimcher, M. J. (1971). *J. Ultrastruct. Res.* **36,** 565–586.
Scott, B. L., and Pease, D. C. (1956). *Anat. Rec.* **126,** 465–495.
Scott, J. E., (1975). *Philos. Trans. R. Soc. London, Ser. B* **271,** 235–242.
Scott, J. H. (1954). *Proc. R. Soc. Med.* **47,** 91–100.
Scott, J. H., and Symons, N. B. B. (1974). "Introduction to Dental Anatomy," 7th ed. Churchill, London.
Scott, W. J., Ritter, E. J., and Wilson, J. G. (1977). *J. Embryol. Exp. Morphol.* **42,** 93–104.
Searls, R. L. (1965a). *Dev. Biol.* **11,** 155–168.
Searls, R. L. (1965b). *Proc. Soc. Exp. Biol. Med.* **118,** 1172–1176.
Searls, R. L. (1967). *J. Exp. Zool.* **166,** 39–50.
Searls. R. L. (1968). *Dev. Biol.* **17,** 382–399.
Searls, R. L. (1971). *Exp. Cell Res.* **64,** 163–169.
Searls, R. L. (1972). *Exp. Cell Res.* **73,** 57–64.
Searls, R. L. (1973a). *Clin. Orthop. Relat. Res.* **96,** 327–344.
Searls, R. L. (1973b). *In* "Developmental Regulation. Aspects of Cell Differentiation" (S. J. Coward, ed.), pp. 219–251. Academic Press, New York.
Searls, R. L. (1976). *J. Embryol. Exp. Morphol.* **35,** 369–381.
Searls, R. L., and Janners, M. Y. (1969). *J. Exp. Zool.* **170,** 365–376.
Searls, R. L., and Janners, M. Y. (1971). *Dev. Biol.* **24,** 198–213.
Searls, R. L., and Zwilling, E. (1964). *Dev. Biol.* **9,** 38–55.
Searls, R. L., Hilfer, S. R., and Mirow, S. M. (1972). *Dev. Biol.* **28,** 123–137.
Sěbek, J., Skalova, J., and Hert, J. (1972). *Folia Morphol.* **20,** 29–37.
Seegmiller, R. E. (1977). *Dev. Biol.* **58,** 164–173.
Seegmiller, R. E., and Fraser, F. C. (1977). *J. Embryol. Exp. Morphol.* **38,** 227–238.
Seegmiller, R. E., and Runner, M. N. (1974). *J. Embryol. Exp. Morphol.* **31,** 305–312.
Seegmiller, R. E., Fraser, F. C., and Sheldon, H. (1971). *J. Cell Biol.* **48,** 580–593.
Seegmiller, R. E., Overman, D. O., and Runner, M. N. (1972a). *Dev. Biol.* **28,** 555–572.
Seegmiller, R. E., Ferguson, C. C., and Sheldon, H. (1972b). *J. Ultrastruct. Res.* **38,** 288–301.
Selleck, D. K., and Elmer, W. A. (1973). *Am. Zool.* **13,** 1313.
Serafini-Fracassini, A., and Smith, J. W. (1974). "The Structure and Biochemistry of Cartilage." Churchill, London.
Servoss, J. M. (1973). *Am. J. Anat.* **136,** 479–486.
Shanfeld, J. L., Shapiro, I., and Davidovitch, Z. (1975). *Anal. Biochem.* **66,** 450–459.
Shanfeld, J. L., Mirin, R., and Davidovitch, Z. (1976). *J. Dent. Res.* **55,** B303.
Shapiro, F., Holtrop, M. E., and Glimcher, M. J. (1977). *J. Bone Joint Surg.* **59A,** 703–723.
Sharpey, W. (1848). *In* "Elements of Anatomy" (J. Quain, ed.), 5th ed., Vol. I. Taylor, Walton, & Maberley, London.
Shellis, R. P., and Miles, A. E. W. (1974). *Proc. R. Soc. London, Ser. B* **185,** 51–72.
Shepard, N., and Mitchell, N. (1977). *Anat. Rec.* **187,** 463–476.

Shepard, T. H. (1971). *J. Embryol. Exp. Morphol.* **25,** 347–363.

Shepard, T. H., Fry, L. R., and Moffett, B. C., Jr. (1969). *Teratology* **2,** 13–22.

Shibata, N., Tatsumi, N., Tanaka, K., Okamura, Y., and Senda, N. (1972). *Biochim. Biophys. Acta* **256,** 565–576.

Shibuya, T. and Kuroda, Y. (1973). *Jpn. J. Genet.* **48,** 197–206.

Shibuya, T. Fujio, Y., and Kondo, K. (1972). *Jpn. J. Genet.* **47,** 23–32.

Shimomura, Y., Wezeman, F. H., and Ray, R. D. (1973). *Clin. Orthop. Relat. Res.* **90,** 246–254.

Shimomura, Y., Yoneda, T., and Suzuki, F. (1975). *Calcif. Tissue Res.* **19,** 179–188.

Shulman, H. J., and Opler, A. (1974). *Biochem. Biophys. Res. Commun.* **59,** 914–919.

Siebenmann, R. E. (1977). *Schweiz. Med. Wochenschr.* **107,** 468–474.

Silberberg, R., and Lesker, P. (1975). *Growth* **39,** 17–34.

Silberberg, R., Hasler, M., and Silberberg, M. (1966). *Acta Anat.* **65,** 275–298.

Silberberg, R., Hasler, M., and Lesker, P. (1976). *Acta Anat.* **96,** 162–175.

Silbermann, M., and Frommer, J. (1972a). *Anat. Rec.* **172,** 659–668.

Silbermann, M., and Frommer, J. (1972b). *Anat. Rec.* **174,** 503–512.

Silbermann, M., and Frommer, J. (1972c). *Am. J. Anat.* **135,** 359–370.

Silbermann, M., and Frommer, J. (1973a). *Histochemie* **36,** 185–192.

Silbermann, M., and Frommer, J. (1973b). *Histochemie* **37,** 365–372.

Silbermann, M., and Frommer, J. (1974). *Acta Anat.* **90,** 330–346.

Silbermann, M., and Kadar, T. (1977). *Acta Anat.* **97,** 459–468.

Silbermann, M., Kadar, T., and Hornung, G. (1977). *Histochemie* **50,** 327–335.

Silberzahn, N. (1968). *C.R. Hebd. Seances Acad. Sci.* **267,** 352–355.

Simmons, D. J. (1971). *Clin. Orthop. Relat. Res.* **76,** 444–480.

Simmons, D. J., Ellsasser, J. C., Cummins, H., and Lesker, P. (1973). *Clin. Orthop. Relat. Res.* **97,** 237–247.

Simmons, N. B., and Marshall, J. H. (1970). *Calcif. Tissue Res.* **5,** 206–221.

Simon, M. R. (1977). *Acta Anat.* **97,** 351–360.

Simon, W. H., Richardson, S., Herman, W., Parsons, J. R., and Lane, J. (1976). *J. Bone Joint Surg., Am. Vol.* **58,** 517–525.

Simons, E. V. (1974). *Acta Morphol. Neerl. Scand.* **12,** 331–344.

Singh, I. J., Tonna, E. A., and Gandel, C. P. (1974). *J. Morphol.* **144,** 421–436.

Slack, J. M. W., and Savage, S. (1978). *Nature* **271,** 760–761.

Slavkin, H. C., ed. (1972). "The Comparative Molecular Biology of Extracellular Matrices." Academic Press, New York.

Slavkin, H. C., and Bavetta, L. A., eds. (1972). "Developmental Aspects of Oral Biology." Academic Press, New York.

Slavkin, H. C., and Greulich, R. C., eds. (1975). "Extracellular Matrix Influences on Gene Expression." Academic Press, New York.

Sledge, C. B. (1973). *Fed. Proc., Fed. Am. Soc. Exp. Biol.* **32,** 1503–1505.

Smith, A. A., Searls, R. L., and Hilfer, S. R. (1975). *Dev. Biol.* **46,** 222–226.

Smith, A. R., and Crawley, A. M. (1977). *J. Embryol. Exp. Morphol.* **37,** 33–48.

Smith, A. R., Lewis, J. H., Crawley, A., and Wolpert, L. (1974). *J. Embryol. Exp. Morphol.* **32,** 375–390.

Smith, D. M., Johnston, C. C., Jr., Severson, A. R., and Bell, N. (1970). *Clin. Res.* **18,** 626.

Smith, D. M., Johnston, C. C., Jr., and Severson, A. R. (1973). *Calcif. Tissue Res.* **11,** 56–69.

Smith, G. N., Jr., Toole, B. P., and Gross, J. (1973). *Am Zool.* **13,** 1350.

Smith, G. N., Jr., Toole, B. P., and Gross, J. (1975). *Dev. Biol.* **43,** 221–232.

Smith, H. G., and McKeown, M. (1974). *J. Anat.* **118,** 543–561.

Smith, H. M. (1947). *Turtox News* **25,** 234–236.

Smith, J. D., and Abramson, M. (1974). *Arch. Otolaryngol.* **99,** 203–205.

Smith, J. W., and Serafini-Fracassini, A. (1967). *J. Cell Sci.* **2,** 129–136.
Smith, P. H. (1972). *Connect. Tissue Res.* **1,** 181–188.
Sofaer, J. A. (1975). *Arch. Oral Biol.* **20,** 57–62.
Sokoloff, L. (1974). *J. Rheumatol.* **1,** 9–16.
Sokoloff, L., Malemud, C. J., and Green, W. T., Jr. (1970). *Arthritis Rheum.* **13,** 118–124.
Sokoloff, L., Malemud, C. J., and Srivastava, V. M. L. (1973). *Fed. Proc., Fed. Am. Soc. Exp. Biol.* **32,** 1499–1502.
Solursh, M., and Karp, G. C. (1975). *J. Exp. Zool.* **191.** 73–84.
Solursh, M., and Meier, S. (1972). *J. Exp. Zool.* **181,** 253–262.
Solursh, M., and Meier, S. (1973). *Dev. Biol.* **30,** 279–289.
Solursh, M., and Meier, S. (1974). *J. Exp. Zool.* **187,** 311–322.
Solursh, M., Meier, S., and Vaerewyck, S. (1973). *Am. Zool.* **13,** 1051–1066.
Solursh, M., Vaerewyck, S. A., and Reiter, R. S. (1974). *Dev. Biol.* **41,** 233–244.
Solursh, M., Ahrens, P. B., and Reiter, R. S. (1978). *In Vitro.* **14,** 51–61.
Soni, N. N., and Malloy, R. B. (1974). *J. Dent. Res.* **53,** 474–480.
Sorgente, N., Kuettner, K. E., Soble, L. W., and Eisenstein, R. (1975). *Lab. Invest.* **32,** 217–222.
Spadaro, J. A. (1977). *Clin. Orthop. Relat. Res.* **122,** 325–332.
Srivastava, V. M. L., Malemud, C. J., and Sokoloff, L. (1974a). *Connect. Tissue Res.* **2,** 127–136.
Srivastava, V. M. L., Malemud, C. J., Hough, A. J., Bland, J. H., and Sokoloff, L. (1974b). *Arthritis Rheum.* **17,** 165–169.
Stanka, P. (1975). *Cell Tissue Res.* **159,** 413–422.
Stark, M., Miller, E. J., and Kühn, K. (1972). *Eur. J. Biochem.* **27,** 192–196.
Stark, R. J., and Searls, R. L. (1973). *Dev. Biol.* **33,** 138–153.
Stark, R. J., and Searls, R. L. (1974). *Dev. Biol.* **38,** 51–63.
Steen, T. P. (1968). *J. Exp. Zool.* **167,** 49–78.
Steen, T. P. (1970). *Am. Zool.* **10,** 119–132.
Steen, T. P. (1973). *Am. Zool.* **13,** 1349.
Steinetz, B. G., and Beach, V. L. (1963). *Endocrinology* **72,** 771–776.
Steinetz, B. G., Manning, J. P., Butler, M., and Beach, V. L. (1965). *Endocrinology* **76,** 876–882.
Steinetz, B. G., Matthews, J. R. Butler, M. C., and Thompson, S. W., (1973). *Am. J. Pathol.* **73,** 735–746.
Stephens, T. D., and Seegmiller, R. E. (1976). *Teratology* **13,** 317–326.
Steven, F. S. (1972). *Clin. Orthop. Relat. Res.* **85,** 257–274.
Stevens, L. C. (1968). *J. Embryol. Exp. Morphol.* **20,** 329–341.
Stewart, P. A., and McCallion, D. J. (1975). *Dev. Biol.* **46,** 383–389.
Stinson, J. C. (1975a). *Med. Hypothesis* **1,** 97–100.
Stinson, J. C. (1975b). *Med. Hypothesis* **1,** 186–190.
Stocum, D. L. (1968). *Dev. Biol* **18,** 441–456.
Stocum, D. L. (1975a). *Dev. Biol.* **45,** 112–136.
Stocum, D. L. (1975b) *Differentiation* **3,** 167–182.
Stocum, D. L., and Dearlove, G. E. (1972). *J. Exp. Zool.* **181,** 49–62.
Storey, E. (1972). *Am. J. Orthod.* **62,** 142–165.
Storey, E. (1973). *Am. J. Orthod.* **64,** 229–247.
Storti, R. V., and Rich, A. (1976). *Proc. Natl. Acad. Sci. U.S.A.* **73,** 2346–2350.
Strom, C. M., and Dorfman, A. (1976a). *Proc. Natl. Acad. Sci. U.S.A.* **73,** 1019–1023,
Strom, C. M., and Dorfman, A. (1976b). *Proc. Natl. Acad. Sci. U.S.A.* **73,** 3428–3432.
Strudel, G. (1953a). *Ann. Sci. Nat., Zool. Biol. Anim.* [11] **15,** 251–329.
Strudel, G. (1953b). *C.R. Seances Soc. Biol. Ser. Fil.* **147,** 132–133.
Strudel, G. (1955). *Arch. Anat. Microsc. Morphol. Exp.* **44,** 209–235.
Strudel, G. (1962). *Dev. Biol.* **4,** 67–86.

Strudel, G. (1963). *J. Embryol. Exp. Morphol.* **11,** 399–412.
Sturdee, A. P., and Connock, M. (1975). *Differentiation* **3,** 43–50.
Stutzmann, J., and Petrovic, A. (1970). *Bull. Assoc. Anat.* **148,** 552–562.
Stutzmann, J., and Petrovic, A. (1974). *Bull. Assoc. Anat.* **58,** No. 163.
Stutzmann, J., and Petrovic, A. (1975a). *Bull. Assoc. Anat.* **59,** 523–534.
Stutzmann, J., and Petrovic, A. (1975b). *C.R. Hebd. Seances Acad. Sci.* **281,** 175–178.
Summerbell, D. (1974a). *J. Embryol. Exp. Morphol.* **32,** 227–238.
Summerbell, D. (1974b). *J. Embryol. Exp. Morphol.* **32,** 651–660.
Summerbell, D. (1976). *J. Embryol. Exp. Morphol.* **35,** 241–260.
Summerbell, D. (1977a). *J. Embryol. Exp. Morphol.* **40,** 1–21.
Summerbell, D. (1977b). *J. Embryol. Exp. Morphol.* **41,** 137–159.
Summerbell, D., and Lewis, J. H. (1975). *J. Embryol. Exp. Morphol.* **33,** 621–643.
Summerbell, D., and Wolpert, L. (1972). *Nature (London) New Biol.* **239,** 24–26.
Summerbell, D., and Wolpert, L. (1973). *Nature (London)* **244,** 228–230.
Summerbell, D., Lewis, J. H., and Wolpert, L. (1973). *Nature (London)* **244,** 492–496.
Sutfin, L. V., Holtrop, M. E., and Ogilvie, R. E. (1971). *Science* **174,** 947–949.
Svajger, A., and Levak-Svajger, B. (1971). *Periodontol. Biol.* **73,** 33–37.
Swett, F. H. (1937). *Q. Rev. Biol.* **12,** 322–339.
Swinyard, C. A., ed. (1969). "Limb Development and Deformity: Problems of Evaluation and Rehabilitation." Thomas, Springfield, Illinois.
Symons, N. B. B. (1951). *Dent. Rec.* **71,** 41–53.
Symons, N. B. B. (1952). *J. Anat.* **86,** 326–332.
Takeichi, M. (1973). *J. Cell Sci.* **13,** 193–204.
Takuma, S. (1962). *J. Dent. Res.* **41,** 883–892.
Talmage, R. V. (1947). *Anat. Rec.* **99,** 91–113.
Tarin, D., and Sturdee, A. P. (1971). *J. Embryol. Exp. Morphol.* **26,** 169–179.
Tarin, D., and Sturdee, A. P. (1974). *J. Embryol. Exp. Morphol.* **31,** 287–303.
Tarlo, L. B. H. (1964). *In* "Bone and Tooth" (H. J. J. Blackwood, ed.), pp. 3–17. Pergamon, Oxford.
Tassava, R. Y., Bennett, L. L., and Zitnik, G. D. (1974). *J. Exp. Zool.* **190,** 111–116.
Tavassoli, M., and Crosby, W. H. (1971). *Proc. Soc. Exp. Biol. Med.* **137,** 641–644.
Telhag, H. (1972). *Clin. Orthop. Relat. Res.* **86,** 224–229.
Telhag, H. (1973). *Acta Orthop. Scand.* **44,** 604–610.
Telhag, H. (1976). *Acta Orthop. Scand.* **47,** 585–587.
Telhag, H., and Havdrup, T. (1975). *Acta Orthop. Scand.* **46,** 185–189.
Ten Cate, A. R. (1972). *J. Anat.* **112,** 401–414.
Ten Cate, A. R. (1975). *Arch. Oral Biol.* **20,** 137–138.
Ten Cate, A. R., and Deporter, D. A. (1975). *Anat. Rec.* **182,** 1–14.
Ten Cate, A. R., and Freeman, E. (1974). *Anat. Rec.* **179,** 543–546.
Ten Cate, A. R., and Syrbu, S. (1974). *J. Anat.* **117,** 351–360.
Ten Cate, A. R., Freeman, E., and Dickinson, J. B. (1974). *J. Dent. Res.* **53,** 245.
Ten Cate, A. R., Deporter, D. A., and Freeman, E. (1976). *Am. J. Orthod.* **69,** 155–168.
Ten Cate, A. R., Freeman, E., and Dickinson, J. B. (1977). *Am. J. Orthod.* **71,** 622–636.
Tenenbaum, H., Thiebold, J., and Bolender, C. (1976). *J. Biol. Buccale* **4,** 261–274.
Thesleff, I., Lehtonen, E., Wartiovaara, J., and Saxén, L. (1977). *Dev. Biol.* **58,** 197–203.
Thompson, D'A. W. (1917). "On Growth and Form." Cambridge Univ. Press, London and New York.
Thonard, J. C., and Wiebkin, O. W. (1973). *J. Periodontol. Res.* **8,** 101–105.
Thorngren, K.-G., and Hansson, L. I. (1973a). *Z. Zellforsch. Mikrosk. Anat.* **142,** 431–442.

Thorngren, K.-G., and Hansson, L. I. (1973b). *Calcif. Tissue Res.* **13,** 113–129.
Thornton, C. S. (1938). *J. Morphol.* **62,** 219–235.
Thornton, C. S. (1942). *J. Exp. Zool.* **89,** 375–390.
Thorogood, P. V., and Craig Gray, J. (1975). *J. Anat.* **120,** 27–47.
Thorogood, P. V., and Hall, B. K. (1976). *J. Embryol. Exp. Morphol.* **36,** 305–313.
Thorogood, P. V., and Hall, B. K. (1977). *Calcif. Tissue Res.* **22,** Suppl., 314–317.
Thorogood, P. V., and Hinchliffe, J. R. (1975). *J. Embryol. Exp. Morphol.* **33,** 581–606.
Thorpe, F. F., and Dorfman, A. (1967). *Curr. Top. Dev. Biol.* **2,** 151–190.
Thyberg, J. (1977). *Histochemie J.* **9,** 259–266.
Thyberg, J., and Friberg, U. (1971). *Z. Zellforsch. Mikrosk. Anat.* **122,** 254–272.
Thyberg, J., Nilsson, S., and Friberg, U. (1973). *Z. Zellforsch. Mikrosk. Anat.* **146,** 83–102.
Tickle, C., Summerbell, D., and Wolpert, L. (1975). *Nature (London)* **254,** 199–202.
Tickle, C., Shellswell, G., Crawley, A.M., and Wolpert, L. (1976). *Nature (London)* **259,** 396–297.
Tilney, L. G., and Mooseker, M. (1971). *Proc. Natl. Acad. Sci. U.S.A.* **68,** 2611–2615.
Toerien, M. J. (1965). *J. Embryol. Exp. Morphol.* **13,** 141–149.
Tonegawa, Y. (1973). *Dev., Growth & Differ.* **15,** 57–72.
Tonna, E. A. (1961). *J. Biophys. Biochem. Cytol.* **9,** 813–824.
Tonna, E. A. (1974). *In* "Humoral Control of Growth and Differentiation" (J. LoBue and A. S. Gordon, eds.), Vol. 1, pp. 275–360. Academic Press, New York.
Tonna, E. A., and Cronkite, E. P. (1962). *Lab. Invest.* **11,** 455–462.
Tonna, E. A., and Pentel, L. (1972). *Lab. Invest.* **27,** 418–426.
Toole, B. P. (1969). *Nature (London).* **222,** 872–873.
Toole, B. P. (1972). *Dev. Biol.* **29,** 321–329.
Toole, B. P. (1973a). *Science* **180,** 302–303.
Toole, B. P. (1973b). *Am. Zool.* **13,** 1061–1067.
Toole, B. P., and Gross, J. (1971). *Dev. Biol.* **25,** 57–77.
Toole, B. P., and Lowther, D. A. (1968a). *Arch. Biochem. Biophys.* **128,** 567–578.
Toole, B. P., and Lowther, D. A. (1968b). *Biochem. J.* **109,** 857–866.
Toole, B. P., and Trelstad, R. L. (1971). *Dev. Biol.* **26,** 28–35.
Toole, B. P., Jackson, G., and Gross, J. (1972a). *Proc. Natl. Acad. Sci. U.S.A.* **69,** 1384–1386.
Toole, B. P., Kang, A. H., Trelstad, R. L., and Gross, J. (1972b). *Biochem. J.* **127,** 715–720.
Torres, A. de L. (1917). *Anat. Rec.* **13,** 443–445.
Travis, D. F. (1968). *In* "Biology of the Mouth" (P. Person, ed.), Publ. No. 89, pp. 237–298. Am. Assoc. Adv. Sci., Washington, D.C.
Travis, D. F., François, C. J., Bonar, L. C., and Glimcher, M. J. (1967). *J. Ultrastruct. Res.* **18,** 519–550.
Trelstad, R. L. (1973). *J. Histochem. Cytochem.* **21,** 521–528.
Trelstad, R. L. (1977). *Dev. Biol.* **59,** 153–163.
Trevisan, R. A., and Scapino, R. P. (1976a). *Acta Anat.* **94,** 40–58.
Trevisan, R. A., and Scapino, R. P. (1976b). *Acta Anat.* **96,** 335–355.
Trueta, J., and Little, K. (1960). *J. Bone Joint Surg., Br. Vol.* **42,** 267–276.
Tschumi, P. A. (1957). *J. Anat.* **91,** 149–173.
Turner, R. S., and Burger, M. M. (1973). *Nature (London)* **244,** 509–510.
Tyler, M. S. (1978). *Anat. Rec.* (in press).
Tyler, M. S., and Hall, B. K. (1977). *Anat. Rec.* **188,** 229–240.
Uhthoff, H. K., and Germain, J.-P. (1977). *Clin. Orthop. Relat. Res.* **123,** 248–252.
Uitto, J., Hoffmann, H.-P., and Prockop, D. J. (1977). *Arch. Biochem. Biophys.* **179,** 654–662.
Umansky, R. (1966). *Dev. Biol.* **13,** 31–56.
Upton, L. G. (1972). *J. Oral Surg.* **30,** 486–490.

Urist, M. R. (1962). *Perspect. Biol. Med.* **6,** 75–115.

Urist, M. R. (1970). *In* "Cell Differentiation" (O. A. Schjeide and J. de Vellis, eds.), pp. 504–528. Van Nostrand-Reinhold, Princeton, New Jersey.

Urist, M. R. (1972). *In* "The Biochemistry and Physiology of Bone" (G. H. Bourne, ed.), 2nd ed., Vol. 2, pp. 155–195. Academic Press, New York.

Uwa, H. (1969). *Dev., Growth & Differ.* **11,** 77–87.

Uwa, H. (1974). *Dev., Growth & Differ.* **16,** 41–54.

Uwa, H., and Nagata, T. (1976). *Dev., Growth & Differ.* **18,** 279–288.

van de Kamp, N. (1968). *J. Exp. Zool.* **169,** 447–462.

van der Strict, O. (1890). *Arch. Biol.* **10,** 1–41.

van Limborgh, J. (1970). *Acta Morphol. Neerl. Scand.* **8,** 143–160.

van Limborgh, J. (1972). *Acta Morphol. Neerl. Scand.* **10,** 37–47.

Vasan, N. S., and Lash, J. W. (1977). *Biochem. J.* **164,** 179–184.

Vasse, J. (1974). *J. Embryol. Exp. Morphol.* **32,** 417–430.

Vasse, J. (1977). *J. Embryol. Exp. Morphol.* **42,** 135–148.

Vaughan, J. M. (1975). "The Physiology of Bone," 2nd ed. Oxford Univ. Press (Clarendon), London and New York.

Vinson, W. C., and Seyer, J. M. (1974). *Biochem. Biophys. Res. Commun.* **58,** 58–65.

Virchow, R. L. K. (1853). *Virchows Arch. Pathol. Anat. Physiol.* **5,** 409.

Vogel, K. G., and Kelley, R. O. (1977). *J. Cell. Physiol.* **92,** 469–480.

von der Mark, H., von der Mark, K., and Gay, S. (1976). *Dev. Biol.* **48,** 237–249.

von der Mark, K., and Bornstein, P. (1973). *J. Biol. Chem.* **248,** 2285–2289.

von der Mark, K., and von der Mark, H. (1977a). *J. Bone Joint Surg.* **59B,** 458–464.

von der Mark, K., and von der Mark, H. (1977b). *J. Cell Biol.* **73,** 736–747.

von der Mark, K., von der Mark, H., and Gay, S. (1976). *Dev. Biol.* **53,** 153–170.

von der Mark, K., Gauss, V., von der Mark, H., and Muller, P. (1977a). *Nature (London)* **267,** 531–532.

von der Mark, K., von der Mark, H., Timpl, R., and Trelstad, R. L. (1977b). *Dev. Biol.* **59,** 75–85.

von Meyer, H. (1867). *Arch. Anat. Physiol., Physiol. Abt.* **34,** 615–642.

Wahl, L. M. (1971), *Anat. Rec.* **169,** 448.

Wahl, L. M., Wahl, S. M., Mergenhagen, S. E., and Martin, G. R. (1975). *Science* **187,** 261–263.

Walder, D. N. (1976). *In* "Diving Medicine" (R. H. Strauss, ed.), pp. 97–108. Grune & Stratton, New York.

Walker, D. G. (1972). *Calcif. Tissue Res.* **9,** 296–309.

Walker, D. G. (1975a). *Science* **190,** 784–785.

Walker, D. G. (1975b). *Science* **190,** 785–787.

Walker, K. V. R., and Kember, N. F. (1972). *Cell Tissue Kinet.* **5,** 401–408.

Wallace, H., Clyman, R. I., and Pierro, L. P. (1969). *J. Exp. Zool.* **172,** 245–252.

Wallace, H., Maden, M., and Wallace, B. M. (1974). *J. Embryol. Exp. Morphol.* **32,** 391–404.

Walton, A. G. (1974). *J. Biomed. Mater. Res.* **8,** 409–426.

Ward, F. O. (1838). "Outlines of Human Osteology." London.

Washburn, S. L. (1947). *Anat. Rec.* **99,** 239–248.

Watabe, N. (1965). *J. Ultrastruct. Res.* **12,** 351–370.

Watanabe, M., and Okada, T. S. (1975). *Dev., Growth & Differ.* **17,** 51–60.

Watson, J., DeHaas, W. G., and Hauser, S. S. (1975). *Nature (London).* **254,** 331–332.

Watterson, R. L., Fowler, I., and Fowler, B. J. (1954). *Am. J. Anat.* **95,** 337–399.

Webber, R. J., Malemud, C. J., and Sokoloff, L. (1977). *Calcif. Tissue Res.* **23,** 61–66.

Webster, D. A., and Gross, J. (1970). *Dev. Biol.* **22,** 157–184.

Wedlock, W. E., and McCallion, D. J. (1969). *Can. J. Zool.* **47,** 142–143.

Weinger, J. M., and Holtrop, M. E. (1974). *Calcif. Tissue Res.* **14,** 15–30.

Weinstein, A., Koldovsk, O., and Segal, S. (1971). *Biochem. J.* **125,** 935–941.
Weiss, P. (1925). *Wilhelm Roux Arch. Entwicklungsmech. Org.* **104,** 359–394.
Weiss, P., and Amprino, R. (1940). *Growth,* **4,** 245–258.
Weiss, P., and Moscona, A. A. (1958). *J. Embryol. Exp. Morphol.* **6,** 238–246.
Wells, C. (1973). *Practitioner* **210,** 384–391.
Weston, J. A. (1963). *Dev. Biol.* **6,** 279–310.
Weston, J. A. (1970). *Adv. Morphog.* **8,** 41–114.
Weston, J. A., and Butler, S. L. (1966). *Dev. Biol.* **14,** 246–266.
White, A. A., III. (1975). *Clin. Orthop. Relat. Res.* **106,** 279–284.
Whitehead, P. E., and McEwan, E. H. (1973). *Can. J. Zool.* **51,** 651–658.
Whitfield, J. F., Rixon, R. H., MacManus, J. P., and Balk, S. D. (1973). *In Vitro* **8,** 257–278.
Whitten, J. M. (1969). *Science* **163,** 1456–1457.
Wiebkin, O. W., and Muir, H. (1973). *In* "Biology of Fibroblast" (E. Kulonen and J. Pikkarainen, eds.), pp. 231–236. Academic Press, New York.
Wiebkin, O. W., and Muir, H. (1975). *Philos. Trans. R. Soc. London, Ser. B* **271,** 283–292.
Wiebkin, O. W., and Muir, H. (1977). *Biochem. J.* **164,** 269–272.
Wigley, C. B. (1975). *Differentiation* **4,** 25–55.
Wilby, O. K., and Ede, D. A. (1975). *J. Theor. Biol.* **52,** 199–218.
Willis, R. A. (1962). "The Borderland of Embryology and Pathology." Butterworth, London.
Willmer, E. N. (1970). "Cytology and Evolution," 2nd ed. Academic Press, New York.
Winterburn, P. J., and Phelps, C. F. (1970). *Nature (London)* **228,** 1311–1312.
Wislocki, G. B., Weatherford, H. L., and Singer, M. (1947). *Anat. Rec.* **99,** 265–284.
Wlodarski, K. (1969). *Exp. Cell Res.* **57,** 446–448.
Wlodarski, K., Hinek, A., and Ostrowski, K. (1970). *Calcif. Tissue Res.* **5,** 70–79.
Wlodarski, K., Poltorak, A., and Koziorowska, J. (1971a). *Calcif. Tissue Res.* **7,** 345–352.
Wlodarski, K., Moskalewski, S., Skarzynska, S., Poltorak, A., and Ostrowski, K. (1971b). *Bull. Acad. Pol. Sci., Ser. Sci. Biol.* **19,** 821–825.
Wlodarski, K., Ostrowski, K., Chtopkiewicz, B., and Koziorowska, J. (1974). *Calcif. Tissue Res.* **16,** 251–256.
Woessner, J. F. (1973). *Clin. Orthop. Relat. Res.* **96,** 310–326.
Wolff, E., and Kieny, M. (1957). *C.R. Hebd. Seances Acad. Sci.* **244,** 1089–1091.
Wolff, E., and Kieny, M. (1963). *Dev. Biol.* **7,** 324–341.
Wolff, J. (1892). "Das Gesetz der Transformation der Knochen." Berlin.
Wolpert, L. (1971). *Curr. Top. Dev. Biol.* **6,** 183–224.
Wong, G. L., and Cohn, D. V. (1974). *Nature (London)* **252,** 713–714.
Wong, G. L., and Cohn, D. V. (1975). *Proc. Natl. Acad. Sci. U.S.A.* **72,** 3167–3171.
Wong, G., Luben, R. A., and Cohn, D. V. (1977). *Science* **197,** 663–665.
Woo, S. L.-Y., Matthews, J. V., Åkeson, W. H., Amiel, D., and Convery, F. R. (1975). *Arthritis Rheum.* **18,** 257–264.
Woolley, D. E., Glanville, R. W., Lindberg, K. A., Bailey, A. J., and Evanson, J. M. (1973). *FEBS Lett.* **34,** 267–269.
Woolley, D. E., Crossley, M. J., and Evanson, J. M. (1977). *Arthritis Rheum.* **20,** 1231–1239.
Wright, D. M., and Moffett, B. C., Jr. (1974). *Am. J. Anat.* **141,** 235–250.
Yagiela, J. A., and Woodbury, D. M. (1977). *Anat. Rec.* **188,** 287–306.
Yasuda, Y. (1973). *Anat. Rec.* **175,** 561–578.
Yasuda, M. (1975a). *J. Embryol. Exp. Morphol.* **33,** 745–756.
Yasuda, M. (1975b). *Proc. Int. Congr. Anat., 10th, 1974* p. 125.
Yeager, V. L., Chiemchanya, S., and Chaiseri, P. (1975). *J. Gerontol.* **30,** 9–14.
Yntema, C. L., and Hammond, W. S. (1954). *J. Comp. Neurol.* **101,** 515–541.
Young, R. W. (1962). *J. Cell Biol.* **14,** 357–370.

Young, R. W. (1963). *Clin. Orthop. Relat. Res.* **26,** 147–160.
Young, R. W. (1964). *In* "Bone Biodynamics" (H. M. Frost, ed.), pp. 117–139. Little, Brown, Boston, Massachusetts.
Zika, J. M., and Klein, L. (1975). *Calcif. Tissue Res.* **18,** 101–111.
Zilliken, F. (1967). *Exp. Biol. Med.* **1,** 199–212.
Zimny, M. L., and Redler, I. (1969). *J. Bone Joint Surg., Am. Vol.* **51,** 1179–1190.
Zimny, M. L., and Redler, I. (1972). *Clin. Orthop. Relat. Res.* **82,** 37–44.
Zwilling, E. (1949). *J. Exp. Zool.* **111,** 175–188.
Zwilling, E. (1955). *J. Exp. Zool.* **128,** 423–441.
Zwilling, E. (1956a). *J. Exp. Zool.* **132,** 173–187.
Zwilling, E. (1956b). *J. Exp. Zool.* **132,** 241–253.
Zwilling, E. (1964). *Dev. Biol.* **9,** 20–37.
Zwilling, E. (1966). *Ann. Med. Exp. Biol. Fenn.* **44,** 134–139.
Zwilling, E. (1968). *Dev. Biol., Suppl.* **2,** 184–207.
Zwilling, E. (1972). *Dev. Biol.,* **28,** 12–17.
Zwilling, E. (1974). *Dev. Biol.,* **39,** 37–48.
Zwilling, E., and Hansborough, L. (1956). *J. Exp. Zool.* **132,** 219–239.

Author Index

Numbers in italics refer to the pages on which complete references are listed.

I

J

K

L

M

S

Subject Index

D

H

N

O

A
B 8
C 9
D 0
E 1
F 2
G 3
H 4
I 5
J 6